GW01238149

Berger Automating with STEP 7 in LAD

Automating with STEP 7 in LAD

SIMATIC S 7-300/400 Programmable Controllers

by Hans Berger

Publicis MCD Verlag

Die Deutsche Bibliothek – CIP-Einheitsaufnahme

Automating with STEP 7 in LAD : SIMATIC S7-300/400
programmable controllers / issued by Siemens-Aktiengesellschaft,
Berlin and Munich. By Hans Berger. – Erlangen ; Munich : Publicis-MCD-Verl.

ISBN 3-89578-094-4

Buch. 1998
Gb.

Diskette. Programming examples. – 1998

The programming examples concentrate on describing the LAD functions and providing SIMATIC S7 users with programming tips for solving specific tasks with this controller.

The programming examples given in the book do not pretend to be complete solutions or to be executable on future STEP 7 releases or S7-300/400 versions. Additional care must be taken in order to comply with the relevant safety regulations.

The author and publisher have taken great care with all texts and illustrations in this book. Nevertheless, errors can never be completely avoided. The publisher and the author accept no liability, regardless of legal basis, for any damage resulting from the use of the programming examples.

ISBN 3-89578-094-4

Issued by Siemens Aktiengesellschaft, Berlin and Munich
Published by Publicis MCD Verlag, Erlangen and Munich

Printed in Germany

Preface

The new SIMATIC basic automation system unites all the subsystems of an automation solution under a uniform system architecture into a homogenous whole from the field level right up to process control. This is achieved with integrated configuring and programming, data management and communications with programmable controllers (SIMATIC S7), automation computers (SIMATIC M7) and control systems (SIMATIC C7). With the programmable controllers, three series cover the entire area of process and production automation: S7-200 as compact controllers ("Micro PLCs"), S7-300 and S7-400 as modularly expandable controllers for the low-end and high-end performance ranges.

STEP 7, a further development of STEP 5, is the programming software for the new SIMATIC. Microsoft Windows 95 or Microsoft Windows NT has been chosen as the operating system, to take advantage of the familiar user interface of standard PCs (windows, mouse operation). STEP 7 complies with DIN EN 6.1131-3, the standard for PLC programming languages. For block programming you can choose between STL (statement list; an Assembler-like language), LAD (ladder logic; a representation similar to relay logic diagrams), FBD (function block diagram) and SCL (a Pascal-like high-level language). Several optional packages supplement these languages: S7-GRAPH (sequential control), S7-HiGraph (programming with state-transition diagrams) and CFC (connecting blocks; similar to function block diagram). The various methods of representation allow every user to select the suitable control function description. This broad adaptability in representing the control task to be solved significantly simplifies working with STEP 7.

This book contains the description of the LAD programming language for S7-300/400. In the first section, the book introduces the S7-300/400 automation system and explains the basic handling of STEP 7. The next section addresses first-time users or users changing from relay contactor controls; the "Basic Functions" of a binary control are described here. The digital functions explain how digital values are combined; for example, basic calculations, comparisons, data type conversion. With LAD, you can control program processing (program flow) and design structured programs. As well as a cyclically processed main program, you can also incorporate event-driven program sections as well as influence the behavior of the controller at startup and in the event of errors/faults. The book concludes with a general overview of the system functions and the function set.

The contents of this book describe Version 4.0 of the STEP 7 programming language.

Erlangen, November 1997

Publicis MCD Verlag

The Contents of the Book at a Glance

Overview of the S7-300/400 automation system

Introduction

1 S7-300/400 Automation System

Design of the automation system

Configuring the station

Addressing modules

Memory areas

Operand areas

2 STEP 7 Programming Software

STEP 7 Basic

Processing a project

Hardware configuration

Communications links

Creating programs

Symbol table

Program test

On-line functions

3 LAD Programming Language

Program processing

Blocks

Editing LAD elements

Data types

PLC functions comparable with a relay logic control

Basic Functions

4 Series and Parallel Circuits

NO Contact and NC Contact

Series Circuits

Parallel Circuits

Negating the Result of the Logic Operation

5 Memory Functions

Single coil

Set and Reset Coil

Memory Box, Midline Outputs

Edge Evaluation

6 Transfer Functions

MOVE Box

SFCs for Data Transfer

7 Timer Functions

Starting a timer with five different characteristics

8 Counter Functions

Counting up, counting down, setting, checking

Handling numbers and digital operands

Digital Functions

9 Comparison Functions

Comparisons according to INT, DINT and REAL

10 Arithmetic Functions

Basic arithmetic functions according to INT, DINT and REAL

11 Math Functions

Trigonometric functions,

Square, square root

Exponentiation, forming logarithms

12 Conversion Functions

Data type conversion

Generating complements

13 Shift Functions

Shift and rotate

14 Word Logic

AND, OR, exclusive OR

Controlling program execution, block functions

Processing the user program

Supplements to LAD, block libraries, overview of LAD functions

Program Flow Control

15 Status Bits

Setting the status bits

Evaluating the status bits

Binary result, EN/ENO

16 Jump Functions

Absolute jump

Jump if RLO = “1”

Jump if RLO = “0”

17 Master Control Relay

MCR dependency

MCR area, MCR zone

18 Block Functions

Block call, block end

Static local data

Addressing data operands

Opening data blocks

19 Block Parameters

Formal parameters, actual parameters

Declaring, initializing and “passing on”

Program Processing

20 Main Program

Program structure

Cycle time, response time

Program functions

Data exchange with system functions

Start information

21 Interrupt Processing

Hardware interrupt

Watchdog interrupt

Time-of-day interrupt

Time-delay interrupt

Handling interrupts

22 Startup Characteristics

Complete restart, restart

STOP, memory reset

Parameterize modules

23 Error Handling

Synchronous errors

Asynchronous errors

Diagnostics

Appendix

24 Supplements to LAD

KNOW_HOW_PROTECT

Indirect addressing

Brief description of “Message Frame Example”

25 Block Libraries

Std OBs
  Organization blocks

Built In
  System blocks

IEC
  Loadable functions

FB Lib 1
  S5 converter functions

FB Lib 2
  TI converter functions

PID Control
  Closed-loop control functions

Net DP
  DP functions

26 Function Overview

Basic functions

Digital functions

Program flow control

The Contents of the Diskette at a Glance

This book contains many illustrations of the representation and use of the LAD programming language. All the program sections shown in the book and any additional examples can be found on the accompanying diskette.

After installing the examples, you can open, view and print; you can also modify them or copy them to other projects or libraries.

An operand field width of 12 has been selected in the LAD Editor for the examples in the book. All examples contain symbols and comments. If there is more comment in a network than is displayed, click on the text field and scroll to the hidden portion of text.

The introduction page of the "Appendix" contains installation instructions for the library stored on the diskette.

Data Types Examples of definition and application		**Program Processing** Examples of SFC calls	
FB 101	Elementary data types	FB 120	Chapter 20: Main Program
FB 102	Complex data types	FB 121	Chapter 21: Interrupt Processing
FB 103	Parameter types	FB 122	Chapter 22: Startup Characteristics
		FB 123	Chapter 23: Error Handling
Basic Functions LAD representation examples		**Conveyor Example** Examples of basic functions and local instances	
FB 104	Chapter 4: Series and Parallel Circuits	FC 11	Belt control
FB 105	Chapter 5: Memory Functions	FC 12	Counter control
FB 106	Chapter 6: Transfer Functions	FB 20	Feed
FB 107	Chapter 7: Timer Functions	FB 21	Conveyor belt
FB 108	Chapter 8: Counter Functions	FB 22	Parts counter
Digital Functions LAD representation examples		**Message Frame Example** Data handling examples	
FB 109	Chapter 9: Comparison Functions	UDT 51	Data structure for the frame header
FB 110	Chapter 10: Arithmetic Functions	UDT 52	Data structure for a message
FB 111	Chapter 11: Math Functions	FC 51	Time-of-day check
FB 112	Chapter 12: Conversion Functions	FC 52	Copy data area with indirect addressing
FB 113	Chapter 13: Shift Functions	FB 51	Prepare message frame
FB 114	Chapter 14: Word Logic	FB 52	Save message frame
Program Flow Control LAD representation examples		**General Examples** Examples of digital functions	
FB 115	Chapter 15: Status Bits	FC 41	Range monitor
FB 116	Chapter 16: Jump Functions	FC 42	Limit value detection
FB 117	Chapter 17: Master Control Relay	FC 43	Compund interest
FB 118	Chapter 18: Block Functions	FC 44	Doublewordwise edge evaluation
FB 119	Chapter 19: Block Parameters		

Contents

The author and publisher are always grateful to hear your responses to the contents of the book.

Publicis MCD Verlag
Postfach 3240
D-91050 Erlangen
Federal Republic of Germany
Fax: ++49 9131/727838
E-mail: publishing-books@publicis-mcd.de

Introduction

This portion of the manual provides an overview of the SIMATIC S7-300/400 programmable controller.

- **The S7-300/400 programmable controller** is of modular design. The modules with which it is configured can be central (in the vicinity of the CPU) or distributed without any special settings or parameter assignments having to be made. In SIMATIC S7 systems, distributed I/O is an integral part of the system. The CPU, with its various memory areas, forms the hardware basis for processing of the user programs. A load memory contains the complete user program; the parts of the program relevant to its execution at any given time are in a work memory whose short access times are the prerequisite for fast program processing.
- **STEP 7** is the programming software for S7-300/400; the automation tool is the SIMATIC Manager. The SIMATIC Manager is a Windows 95/NT application, and contains all functions needed to set up a project. When necessary, the SIMATIC Manager starts additional tools, for example to configure stations, initialize modules, and to write and test programs.
- You formulate your automation solution in the **LAD programming language**. The LAD program is structured, that is to say it consists of blocks with defined functions which are composed of networks or rungs. Different priority classes allow a graduated interruptibility of the user program currently executing. STEP 7 works with variables of different data types, from binary variables (data type BOOL) to digital variables (for instance of data type INT or REAL for computing tasks) to complex data types such as arrays or structures (combining of variables of different data types to form a single variable).

The first chapter contains an overview of the hardware in an S7-300/400 **programmable controller**, the second chapter an overview of the STEP 7 programming software. The basis for the description is the function scope for STEP 7 Version 4.0.

Chapter 3, "The LAD Programming Language", serves as an introduction to the most important LAD elements. A more detailed approach follows in the subsequent chapters of the manual. The descriptions of all LAD functions are enhanced by brief examples. At the end of some chapters you will find extensive examples, the intention of which is to show the use of the LAD language in a larger context.

1 SIMATIC S7-300/400 Programmable Controller
Structure of the programmable controller; configuring a station (introduction); addressing modules; CPU memory areas; process images, bit memory and temporary local data

2 STEP 7 Programming Software
SIMATIC Manager; processing a project; configuring a station; configuring communication links; writing programs (symbol table, editor); testing programs

3 LAD Programming Language
Program processing with priority classes; program blocks; programming blocks; addressing variables; data types

1 SIMATIC S7-300/400 Programmable Controller

1.1 Structure of the Programmable Controller

1.1.1 Components

The SIMATIC S7-300/400 is a modular programmable controller comprising the following components:

- ▷ Racks
  accommodate the modules and connect them to each other
- ▷ Power supply (PS);
  Provides the internal supply voltages
- ▷ Central processing unit (CPU)
  Stores and processes the user program
- ▷ Interface modules (IMs);
  Connect the racks to one another
- ▷ Signal modules (SMs);
  Adapt the signals from the system to the internal signal level or control actuators via digital and analog signals
- ▷ Function modules (FMs);
  Execute complex or time-critical processes independently of the CPU
- ▷ Communications processors (CPs)
  Interface the programmable controllers to each other or to other devices via serial links

A programmable controller (or station) may consist of several racks, which are linked to one another via bus cables. Figure 1.1 shows the structure of a programmable controller using an S7-300 station as example.

In S7-314/315 systems, you can connect up to three additional racks to the controller rack at a maximum distance of 10 meters. In S7-400 systems, depending on the interface modules, you can connect additional local racks up to three meters, remote racks up to 100 meters away from the controller rack (not to be confused with

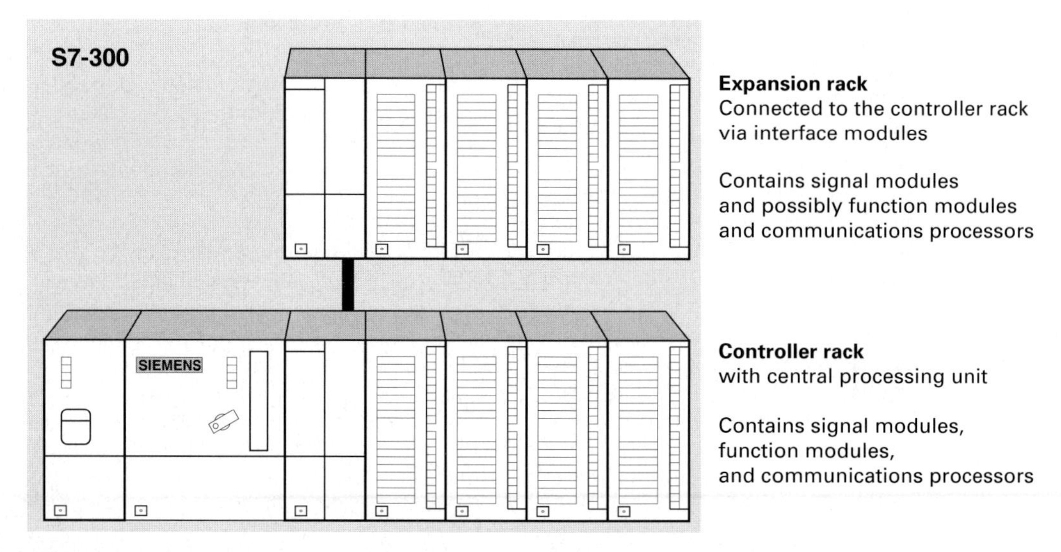

Figure 1.1 Modular Structure of a SIMATIC S7-300 Programmable Controller

distributed I/O). An interface module in the controller rack can control as many as four expansion racks. It is possible to connect up to 22 expansion racks in total.

The racks connect the modules with two busses: the I/O bus (or P bus) and the communication bus (or K bus). The I/O bus is designed for the high-speed interchange of input and output signals, the communication bus for the exchange of large amounts of data. The communication bus connects the CPU and the programming device interface with function modules and communications processors.

A special feature in S7-400 systems is the segmented rack. This rack can accommodate two CPUs with a shared power supply while keeping them functionally separate. The two CPUs can interchange data with one another via the K bus, but have completely separate P busses for their own signal modules.

In an S7-400 system, as many as four specially designed CPUs in a suitable rack can take part in multi-processor operation. Each module in this station is assigned to only one CPU, both with its address and its interrupts.

The IM 463-2 interface module allows you to connect S5 expansion units (EU 183U, EU 185U, EU 186U as well as ER 701-2 and ER 701-3) to an S7-400, and also allows centralized expansion of the expansion units. An Im 314 in the S5 expansion unit handles the link. You can operate all analog and digital modules allowed in these expansion units. An S7-400 can accommodate as many as four IM 463-2 interface modules; as many as four S5 expansion units can be connected in a distributed configuration to each of an IM 463-2's interfaces.

1.1.2 Distributed I/O

The term distributed I/O is understood to mean modules that are connected to a DP master module via PROFIBUS-DP. PROFIBUS-DP is a manufacturer-independent standard to EN 50170 Volume 2 for the connection of DP standard slaves.

You can set the baud rate for PROFIBUS-DP to between 9.6 kbaud and 12 Mbaud, depending on the permissible line length in a segment (subnetwork). Segments are connected using RS 485 repeaters which may be up to 1000 meters apart. The physical medium is either a shielded two-wire cable in a network extending over a maximum of 9.6 kilometers or a fiber-optic cable extending over no more than 23.8 kilometers.

A DP master (such as the CPU 414-2 DP or the CP 342-5 DP) controls as many as 31 DP slaves in a segment or up to 63 throughout the entire network. You can also connect programming devices, operator interface systems, ET 200's or SIMATIC S5 DP slaves to the PROFIBUS-DP network.

Figure 1.2 shows an example for networking with distributed I/O. A CPU 414-2DP, together with a PG 740, is DP master, and controls the ET 200B, ET 200C, ET 200M and OP 25 distributed I/O stations. The ET 200M consists of an S7-300 rack containing the appropriate modules, whereby an IM 513 PROFIBUS interface module replaces the CPU. The DP/AS Interface Link allows operation of the AS-Interface bus (AS-Interface is short for Actuator Sensor Interface) as subnetwork on PROFIBUS-DP.

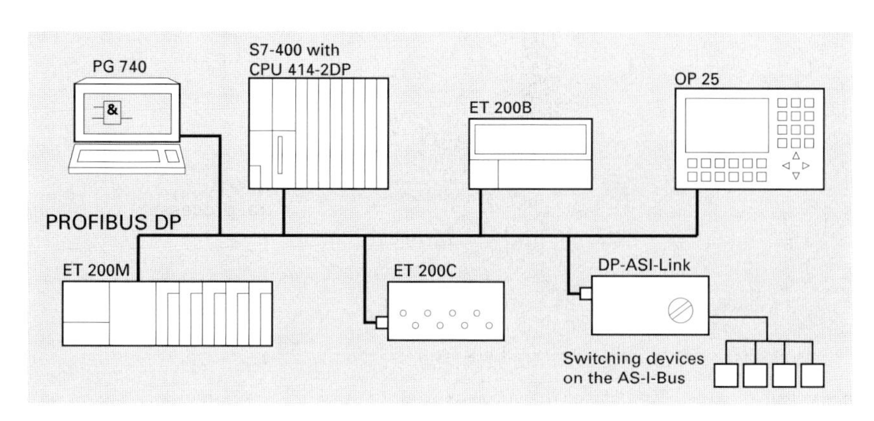

Figure 1.2 Connection to distributed I/O

1.1.3 MPI Network

Every central processing unit has an "interface with multipoint capability" (multipoint interface, or MPI). The MPI gives a CPU network capability without the need for additional modules. An MPI network may comprise as many as 32 nodes; these may be

- ▷ programming devices (PGs, PCs with appropriate interface card),
- ▷ operator panels (OPs),
- ▷ appropriately designed function modules (FMs) or communications processors (CPs) and
- ▷ additional CPUs in other programmable controllers.

This allows you to use only a single programming device as operator interface to all connected CPUs, and also allows low-overhead interchanging of data between the CPUs' user programs via global data communication (which is integrated in the CPUs).

The transmission rate on the MPI bus is permanently set to 187.5 kbaud. The cable in one segment (subnetwork) may be up to 50 meters long, but can be increased to a total of 1000 meters through the use of RS 485 repeaters. The ends of the MPI bus must be terminated and biased.

Every node in the network is identified by an MPI address. A node's MPI address is preset at the factory (PG = 0, OP = 1, CPU = 2). This allows you to put a programmable controller with only one CPU into operation with one PG and one OP without setting any addresses.

Within a segment, nodes connected to an MPI network must have different MPI addresses. The MPI address is specifiable. The entire network may comprise up to 126 nodes, a subnetwork up to 32 nodes.

Figure 1.3 shows an example of an MPI network. Two S7-300s, two S7-400s, one OP 25 and two PG 740s are connected to the MPI bus, which consists of two subnetworks connected via RS 485 repeaters, whereby one PG is loosely connected via a spur line to a bus connector with PG socket.

1.1.4 SIMATIC NET

SIMATIC NET is the new product designation for networks and network components (formerly SINEC, Siemens Network and Communication).

You can connect the S7-300/400 programmable controllers via communications processors to the following bus systems:

- *Industrial Ethernet* is the network for the management level and the cell level defined by international standard IEEE 802.3, with emphasis on the industrial area (former product designation: SINEC H1). SIMATIC S7

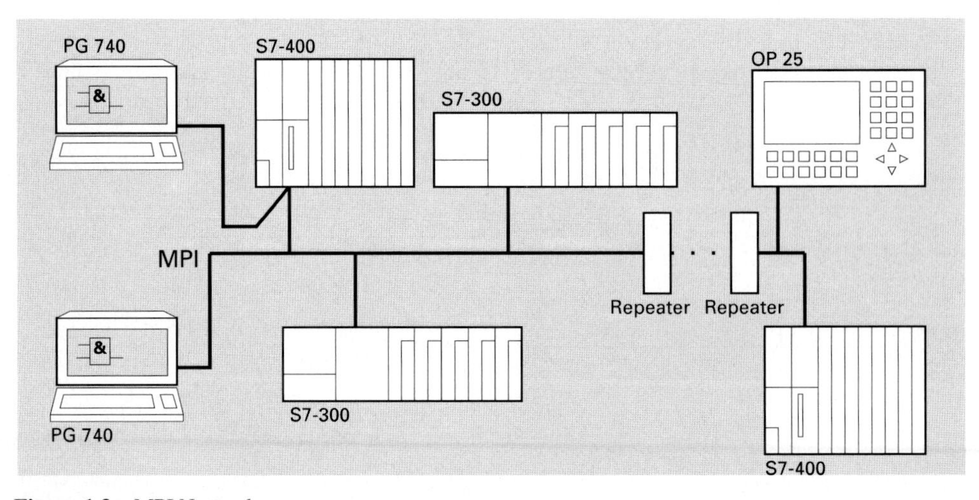

Figure 1.3 MPI Network

provides communications processors CP 343-1 and CP 343-1 TCP (S7-300) as well as CP 443-1 and CP 443-1 TCP (S7-400).

- *PROFIBUS* stands for 'Process Field Bus', and is a manufacturer-independent standard to DIN 19245 for the networking of field devices (former product designation: SINEC L2). PROFIBUS provides two protocols: PROFIBUS-FMS for the transfer of large amounts of data and PROFIBUS-DP for distributed I/O. Communications processors CP 342-5 DP and CP 342-5 FMS (S7-300) and CP 443-5 (S7-400) are available for interfacing to SIMATIC S7.
- The *AS-Interface* is a networking system for binary sensors and actuators at the lowest field level (former product designation: SINEC S1). The AS Interface is connected to an S7-300 via a CP 342-2 communications module. The DP/AS-Interface Link connects the AS-Interface to PROFIBUS-DP.
- A *point-to-point* link is used to connect devices with a serial interface to the programmable controller, for instance bar code readers, printers, or SIMATIC S5 programmable controllers. For a point-to-point link you need a CP 340 (S7-300) or a CP 441-1 or CP 441-2 (S7-400) communications processor with one or two interfaces.

1.2 Configuring a Station

1.2.1 Signal Path

When you wire your machine or plant, you determine which signals are connected where on the programmable controller (Figure 1.4). An input signal, for example the signal from mo-

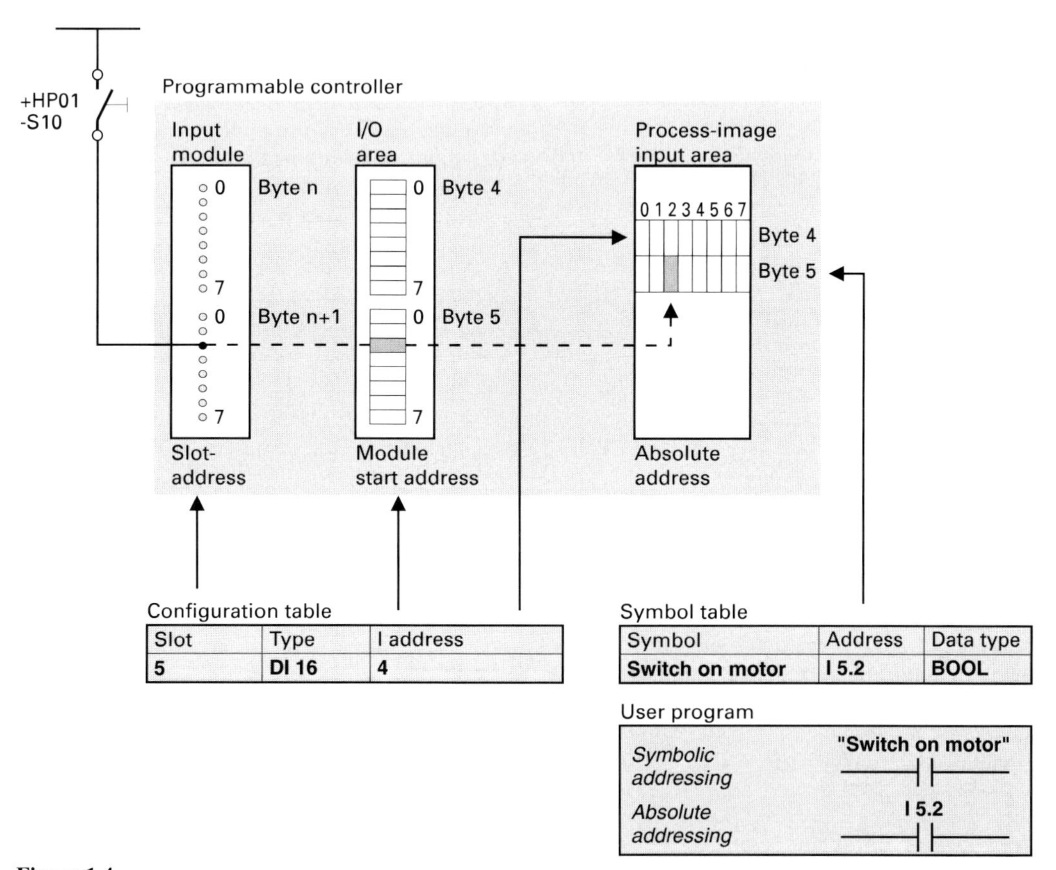

Figure 1.4
Correlation between module address, absolute address and symbolic address (path of a signal from sensor to scanning in the program)

mentary-contact switch +HP01-S10, the one for "Switch motor on", is run to an input module, where it is connected to a specific terminal. This terminal has an "address" called the I/O address (for instance byte 5, bit 2). The CPU copies the signal to the process-image input table, where it is then addressed as input (I 5.2, for example). The term "I 5.2" is the *absolute address*. You can now give this input a name by assigning a symbol to the absolute address in the symbol table which corresponds to this input signal (such as "Switch motor on"). The term "Switch motor on" is the *symbolic address*.

1.2.2 Locating Modules

In a programmable controller (that is, in an S7 station), every slot has a fixed address. This slot address is dependent on the number of the rack and the number of the slot. A module is unambiguously identifiable via the slot address.

STEP 7 provides *configuration tables* for locating modules, one table for each rack. You can generate the configuration tables during the planning phase or wait until the hardware has been set up. STEP 7 provides you with a catalog of racks, modules and interface modules available in the system. Taking the rules applying to module slots into account, you simply transfer components from this catalog to the relevant line in the configuration table (which corresponds to a particular slot).

All settings on the modules are done at the software level; SIMATIC S7 modules have no switches or jumpers to set.

1.2.3 Addressing Modules

In addition to the slot address, which defines the slot, each module has a start address which defines the location in the logical address space (in the I/O addressing range).

The module start address is determinative for the referencing of the input and output signals by the program. In the case of digital modules, the individual signals (bits) are bundled into groups of eight called "bytes". There are modules with one, two or four bytes. These bytes have the relative addresses 0, 1, 2 and 3; addressing of the bytes begins at the module start address. Example: In the case of a digital module with four bytes and the start address 8, the individual bytes are referenced by addresses 8, 9, 10 and 11. In the case of analog modules, the individual analog signals (voltages, currents) are called "channels", each of which reserves two bytes. Analog modules are available with two, four, eight and 16 channels, corresponding to four, eight, 16 or 32 bytes address area.

On power-up (if there is no setpoint configuration), the CPU defaults to a slot-oriented module start address which depends on the module type, the slot, and the rack. This module start address corresponds to (relative) byte 0. You can view this address in the configuration table.

In S7-315 DP and S7-400 systems, you can change this address. You have the option of assigning the start addresses of the modules in your hardware configuration yourself. In a raster of 16 (S7-300) or four (S7-400), you can assign each slot an arbitrary address. You also have the option of assigning different start addresses for inputs and outputs on a hybrid digital or analog module. FMs and CPs normally reserve the same start address for inputs and outputs.

The advantages of self-assigned addresses are most apparent when it comes to distributed I/O. The counting of module addresses according to slot serves little purpose, since both digital and analog modules can be plugged into a distributed station. Default assignments thus do not guarantee that all digital modules lead to the process image (which always begins at address 0 and ends at a maximum address specific to the relevant CPU), nor is it guaranteed that the analog modules do not lead to the process image (which one usually wishes to avoid).

Modules which are nodes in an MPI network (CPUs, FMs and CPs) also have an MPI address.This address is decisive for global communication and for the link to programming devices and operator interface systems. Each module on the bus also has a node address, which you set when you initialize the module.

1.2.4 Initializing Modules

When you initialize the modules, you pass to them settings and data with which they are to work on system startup. By selecting a module from a configuration table, you screen a dialog box in which you can define the module's properties (for instance limit values for an analog values, enabling of diagnostic or process interrupts).

You also set the properties of a central processing unit in this way (for example clock memory, retentivity, priority classes), thereby affecting your program's performance.

The programmed module properties are stored in system data blocks. You can transfer these blocks to the CPU, or have them automatically transferred together with your program. The CPU accepts the modified parameters immediately, and forwards them to the modules on the next system startup.

Some modules can be initialized with system functions via data record transfer at program runtime.

1.3 Addressing Modules

In SIMATIC S7 systems, each module has two address areas: a *useful data area*, which can be directly addressed, for instance with MOVE box, and a *system data* area for transferring data records.

When modules are addressed, it makes no difference whether they are in racks with centralized configuration or used as distributed I/O. All modules have the same (logical) address capacity.

CPUs can also interchange data when they are linked to one another via the MPI bus. This so-called *global data* communication is also designed for small amounts of data.

1.3.1 Useful Data Transfer

Useful data

A module's useful data properties depend on the module type. In the case of signal modules, they are either digital or analog input/output signals, in the case of function modules and communications processors, they might, for example, be control or status information. The amount of useful data is module-specific. There are modules which reserve one, two, four or more bytes in this area. Addressing always beging at relative byte 0. The address of byte 0 is the module start address; it is stipulated in the configuration table.

The useful data represent the I/O address area, subdivided, depending on the direction of transfer, into peripheral inputs (PIs) and peripheral outputs (PQs). If the useful data are in the process-image area, the CPU automatically handles the transfers to update the process images.

Peripheral inputs

You use the peripheral input (PI) address area when you read from the useful data area on an input module. Part of the PI address area leads to the process image. This part always begins at I/O address 0; the length of the area is CPU-specific.

With a Direct I/O Read operation, you can address the modules whose interfaces do not lead to the process-image input area (for instance analog input modules). The signal states of modules which lead to the process-image input area also cannot be read with a Direct Read operation. The momentary signal states of the input bis are then scanned Note that this signal state may differ from the relevant inputs in the process image! (the process-image input area is updated at the beginning of the program scan.)

Peripheral inputs may reserve the same absolute addresses as peripheral outputs.

Peripheral outputs

You use the peripheral output (PQ) address area when you write values to the useful data area on an output module. Part of the PQ address area leads to the process image. This part always begins at I/O address 0; the length of the area is CPU-specific.

With a Direct I/O Write operation, you can address modules whose interfaces do not lead to the process-image output area (such as analog output modules). The signal states of modules controlled by the process-image output area can also be directly affected. The signal states of the output bits then change immediately. Note that a Direct I/O Write operation also changes the signal states of the relevant modules in the process-image output area! In this way, there is no difference between process-image output area and the signal states on the output modules.

Peripheral outputs can reserve the same absolute addresses as peripheral inputs.

Substitute value strategy

In STOP mode or during startup, the CPU disables the output modules. In this case, zero or, if the module is designed accordingly, a specified substitute value, is output.

When no further values can be read from an input module or when an attempt is made to read from a non-existent module, the CPU invokes OB 122, the organization block for synchronization errors. In this OB, you can return a substitute value in accumulator 1 using system function SFC 44 REPL_VAL instead of returning the value (not) read.

1.3.2 Data Record Transfers

Data records are used to exchange data via a module's system data area. These data records may contains information such as parameter assignment data or diagnostic info, but also user data. The CPU automatically transfers data records containing parameter assignment data or diagnostic data when required. System functions allow you to transfer data records yourself. The data records are identified by number. Depending on the type of record involved, the information may comprise up to 240 bytes.

The system functions for data transfer are described in detail in section 22.5, "Initializing Modules".

1.3.3 GD Communication

The CPUs in an MPI network can exchange data amongst themselves. This capability, called global data communication, is restricted to small amounts of data. Global data (GD) are

▷ Inputs and outputs (process images)

▷ Bit memory

▷ Data in data blocks

▷ Timers and counters (as Send data).

The data are exchanged in the form of data packets (GD packets) between CPUs, which are combined into GD circles. A GD circle may be

- a bilateral connection between two CPUs (each of which can both send and receive a GD packet)
- a unilateral connection in which a single CPU sends a GD packet to several other CPUs, which receive that packet.

A CPU may belong to several GD circles. The possible number of GD circles and the size of the GD packets is CPU-specific.

Please note that a receiving CPU does not acknowledge the receipt of global data. The sender thus receives no information of any kind to indicate whether data was received or which CPU received that data. You do, however, have the option of viewing the status of communications between two CPUs as well as the status of all of the GD circles to which a CPU belongs.

Caution: Handle GD communication for each CPU as carefully as you do your CPU programs. Connections which fail to be established or are interrupted can produce grave problems. An interrupted connection allows the values last transferred to stand.

Global data communication requires a considerable portion of the processing time in the CPU operating system. For this reason, you can define a "scan rate" in the global data table which specifies after how many program cycles the data are to be sent or received. Because the data are not updated in every program cycler, you should not use this form of communication to transfer time-critical data.

Global data are sent and received asynchronously between sender and receiver at the scan cycle checkpoint (after cyclic program scanning and before a new program cycle begins). The S7-400 CPUs also allow event-driven global data communication using system functions SFC 60 GD_SND and SFC 61 GD_RCV.

1.4 CPU Memory Areas

1.4.1 User Memory

Figure 1.5 shows the CPU memory areas important to your program. The user program itself is in two areas, namely load memory and work memory.

Load memory can be a plug-in memory card or integrated memory. The entire user program, including module configuration and module pa-

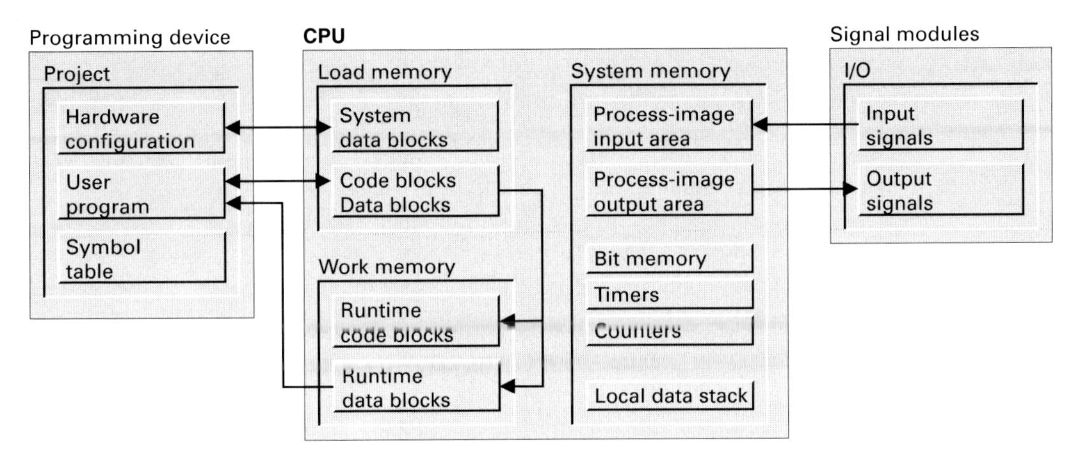

Figure 1.5 CPU memory areas

rameters, are in load memory. In future, load memory should also be capable of holding all symbols.

Work memory is designed in the form of integrated high-speed RAM. Work memory contains the relevant portions of the user program; these are essentially the program code and the user data. "Relevant" is a characteristic of the existing objects, and does not mean that a particular code block will necessarily be called and processed.

The operating system copies the relevant parts of a user program from a plug-in memory card to work memory when the supply voltage is switched on. All blocks reloaded from the programming device are updated accordingly in both load and work memory. Changes made to data blocks by the user program, however, affect only work memory; on the next non-battery-backed power-up, the (old) data block in load memory overwrites the (new) data block in work memory. Data blocks generated at program runtime are in work memory only.

1.4.2 Memory Card

On most CPUs, load memory can be expanded with a plug-in memory card. There are two types of memory card: RAM cards and Flash EPROM cards.

If you want to expand load memory only, use a RAM card. A RAM card allows you to modify individual blocks on-line.

If you want to protect your user program, including configuration data and module parameters, against power failure, use a Flash EPROM card. In this case, load the entire program off-line onto the Flash EPROM card with the card plugged into the programming device. On S7-400 and 315 CPUs (with 5 V Flash EPROM card), you can also load the program on-line with the memory card plugged into the CPU.

1.4.3 System Memory

System memory contains the variables which you address in your program. The variables (addresses) are combined into areas (address areas) containing a CPU-specific number of addresses. Addresses may be, for example, inputs used to scan the signal states of momentary-contact switches and limit switches, and outputs, which you can use to control contactors and lamps. The system memory on a CPU contains the following address areas:

▷ Inputs (I)
Inputs are an image ("process image") of the digital input modules.

▷ Outputs (Q)
Outputs are an image ("process image") of the digital output modules.

▷ Bit memory (M)
Stores of information addressable throughout the whole program.

▷ Timers (T)
Timers are locations used to implement waiting and monitoring times.

▷ Counters (C)
Counters are software-level locations which can be used for up and down counting.

▷ Temporary local data (L)
Locations which serve as dynamic intermediate buffers during block processing. The temporary local data are located in the L stack, which the CPU reserves dynamically during program scanning.

The letters enclosed in parentheses represent the abbreviations to be used for the different addresses when writing programs. You may also assign a symbol to each variable and then use the symbol in place of the address identifier.

1.5 Address Areas in System Memory

The address areas in the CPU's system memory are the process image with the inputs and outputs, the bit memory, the temporary local data, and the timers and counters. The timers and counters are described in detail in the corresponding chapters in the next part of the book.

1.5.1 Process Image

The process image contains the image of the digital input and digital output modules, and is thus subdivided into process-image input area and process-image output area. The process-image input area is addressed via the address area for inputs (I), the process-image output area via the address area for outputs (Q). As a rule, the machine or process is controlled via the inputs and outputs.

When you address the inputs and outputs in your program, an S7-400 system must be configured with the appropriate input/output modules. In S7-300 systems, the areas of the process image to which no modules are assigned may be used as additional memory in a manner similar to the use of bit memory.

Inputs

An input is an image of the corresponding bit on a digital input module. Scanning an input is the same as scanning the bit on the module itself. Prior to program scanning in every program cycle, the CPU's operating system copies the signal state from the module to the process-image input area. The use of a process-image input area has many advantages:

- Inputs can be scanned and linked bit by bit (I/O bits cannot be directly addressed).
- Scanning an input is much faster than addressing an input module (for example, there is no transient recovery time on the I/O bus, and the system memory response times are shorter than the module's response times). The program is therefore scanned that much more quickly.
- The signal state of an input is the same throughout the entire program cycle (there is data consistency throughout a program cycle). When a bit on an input module changes, the change in the signal state is transferred to the input at the start of the next program cycle.
- Inputs can also be set and reset because they are located in random access memory. Digital input modules can only be read. Inputs can be set during debugging or start-up to simulate sensor states, thus simplifying program testing.

In opposition to these advantages stands an increased program response time (also refer to section 20.2.4, "Response Time").

Outputs

An output is an image of the corresponding bit on a digital output module. Setting an output is the same as setting the bit on the output module itself. The CPU's operating system copies the signal state from the process-image output area to the module. The use of a process-image output area has many advantages:

- Outputs can be set and reset bit by bit (direct addressing of I/O bits is not possible).
- Setting an output is much faster than addressing an output module (for example, there is no transient recovery time on the I/O but, and the system memory response times are shorter than the module response times). The program is therefore scanned that much more quickly.
- A multiple signal state change at an output during a program cycle does not affect the bit on the output module. It is the signal state of the output at the end of the program cycle which is transferred to the module.

- Outputs can also be scanned because they are located in random access memory. While it is possible to write to digital output modules, it is not possible to read them. The scanning and linking of the outputs makes additional storage of the output bit to be scanned unnecessary.

These advantages are offset by an increased program response time. A detailed description of how a programmable controller's response time is put together can be found in section 20.2.4, "Response Time".

1.5.2 Bit Memory

The area called bit memory hold what could be regarded as the controller's "auxiliary contactors". Bit memory is used primarily for storing binary signal states. The bits in this area can be treated as outputs, but are not "externalized". Bit memory is located in the CPU's system memory area, and is therefore available at all times. The number of bits in bit memory is CPU-specific.

Bit memory is used to store intermediate results which are valid beyond block boundaries and are processed in more than one block. *Temporary local data* (available in all blocks but valid for the current block call only) and *static local data* (available only in function blocks but valid over multiple block calls) are provided for storing intermediate results within a block.

Retentive bit memory

Part of bit memory may be designated "retentive", which means that the bits in that part of bit memory retain their signal states even under off-circuit conditions. Retentivity always begins with memory byte 0 and ends at the designated location. Retentivity is set when the CPU is initialized.

Clock memory

Many procedures in the controller require a periodic signal. Such a signal can be implemented using timers (clock pulse generator), watchdog interrupts (time-controlled program scanning), or simply by using clock memory.

Clock memory consists of bits whose signal states change periodically with a mark-to-space ratio of 1:1. The bits are combined into a byte,

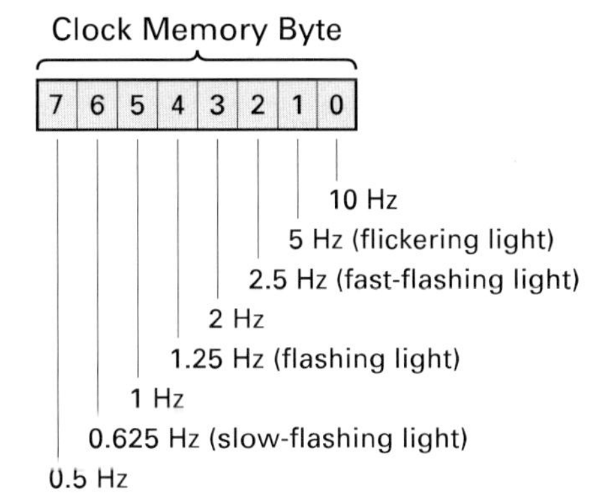

Figure 1.6 Contents of the Clock Memory Byte

and correspond to fixed frequencies (Figure 1.6). You specify the number clock memory bits when you initialize the CPU. Please note that the updating of clock memory is asynchronous to the scanning of the main program.

1.5.3 Temporary Local Data

The temporary local data are stored in the local data stack (L), which is part of system memory. The CPU's operating system makes the temporary local data for each code block available as that block is called. The temporary local data may be used to store and re-use results determined when the block in question was processed.

The amount of temporary local data a block requires is in the block header. The operating system thus learns exactly how many bytes have to be reserved in the L stack as soon as the block is called. You, too, can find out how many local data bytes the block needs by looking at the block header.

When the block terminates, the L stack is made available to other blocks. When a block is called, the values in the L stack are virtually arbitrary, and are available to that block only while it is executing. It is for this reason that this area is referred to as the "temporary" local data area.

The total size of the L stack depends on the CPU. The number of temporary local data bytes available in a priority class (in an organization block's program) is also fixed. In an S7-300 system, the number is preset (on the CPU 314, for instance, 256 bytes are available for each priority class); in an S7-400 system, you can adapt the number of temporary local data bytes to suit your needs when you initialize the CPU. These bytes must be shared by the blocks called in the relevant organization block, as well as by the blocks called by these blocks. In this conjunction, please note that the editor also uses temporary local data, for instance when passing block parameters.

The start information for an organization block comprises 20 bytes. These 20 bytes must be available is every priority class used. If you program an evaluation routine for synchronization errors (programming and access errors), you must provide 20 additional bytes for the start information for these error organization blocks.

You must specify the number of temporary local data bytes, as well as the name and data type, in the declaration part of the relevant block. All operations which apply to bit memory are also permitted for temporary local data.

2 STEP 7 Programming Software

2.1 STEP 7 Basic Package

This chapter describes the STEP 7 basic package, Version 4.0. While the first chapter presented an overview of the characteristics of the programmable controller, this chapter tells you how to set these characteristics.

The basic package contains the statement list (STL), ladder logic (LAD) and function block diagram (FBD) programming languages. In addition to the basic package, option packages such as S7-SCL (Structured Control Language), S7-GRAPH (sequence planning) and S7-HiGraph (state-transition diagram) are also available.

2.1.1 Installation

STEP 7 V4.0 is a 32-bit application requiring Microsoft Windows 95 or Microsoft Windows NT as the operating system. To work with the STEP 7 software under Windows 95, you require a programming device (PG) or a PC with an 80486 processor or higher and at least 16 MB or RAM, with 32 MB recommended. For Windows NT, you require a Pentium processor and at least 32 MB of RAM; you must have administration authorization to install STEP 7 under Windows NT. A swap-out file is also needed. This file is approximately 128 MB minus main memory; for example, if the main memory configuration is 16 MB, the swap-out file would comprise about 112 MB. You should reserve around 50 MB for your user data, and between 10 and 20 MB for each additional option package. The memory requirements may increase for certain operations, such as copying a project. If there is insufficient space for the swap-out file, errors such as program crashes may occur.

Windows 95/NT's SETUP program, which is on the first diskette, is used for installation. On the PG STEP 7 is already factory-installed.

An MPI interface is needed for the on-line connection. The programming devices have the multipoint interface already built in, but PCs must be retrofitted with an MPI module. If you want to use PC memory cards, you will need a prommer.

STEP 7 V4 has multi-user capability, that is, a project that is stored, say, on a central server can be edited simultaneously from several workstations. You make the necessary settings in the Windows Control Panel with the "SIMATIC Workstation" program. In the dialog box that appears, you can parameterize the workstation as a single-user system or a mult-user system with the protocols used.

2.1.2 Authorization

An authorization (right of use) is required to operate STEP 7. The authorization is provided on diskette. Before using STEP 7, you must copy this authorization from the diskette to the hard disk. You may also transfer the authorization to another device by copying it back to the (original) authorization diskette, then transferring it to the new device.

Should you lose your authorization for some reason, such as a hard disk defect, you can use the emergency license, which is also on the authorization diskette and is good for a limited time only, until you are able to obtain a replacement authorization.

2.1.3 SIMATIC Manager

The SIMATIC Manager is the main tool in STEP 7; you will find its icon in Windows.

The SIMATIC Manager is started by double-clicking on its icon. Programming begins with opening or creating a "project". The projects COM_ SFB, S7_MIX and S7_ZEBRA are provided as learning material. When you open project S7_ZEBRA with FILE → OPEN, you will see the split project window: at the left the structure of the open object (the object hierarchy), at the right the selected object (Figure 2.1). Clicking on the box containing a plus sign in the left window displays additional levels of the structure; selecting an object in the left half of the window displays its contents in the right half of the window.

Under the SIMATIC Manager, you work with the objects in the STEP 7 world. These "logical" objects correspond to "real" objects in your plant. A project contains the entire plant, a station corresponds to a programmable controller. A project may contain several stations connected to one another, for example, via an MPI subnet. A station contains a CPU, and the CPU contains a program, in our case an S7 program. This program, in turn, is a "container" for other objects, such as the object Blocks, which contains, among other things, the compiled blocks.

The STEP 7 objects are connected to one another via a tree structure. Figure 2.2 shows the most important parts of the tree structure (the "main branch", as it were) when you are working with the STEP 7 basic package for S7 applications in off-line view. The objects shown in bold type are containers for other objects. All objects in the Figure are available to you in the off-line view. These are the objects that are on the programming device's hard disk. If your programming device is on-line on a CPU (normally a PLC target system), you can switch to the on-line view by selecting VIEW → ONLINE. This option displays yet another project window containing the objects on the destination device; the objects shown in italics in the Figure are then no longer included. The title bar of the active project window shows you whether you are currently working on-line or off-line.

Select OPTIONS → CUSTOMIZE to change the SIMATIC Manager's basic settings, such as the session language, the archive program and the storage location for projects and libraries, and configuring the archive program.

The following applies for the general editing of objects:

To *select an object* means to click on it once with the mouse so that it is highlighted (this is possible in both halves of the project window).

To *name an object* means to click on the name of the selected object (a frame will appear around the name and you can change the name in the window) or select the menu item EDIT → OBJECT PROPERTIES and change the name in the dialog box.

To *open an object*, double-click on that object. If the object is a container for other objects, the SIMATIC Manager displays the contents of the object in the right half of the window. If the object is on the lowest hierarchical level, the SIMATIC Manager starts the appropriate tool for editing the object (for instance, double-clicking on a block starts the editor, allowing the block to be edited).

In this book, the menu items in the standard menu bar at the top of the window are described

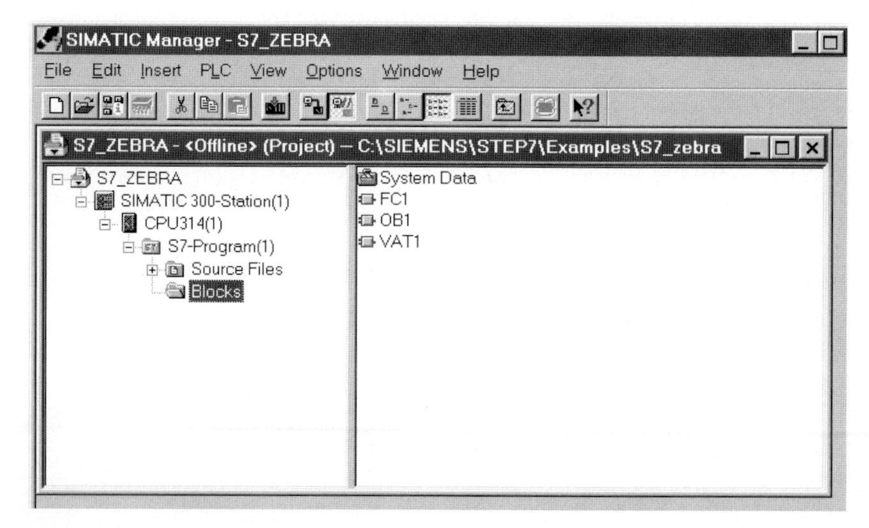

Figure 2.1
SIMATIC Manager Example

as *operator sequences*. Programmers experienced in the use of the operator interface use the icons from the toolbar. The use of the *right mouse button* is very effective. Clicking on an object once with the right mouse button screens a menu showing the current editing options.

2.1.4 Projects and Libraries

In STEP 7, the "main objects" at the top of the object hierarchy are projects and libraries.

Projects are used for the systematic storing of data and programs needed for solving an automation task. Essentially, these are

- ▷ the hardware configuration data,
- ▷ the parameterization data for the modules,
- ▷ the configuring data for communication via networks,
- ▷ the programs (code and data, symbols, sources).

The objects in a project are arranged hierarchically. The opening of a project is the first step in editing all (subordinate) objects which that object contains. The following sections discuss how to edit these objects.

Libraries are used for storing reusable program components. Libraries are organized hierarchically. They may contain STEP 7 programs which in turn may contain a user program (a container for compiled blocks), a container for source programs, and a symbol table. With the exception of on-line connections (no testing

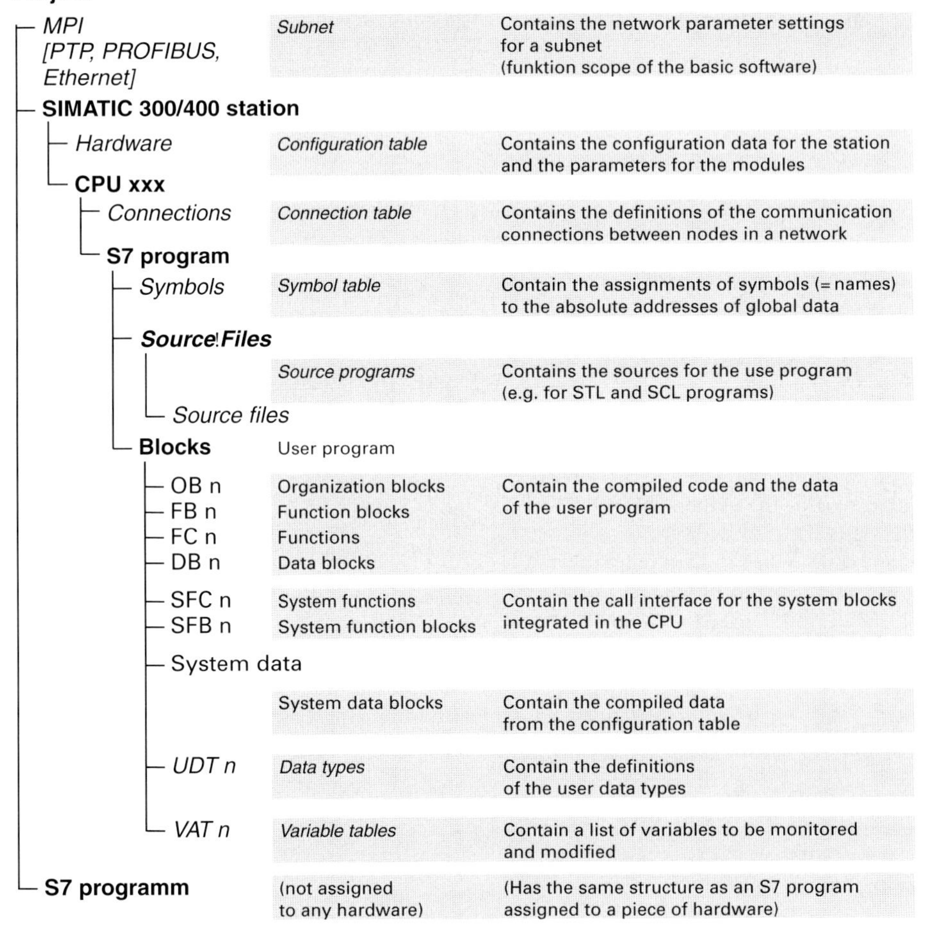

Figure 2.2 Object Hierarchy in a STEP 7 Project

possible), the creation of a program or program or program section in a library provides the same functionality as in an object.

As supplied, STEP 7 V4 provides the *StdLib30* library containing the following programs:

▷ Built In
Contains the call interfaces of the system blocks for off-line programming integrated in the CPU

▷ FB Lib 1
Contains loadable functions for the S5/S7 converter (replacement of S5 standard function blocks in conjunction with program conversion)

▷ FB Lib 2
Contains additional loadable functions and function blocks

▷ IEC
Contains loadable functions for editing variables with complex data types (STRING, DATE_AND_TIME)

▷ Net DP
Contains loadable functions for controlling CP modules

▷ PID Control
Contains loadable function blocks for closed-loop control

▷ Std OBs
Contains the templates for the organization blocks (essentially the variable declaration for the start information)

You will find an overview of the contents of these libraries in Chapter 25, "Block Libraries". Should you, for example, purchase an S7 module with standard blocks, the associated installation program installs the standard blocks as a library on the hard disk. You can then copy these blocks from the library to your project. A library is opened with FILE → OPEN, and can then be edited in the same way as a project. You can also create your own libraries.

The menu item FILE → NEW... generates a new object at the top of the object hierarchy (project, library). The location in the directory structure where the SIMATIC Manager is to create a project or library must be specified under the menu item OPTIONS → CUSTOMIZE or in the New dialog box.

When you create a project, you can select the project type "Project 2.x". You can also edit projects of this type with STEP 7 V2 (see also Section 2.2.3 "Project Versions").

The INSERT menu is used to add new objects to existing ones (such as adding a new block to a program). Before doing so, however, you must first select the object container in which you want to insert the new object from the left half of the SIMATIC Manager window.

2.1.5 On-line Help

The SIMATIC Manager's on-line help provides information you need during your programming session without the need to refer to hardcopy manuals. You can select the topics you need information on by selecting the HELP menu. The on-line help option GETTING STARTED, for instance, provides a brief summary on how to use the SIMATIC Manager.

The on-line help also provides context-dependent help, that is to say, you can call up information about an object selected with your mouse or the current error message when you press F1. In the toolbar you will see a button with an arrow and a question mark. When you click on this button, the mouse pointer is also given a question mark. With this "help" mouse pointer, you can now click on an object on the screen, such as an icon or menu item, and you will receive the on-line help associated with that object.

2.2 Editing Projects

When you set up a project, you create "containers" for the resulting data, then you generate the data and fill these containers. Normally, you create a project with the relevant hardware, configure the hardware, or at least the CPU, and receive in return containers for the user program. However, you can also put an S7 program directly into the project container without involving any hardware at all. Note that initializing of the modules (address modifications, CPU settings, configuring connections) is possible with the Hardware Configuration only.

We strongly recommend that the entire project editing process be carried out using the SIMATIC Manager. The creating, copying or deleting of directories or files as well as the changing of names (!) with the Windows Explorer within the structure of a project can cause problems with the SIMATIC Manager.

2.2.1 Creating Projects

From STEP 7 V3.2, the *STEP 7 Assistant* helps you in creating a new project. You specify the CPU used and the assistant creates for you a project with an S7 station and the selected CPU as well as an S7 program container, a source container and a block container with the selected organization blocks. You will find general information on operator entries for object editing in Section 2.1.3, "SIMATIC Manager".

- Creating a new project
Select FILE → NEW, enter a name in the dialog box, change the project type and storage location if necessary, and confirm with "OK" or RETURN.

- Inserting a new station in the project
Select the project and insert a station with INSERT → STATION → SIMATIC 300 STATION (in this case an S7-300).

- Configuring a station
Click on the plus box next to the project in the left half of the project window and select the station; the SIMATIC Manager displays the Hardware object in the right half of the window. Double-clicking on *Hardware* starts the Hardware Configuration, with which you edit the configuration tables. If the module catalog is not on the screen, screen it with VIEW → CATALOG.

You begin configuring by selecting the rail with the mouse, for instance under "SIMATIC 300" and "RACK 300", "holding" it, dragging it to the free portion in the upper half of the station window, and "letting it go" (drag & drop). You then see a table representing the slots on the rail.

Next, select the required modules from the module catalog and, using the procedure described above, drag and drop them in the appropriate slots. To enable further editing of the project structure, a station requires at least one CPU, for instance the CPU 314 in slot 2. You can add all other modules later. Editing of the hardware configuration is discussed in detail in Section 2.3, "Configuring Stations".

Store and compile the station, then close and return to the SIMATIC Manager. In addition to the hardware configuration, the open station now also shows the CPU.

When it configures the CPU, the SIMATIC Manager also creates an S7 program with all objects. The project structure is now complete.

- Viewing the contents of the S7 program
Open the CPU; in the right half of the project window you will see the symbols for the S7 program and for the connection table.

Open the S7 program; the SIMATIC Manager displays the symbols for the compiled user program (*Blocks*), the container for the source programs, and the symbol table in the right half of the window.

Open the user program (*blocks*); the SIMATIC Manager displays the symbols for the compiled configuration data (*System data*) and an empty organization block for the main program (OB 1) in the right half of the window.

- Editing user program objects
We have now arrived at the lowest level of the object hierarchy. The first time OB 1 is opened, the window with the object properties is displayed and the editor needed to edit the program in the organization block is opened. You add another empty block for incremental editing by opening INSERT → S7 BLOCK →... (*Blocks* must be highlighted) and selecting the required block type from the list provided.

When opened, the *System data* object shows a list of available system data blocks. You receive the compiled configuration data. These system data blocks are edited via the *Hardware* object in container *Station*. You can transfer *System data* to the CPU with PLC → DOWNLOAD and initialize the CPU.

The object container *Source Files* is empty. It is not required for writing LAD programs, and may be deleted.

- Module-independent S7 programs
If you wish, you can create a program without having first configured a station. To do so, you must generate the container for your program yourself. Mark the project and generate an S7 program with INSERT → PROGRAM → S7 PROGRAM. Under this S7 program, the SIMATIC Manager creates the object containers *Sources* and *Blocks*. *Blocks* contains an empty OB 1. You can now enter your program.

A module-independent program lies directly under the project. If you want to test this program on-line, generate a new window to display the on-line view of the project with VIEW → ONLINE while the project (window) is open. If your programming device has a connection to only one CPU, all blocks available in the CPU are displayed in the on-line object *Blocks*. If more than one node is possible, select the on-line S7 program, then menu item EDIT → OBJECT PROPERTIES, and set the "geographical address" (rack and slot) of the CPU in the *Address* tab.

• S7 program in a library

You can also create a program under a library, for instance if you want to use it more than once. In this way, the standard program is always available and you can copy it entire or in part into your current program. Please note that you cannot establish on-line connections in a library, which means that you can test a program only within a project.

You cannot copy complete libraries. It is, however, possible to transfer the S7 programs and blocks in a library to other libraries or projects, even from version 2 libraries to version 3 libraries and projects (see also Section 2.2.3, "Project Versions").

2.2.2 Rearranging, Managing and Archiving

With FILE → REARRANGE... you can reduce the memory requirements for a project or library. When it executes this menu item, the SIMATIC Manager eliminates the gaps created by deletions.

The SIMATIC Manager manages projects and libraries in project lists and library lists. When you execute FILE → MANAGE, the SIMATIC Manager shows you a list of all known projects, with name and path. You can then delete projects you no longer want to display from the list ("conceal") or include new projects in the project list ("show"). You manage libraries in the same way.

You can also archive a project or library (FILE → ARCHIVE...). In this case, the SIMATIC Manager stores the selected object in an archive file in compressed form. In order to archive a project or library, you need an archive program.

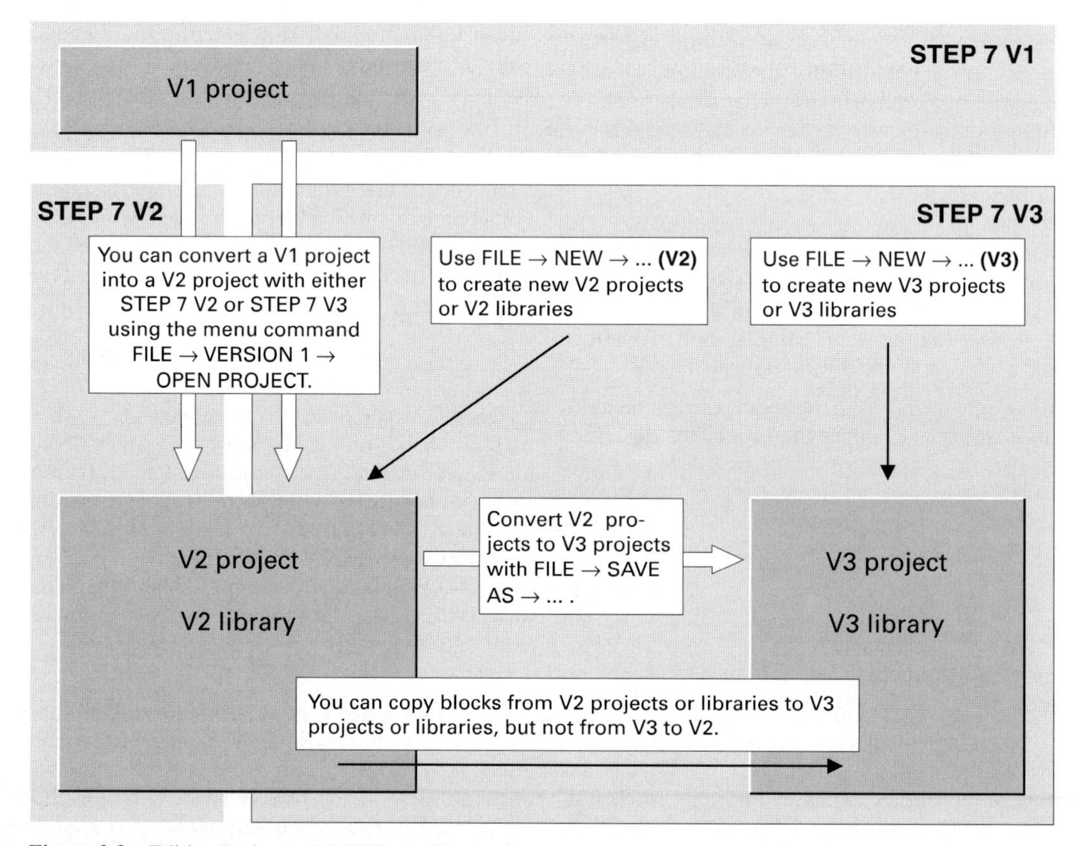

Figure 2.3 Editing Projects with Different Versions

STEP 7 contains the archive program ARJ, but you may also use other archive programs (*winzip* from version 6.0, *pkzip* from version 2.04g, *arj* from version 2.4.1a, or *lha* from version 2.13). You may create the archive file on hard disk or diskette(s). You can choose between archiving all files or only those which have been modified (reset the "archive bit" and select "incremental archiving"). You make the appropriate settings for archiving with OPTIONS → CUSTOMIZE on the "Archive" tab.

You can decompress an archived object for editing with FILE → RETRIEVE.

2.2.3 Project Versions

Since STEP 7 V4 has become available, there are three different versions of SIMATIC projects. STEP 7 V1 creates version 1 projects, STEP 7 V2 version 2 projects, and STEP 7 V3/V4 can be used to create and edit both version 2 and version 3 projects.

If you have a version 1 project, you can convert it into a version 2 project with FILE → OPEN VERSION 1 PROJECT... The project structure with the programs, the compiled version 1 blocks, the STL source programs, the symbol table and the hardware configuration remain unchanged.

Version 2 projects can be created and edited with STEP 7 V3 (Figure 2.3). However, version 2 products still have the function scope of STEP 7 V2; option packages available only in STEP 7 V3 or STEP 7 V4 cannot be used on these projects. The creation of a version 2 project serves a practical purpose when you want to edit it and run it as a V2 project. If you want to convert a version 2 project into a version 3 project, select the menu item FILE → SAVE AS ... and then, in the window that appears, select the option "With Consistency Check". You then define the folder (directory), the project name and the project type ("Project" stands for Version 3 project). Version 3 projects can no longer be saved as Version 2 projects.

Program development with STEP 7 is upwards-compatible, so that you can transfer blocks from version 2 projects and libraries to version 3 projects and libraries. However, the reverse, that is, the transfer of blocks from version 3 objects to version 2 objects, is not possible. For this reason, STEP 7 V4 includes the version 2 library *stdlibs (V2)* in order to make it possible to copy standard blocks from this library to version 2 projects.

2.3 Configuring Stations

You plan and define the hardware complement of your programmable controller using the Hardware Configuration tool. Configuring is done off-line without any connection to the CPU. This same tool allows you to address and initialize the modules

You begin configuring your hardware by double-clicking on the *Hardware* object in container *SIMATIC 300/400-Station*. You must make the basic hardware configuration settings with OPTIONS → CUSTOMIZE.

When configuring has been completed, STATION → CONSISTENCY CHECK will show you whether your entries were free of errors. STATION → SAVE stores the configuration tables with all parameter assignment data in your project on the hard disk.

STATION → SAVE AND COMPILE not only saves but also compiles the configuration tables and stores the compiled data in the *System data* object in the off-line container *Blocks*. After compiling, you can transfer the configuration data to a CPU with PLC → DOWNLOAD. The object *System data* in on-line container *Blocks* represents the current configuration data on the CPU. You can "return" these data to the hard disk with PLC → UPLOAD.

Configuration table

The Hardware Configuration tool works with tables, each of which represents a rack, a module or a DP station. The configuration tables show the slots and the modules plugged into them, or the properties of the module, such as the addresses and the order number. You can generate the configuration tables in the planning phase or wait until the hardware has actually been installed.

2.3.1 Arranging Modules

After it has been opened, the Hardware Configuration displays the station window and module catalog. You can screen and blank out the module catalog with VIEW → CATALOG. You can enlarge or maximize the station window for easier editing. In the upper half, it shows the module racks in the form of tables and the DP sta-

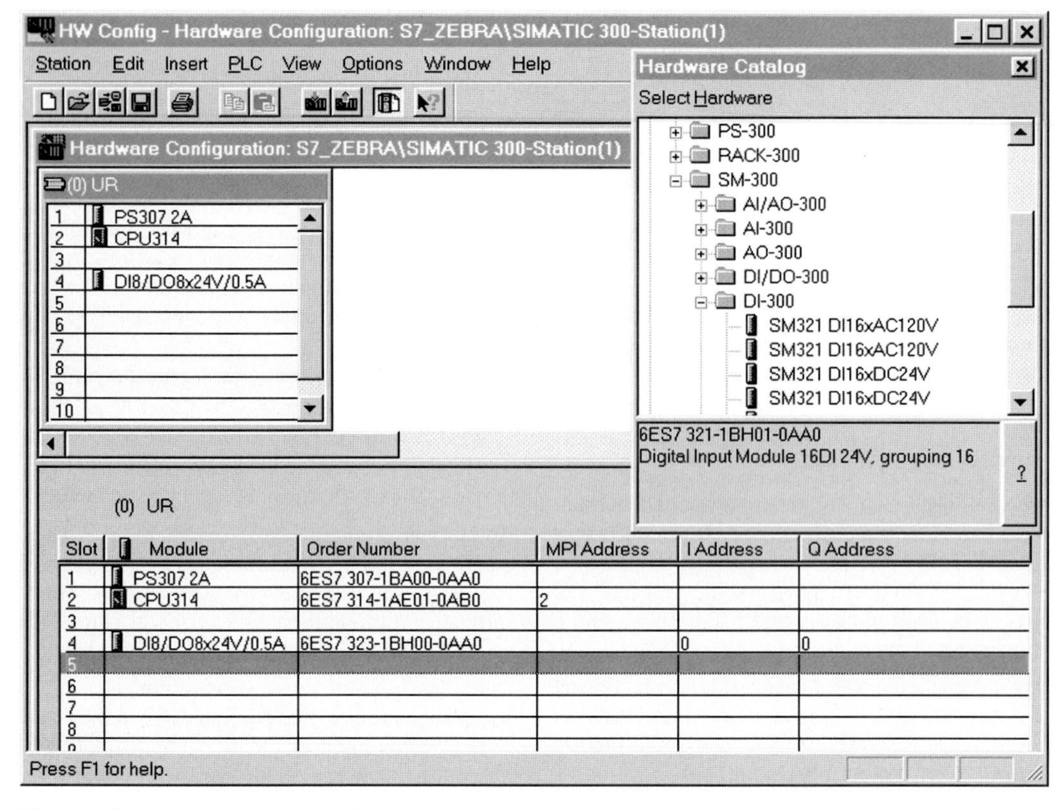

Figure 2.4 Sample Hardware Configuration

tions in the form symbols. If there are several racks, you will see here the connection between the interface modules and if PROFIBUS is used, you will see the structure of the DP master system. The lower half of the station window shows the detailed view of the rack or DP slave marked in the upper half.

You begin configuring by selecting and "holding" the rail from the module catalog, for instance under "SIMATIC 300" or "RACK 300", with the mouse, dragging it to the upper half of the station window, and dropping it anywhere in that window. An empty configuration table is screened for the controller rack. Next, select the required modules from the module catalog and, in the manner described above, drag and drop them in the appropriate slots. A No Parking symbol tells you cannot drop the selected module at the intended slot.

You can generate the configuration table for another rack by dragging the selected rack from the catalog and dropping it in the station window. In S7-400 systems, a non-interconnected rack (or more precisely: the relevant receive interface module) is assigned an interface via the "Link" tab in the Properties window of a Send IM (select module and EDIT → OBJECT PROPERTIES).

The arrangement of distributed I/O stations is described in Section 2.4.2 "Configuring DP Communication".

2.3.2 Addressing Modules

When arranging modules, the Hardware Configuration tool automatically assigns a module start address. You can view this address in the lower half of the station window in the object properties for the relevant modules. In the case of the S7-400 and the CPU 315-2DP, you can change the module addresses. When doing so, please observe the addressing rules for S7-300 and S7-400 systems as well as the addressing capacity of the individual modules.

There are modules which have both inputs and outputs for which you can (theoretically) reserve different start addresses. However, please note carefully the special information provided in the product manuals; the large majority of function and communications modules require the same start address for inputs and outputs.

Modules on the MPI bus or communications bus have an MPI address. You may also change this address. Note, however, that the new MPI address goes into force as soon as the configuration data are transferred to the CPU.

2.3.3 Parameterizing Modules

When you parameterize a module, you define its properties. It is necessary to parameterize a module only when you want to change the default parameters. A requirement for parameterization is that the module is located in a configuration table.

Double-click on the module in the configuration table or select the module and select EDIT → OBJECT PROPERTIES. Several tabs with the specifiable parameters for this module are displayed in the dialog box. When you use this method to parameterize a CPU, you are specifying the run characteristics of your user program.

Some modules allow you to set their parameters at runtime via the user program with system functions SFC 55 WR_PARM, SFC 56 WR_DPARM and SFC 57 PARM_MOD.

2.4 Configuring Communication Connections

A subnet is a homogeneous network of modules with communications capabilities which has no gateway. With STEP 7, you can insert the objects for MPI, Industrial Ethernet, PROFIBUS and PTP subnets in one project.

The SIMATIC Manager generates the object for an MPI subnet automatically when the project is created. You insert the objects for the other subnets by marking the project and selecting INSERT → SUBNET.

2.4.1 Networking Modules with MPI

You specify the nodes for the MPI subnet when you set the module properties. Select a CPU in the configuration table and open it with EDIT → OBJECT PROPERTIES. A dialog box is then displayed. The "General" tab shown in this dialog box contains the "MPI ..." button; at the right of the button is the MPI address. When you click on this button, another dialog box is displayed. You must then select the "Node is connected to the selected network" option from the "Network connection" tab.

Take this opportunity to also set the MPI address you have chosen for this CPU. Note that FM or CP modules with MPI interface that are located in an S7-300 station are automatically assigned an address derived from that CPU's MPI address. The FM or CP module whose slot is closest to that of the CPU is automatically assigned the MPI address which is higher by one than that of the CPU, the next module is assigned the next higher address, and so on. For example, if you set the CPU's MPI address to 5, the nearest FM or CP module is assigned MPI address 6, the next the MPI address 7, and so on. You must take this automatic address allocation into account when you choose the MPI addresses for the CPUs.

The highest MPI address must be higher than or equal to the highest assigned MPI address in the subnet (remember to take automatic address allocation for FM and CP modules into account!). It must have the same value for all nodes in the subnet.

A suggestion: Give the CPUs in the various stations different names ("IDs"). The default name for all stations is "CPUxxx(1)", so that they can be distinguished from one another in the subnet by their MPI address only. If you do not want to assign a name of your own choosing, you can change the default from "CPUxxx(1)" to "CPUxxx(n)", where "n" is the MPI address.

When you assign the MPI address, you should also consider the possibility that, at some later date, you might want to connect a PG or OP to the MPI network for the purpose of service or maintenance. Permanently installed PGs or OPs should be connected directly to the MPI network, an MPI connector with PG socket is available for devices which can be plugged in via a spur. A tip: Reserve address 0 for a service PG, 1 for a service OP, and 2 for a replacement CPU (this corresponds to the preset addresses).

2.4.2 Configuring DP Communication

Distributed I/O consists of stations interfaced to one another via the PROFIBUS-DP bus system.

Table 2.1 Example of a GD Table

GD Identifier	Station1\ CPU314(1)	Station2\ CPU315(2)	Station3\ CPU413(3)	Station4\ CPU416(4)	Station5\ CPU414(5)
GD	>MD16	MD116	MD126	MD126	
GD		>DB20.DBW14:8			MW124:8
GD		DB21.DBW28		>T28	
GD	DB18.DBW20:4		>C10:4		
GD		>ID24		MD24	MD24

A "DP master" connects the PROFIBUS subnet with the CPU's P bus. The stations ("nodes") on the bus are distinguished from one another by the PROFIBUS address ("node number"). A station can have the same characteristics as a module (a "compact" DP slave, such as an ET 200B) or consist of several modules (a "modular" DP slave, such as an ET200M with up to eight S7-300 modules).

Configuring must begin with the DP master module, which you select from the module catalog and place in the configuration table. It is possible that you have already selected a CPU with DP interface. The next line down shows the DP master with one connection to a DP master system in the station window (broken black-and-white line). Using your mouse, drag the stations you select from the module catalog under "PROFIBUS-DP" to this master system symbol. As you drag and drop a station, that station's properties are displayed, and you can set the PROFIBUS address for that station. The DP slave then appears as a symbol in the upper half of the station window and a configuration table for this station appears in the lower half of the window. If you selected a modular station, you can now locate additional modules in the station. You can set the addresses via the object properties of the module.

You can "reinstall" DP slaves that are not in the module catalog. To do so, you need the type file (GSD file, device master data file) for that slave, which you must copy to directory ... \STEP7\S7DATA\GSD. Use OPTIONS → UPDATE GSD FILE to acquaint STEP 7 with the new type file. The new DP slave will be displayed in the module catalog under *PROFIBUS-DP* in container *Additional field devices* following a complete restart of the Hardware Configuration.

Essentially, the distributed I/O modules can be addressed in the same way as the centralized I/O (modules in the racks). All modules reserve addresses in the CPU's logical address area (I/O area P). The address assignments (user data assignments) for "centralized" and "distributed" modules must not overlap. The addressing capacity depends on the DP master. You may change the STEP 7 default PROFIBUS address for the DP slave as well as the default input and output addresses.

The "distributed" modules are initialized in the same manner as the "centralized" modules. Double-clicking on the symbol in the upper half of the station window opens a dialog box with one or more tabs in which you set the required module properties. STEP 7 assigns a *diagnostic address,* which you need in order to check a station's diagnostic data. Distributed modules with the appropriate capability can also be initialized at runtime via system functions. Some DP masters enable event-driven synchronization of DP slaves with the commands SYNC and FREEZE.

2.4.3 Configuring GD Communication

Global data communication (GD communication) is used for the exchange of small volumes of data between CPUs in an MPI subnet. The prerequisite for the configuring of GD communication is an established subnet (see Section 2.4.1, "Networking Modules with MPI"). In GD communication, the data are normally exchanged cyclically, S7-400 systems also enable additional or exclusive event-driven updating of GD data.

GD communication is configured by making entries in a table. After selecting the MPI subnet in the SIMATIC Manager, select the menu command OPTIONS → DEFINE GLOBAL DATA. An empty GD table is displayed. Select a column, then select EDIT → ASSIGN CPU. A project window is opened. Select the station in the left half and the CPU in the right half of this window. "OK" places the CPU in the GD table. Repeat this procedure for each CPU to be configured for GD communication. A GD table may comprise as many as 15 CPU columns.

To configure data interchange between CPUs, select the first location under the CPU that is to be the sender and enter the address whose value is to be transferred (confirm with RETURN). With EDIT → SENDER, define this value as send value, discernible because of ">" and highlighting. In the same line, enter the operand that is to receive the send value (this is preceded by "receiver"). Timers and counters may be used as sender only, the receiver must be a word operand for each timer or counter. A line may contain several receivers but only one sender (Table 2.1).

After completing your entries, compile the GD table with GD TABLE → COMPILE. Following error-free compilation, STEP 7 fills in the "GD Identifier" (for example, GD 2.1.3 would mean GD circle 2, GD packet 1, GD element 3).

After compiling, you can enter the addresses for the communication status in the table with VIEW → GD STATUS. The group status (GST) shows the status of all communication links in the table. The status (GDS) shows the status of one communication link (the status of a GD send packet). In both cases, status information is contained in a doubleword.

The menu command VIEW → SCAN RATES allows you to specify the scan rates for each GD packet and each CPU yourself. The scan rate (SR) specifies after how many CPU cycles the global data are to be updated. The default scan rate is set so that, in the case of a "clean" CPU (a CPU that has no user program), the GD packets will be sent and received approximately every ten milliseconds.

Note that the lower the scan rates, the greater a CPU's "communication load". To keep the communication load within a manageable range, set the scan rate in the sending CPU so that the product of scan rate and cycle time is greater than 60 ms for the S7-300 and greater than 10 ms for the S7-400. The product of these two values must be lower in the receiving CPU than in the sending CPU to ensure that no GD packets will be lost. On S7-400 CPUs, you must disable cyclic data interchange by setting the scan factor to zero when you want to use SFCs and event-controlled GD communication only.

Following a recompilation, use menu command PLC → DOWNLOAD to transfer the GD table to the CPUs, where it immediately comes into force.

2.4.4 Configuring SFB Communication

You use special system blocks for communications between nodes in a network. You can choose between the following:

- *Unconfigured connection* between SIMATIC S7 devices where the connection to the partner is built up at runtime if required.
- *Configured connection* where the connection properties between any two partners is fixed in a connection table.

The system blocks for these communications connections are introduced in Chapter 20 "Main Program".

A configured connection requires special system function blocks SFBs that you call in the CPU program in each case at the locations where you want to send or receive data. You can also control the partner device with these blocks, for example, set it to STOP or initiate a complete restart. The requirement for SFB communication is that a subnet is configured (MPI, PROFIBUS or Ethernet).

Table 2.2 Sample Connection Table

Local ID (HEX)	Partner ID (HEX)	Partner	Type	Active Connection Buildup	Send Status Messages
1	1	Station4 / CPU416(4)	S7 Connection	Yes	No
2		Station2 / CPU315(2)	S7 Connection	Yes	No
3	1	Station5 / CPU414(5)	S7 Connection	Yes	No

Connection table

For communication via configured connections, you require the connection table (Table 2.2). This table is represented by the *Connections* object in container *CPU*. To configure SFB communication, open *Connections*. A table is then screened, and you can make your entries. You must create a connection table for each "active" CPU.

In dialog boxes "Station" and "Module", select the CPU for which you want to configure the connection. You must first have set up the stations and modules with the Hardware Configuration. When you select the menu command INSERT → CONNECTION, you screen a dialog box in which you can define the partner CPU and the type of connection (protocol).

The number of possible connections is CPU-specific. STEP 7 establishes a connection ID for each connection and for each partner. You need this ID whenever you use the communication blocks in your program.

You specify the connection setup in the dialog box screened when you select EDIT → OBJECT PROPERTIES with the connection line marked. You can change the connection partner with EDIT → PARTNER. To activate the connections, you must load the connection table, following compilation and saving, into the PLC (all connection tables into all "active" CPUs).

2.4.5 Network Configuration

NETPRO can be used to configure subnets graphically in STEP 7 V4. You reach NETPRO by opening a subnet in the project window of the SIMATIC Manager.

The window for network configuration displays all previously created subnets and stations (nodes) in the project with the configured connections. A second window shows the network object catalog with a selection of the available SIMATIC stations, subnets and DP slaves. You can toggle the catalog on and off with VIEW → CATALOG. You can adapt and improve the readability of the representation using VIEW → ZOOM IN, VIEW → ZOOM OUT and VIEW → ZOOM FACTOR. If you have already configured DP slaves, these are displayed with VIEW → DP SLAVES.

You start the network configuration by selecting a subnet that you mark in the catalog and then hold and drag to the network window using the mouse. You proceed in the same way with the desired stations, initially without a connection to the subnet. Double-click on a station to open the Hardware Configuration so that you can configure the station or at least the module with network connection. Save the station and return to the network configuration. The interface of this module is represented in the network configuration as a small box under the module view. Click on this box, hold it and drag it to the relevant subnet. Proceed as follows with all other nodes.

After creating the graphical view, parameterize the subnets: Mark a subnet and select EDIT → OBJECT PROPERTIES. In the subsequent dialog box, you can specify the configuration, the lines and the bus parameters in the "Network Settings" tab. When the network connection of a node is marked, you can define the network properties of the node with EDIT → OBJECT PROPERTIES. You define the module properties of the nodes in a similar way (the same operator input as for the Hardware Configuration).

You can highlight the node assignments of a master system graphically using VIEW → HIGHLIGHT → MASTER SYSTEM. If you mark the PROFIBUS connection and select EDIT → MASTER SYSTEM, you can configure SYNC/FREEZE groups, provided you have a suitable DP master.

You can also configure global data communications out of NETPRO: Mark the MPI subnet and select OPTIONS → DEFINE GLOBAL DATA. Editing of the connection table is also possible: Mark a CPU and select OPTIONS → CONFIGURE CONNECTIONS.

You can test a configuration for contradictions with NETWORK → CONSISTENCY CHECK. You conclude the network configuration with NETWORK → SAVE.

2.5 Developing Programs

An S7 program may be allocated in the project hierarchy of a module or generated on a module-independent basis. In the LAD programming language, you write programs incrementally, that is, input is checked immediately and compiled (translated) when stored. If you want to program with symbols, which is strongly recom-

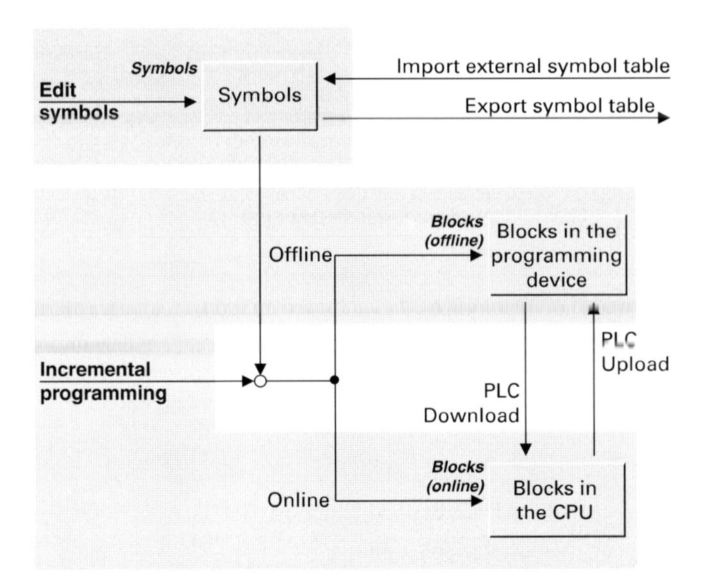

Figure 2.5
Writing Programs with the Incremental Editor

mended, you must first generate a symbol table, which you can update while you are programming.

Incremental programming is used both off-line and on-line for the compiled blocks in the *Blocks* container (Figure 2.5). Transfer these blocks to the CPU with PLC → DOWNLOAD and back to the hard disk with PLC → UPLOAD.

2.5.1 Absolute Addressing of Variables

Variables of elementary data type can be referenced by absolute addresses. These include 1-bit variables (data type BOOL), 8-bit variables (data type BYTE or CHAR), 16-bit variables (data type WORD, INT, S5TIME, DATE), and 32-bit variables (data type DWORD, DINT, REAL, TIME, TOD).

You will find more detailed information on these data types in Section 3.4, "Data Types".

Binary addresses have an address identifier, a byte number, and – separated by a period – a bit number. Numbering of the bytes begins at zero for each address area. The upper limit is CPU-specific. Bits are numbered from 0 to 7. Examples:

I 1.0	Input bit no. 0 in byte no. 1
Q 16.4	Output bit no. 4 in byte no. 16
M 56.7	Memory bit no. 7 in byte no. 56

Digital addresses can be byte address, word addresses or doubleword addresses. These addresses reference whole bytes.

The absolute address of a byte consists of the address identifier and the number of the byte containing the variable. The address identifier is supplemented by a B. Examples:

IB 2	Input byte no. 2
QB 18	Output byte no. 18
MB 58	Memory byte no. 58

Word operands comprise two bytes (a word). The absolute address of a word operand consists of the address identifier and the number of the low-order byte of the word containing the variable. The address identifier is supplemented by a W. Examples:

IW 4	Input word no. 4; contains bytes 4 and 5
QW 20	Output word no. 20; contains bytes 20 and 21
MW 60	Memory word no. 60; contains bytes 60 and 61

Doubleword addresses consist of four bytes (a doubleword). The absolute address of a doubleword comprises the operand identifier and the number of the low-order byte of the word containing the variable. The operand identifier is supplemented by a D. Examples:

ID 8	Input doubleword no. 8; contains bytes 8, 9, 10 and 11

Table 2.3 Sample Symbol Table

Symbol	Address	Data Type	Comment
Main program	OB 1	OB 1	Program for cyclic scanning
Receive mailbox	DB 10	DB 10	Data block for receiving frames
Start signal	I 1.0	BOOL	Start conveyor belt
+HS13-S104	I 1.1	BOOL	Limit switch upper bay 13
Quantity	MW 20	INT	Production per batch
Start-up time B3	T 1	TIMER	Start monitor for belt 3
Temp_Mi7	MD 24	REAL	Temperature in mixer 7

QD 24 Output doubleword no. 24; contains bytes 24, 25, 26 and 27
MD 62 Memory doubleword no. 62; contains bytes 62, 63, 64 and 65

A *data address* is complete in itself, including the relevant data block. Examples:

DB 10.DBX 2.0 Data bit 2.0 in data block DB 10
DB 11.DBB 14 Data byte 14 in data block DB 11
DB 20.DBW 20 Data word 20 in data block DB 20
DB 22.DBD 10 Data doubleword 10 in data block DB 22

You will find additional information on the addressing of data in Section 18.2.2, "Accessing Data Addresses".

2.5.2 Symbolic Addressing of Variables

Symbolic addressing uses a name (called a symbol) in place of an absolute address. You yourself choose this name. A distinction is made between upper and lower case. The use of keywords as symbols is not permitted.

The name, or symbol, must be allocated to an absolute address. A distinction is made between global symbols and symbols that are local to a block.

Global symbols

A global symbol is known throughout the user program, and has the same meaning in all blocks. You define global symbols in the *Symbols* table in container *S7 Program.* You may assign names to the following objects:

▷ Inputs I, outputs Q, peripheral inputs PI and peripheral outputs PQ

▷ Memories M, timers T and counters C

▷ Data blocks DB and code blocks OB, FB, FC, SFB and SFC

▷ User-defined data types UDT

▷ Variable tables VAT

Global symbols may not exceed 24 characters. A global symbol may also include spaces, special characters, and country-specific characters such as umlauts. Exceptions to this rule are the characters 00_{hex} and FF_{hex} and quotation marks ("). When using symbols containing special characters, you must put the symbols in quotation marks in the program. In compiled blocks, the editor always shows global symbols in quotation marks.

Data addresses in data blocks are considered to be local addresses; the associated symbols are defined in the declaration section of the data block in the case of global data blocks and in the declaration section of the function block in the case of instance data blocks.

Symbol table

When an S7 program is created, the SIMATIC Manager also generates an empty symbol table called *Symbols.* Simply open this table and you can define the global symbols and allocate them to absolute addresses (Table 2.3). Definition of a symbol also includes the data type, which defines specific properties of the data behind the symbol, most importantly the form in which the data is represented. The data type BOOL, for instance, identifies a binary variable, and the data type INT a digital variable whole contents represent a 16-bit integer number.

Symbols used for incremental program input must already be allocated to absolute addresses. You can open the symbol table and enter new symbols or correct existing symbols during input with the incremental editor by selecting OPTIONS → SYMBOL TABLE. You can then resume program input with the new or modified symbol.

Symbol tables can be imported and exported. "Exported" means that a file will be created with the contents of your symbol table. You may choose between a pure ASCII file (file ends in *.asc), a sequential allocation list (*.seq), System Data Format (*.sdf for Microsoft Access), and Data Interchange Format (*.dif for Microsoft Excel). You can edit the exported file with the suitable editor. You can also import symbol tables in these formats.

Local symbols

The names for local data are defined in the declaration section of the relevant block. Only letters, digits and the underscore character are permitted. A local symbol begins with a letter and may not exceed 24 characters.

Local symbols are valid only within a block. The same symbol (same variable name) may be used in a different context in another block. The editor displays local symbols preceded by a "#" sign. If the editor cannot distinguish between a local symbol and an address, you must also type a leading "#" when you input the symbol.

Local symbols are available only in the PG database (in the off-line *Blocks* container). If this information is missing following recompilation, the editor uses a replacement symbol.

Not only simple variables can be addressed symbolically, but also data (see Section 18.2.2, "Accessing Data Operands") and components of arrays and structures (see Section 3.4.3, "Complex Data Types").

2.5.3 Programming

Editor

STEP 7 provides you with an editor for the FBD, LAD and STL programming languages. You enter the editor's environment when you open a block in the SIMATIC Manager, for instance by double-clicking on the automatically generated symbol for organization block OB 1 or via the Windows taskbar with START → SIMATIC → STEP 7 → PROGRAM S7 BLOCKS LAD, STL, FBD.

You can adapt the editor's features to suit your requirements with OPTIONS → CUSTOMIZE. Select the font for the text and the programming language (STL, LAD, FBD) from the "Editor" tab. When you generate a new block with the editor, the language in which you generate it will be taken over as the programming language. With the selection of symbols and comments you stipulate the defaults with which a block will be opened.

In the "LAD/FBD" tab you define the print layout and the representation mode for program elements, the size of the address fields (12 in this book), and the colors and line types for element selection and status.

Generating blocks

You begin block programming by opening a block (double-clicking on the block in the project window or selecting FILE → OPEN in the editor). If there is no such block, you can generate one as follows;

- In the SIMATIC Manager:
Select the *Blocks* object in the left half of the project window and generate a new block with INSERT → S7 BLOCK ... . You will see a dialog box with the block header (number of the block, programming language, block attributes). Select the "LAD" programming language. You can enter the remaining block attributes later.

- In the editor:
FILE → NEW displays a dialog box in which you stipulate which block is to be generated in which user program. Once the dialog box has been closed, you can program the block contents.

With the block open, program any subsequent additions to the block header in the editor with FILE → PROPERTIES. Change the programming language by changing the representation mode with VIEW → LAD/STL/FBD and saving the block.

Programming blocks

When a code block is opened, three windows are displayed: the program, the variable declaration, and the program element catalog. Fully visible is the program section with – depending on the editor defaults – the fields for the block title and the block comment, as well as the fields for the network title and network comment for

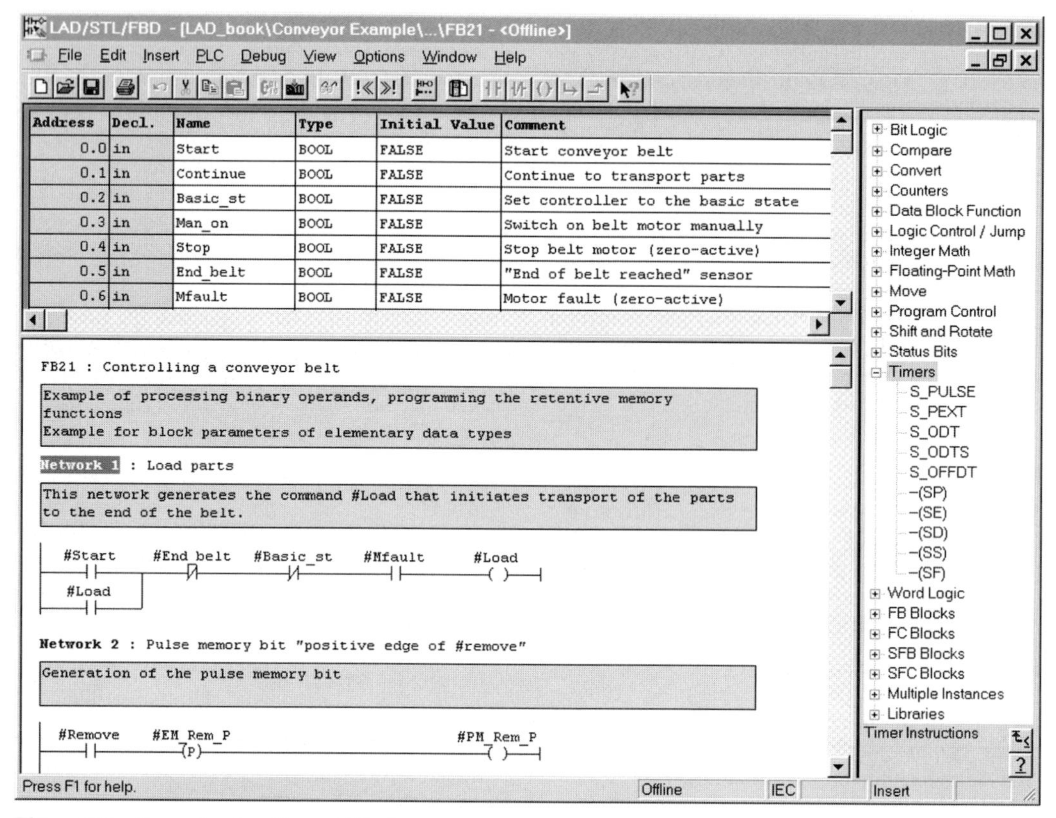

Figure 2.6 Example of an Opened Block

the first network and the program input field (Figure 2.6). In the program section of a code block, you control the display of comments and symbols with menu commands View → Comment, View → Symbolic Representation and View → Symbol Information. You can change the size with View → Zoom In, View → Zoom Out, and View → Zoom Factor. Program input is discussed in detail in Section 3.3, "Editing LAD Elements", and in Section 3.2.6, "Programming Data Blocks".

The table for the variable declaration is located in the window above the program section. If it is not visible, place the mouse pointer on the top boundary of the program section, click the left mouse button until the mouse pointer changes its form, and drag it down. You will see the variable declaration table in which you are to define the local variables (for details, please refer to Section 3.2.4, "Variable Declaration").

The program element catalog is at the right next to the program section, and is visible only as a narrow strip. Enlarge the cutout by placing the mouse pointer on the right margin of the program section, pressing the left mouse button when the mouse pointer changes it form, and dragging toward the left. If the program element catalog is not visible, screen it with View → Catalog. The use of the catalog is described in detail in section 3.3, "Editing LAD Elements".

From the editor you can create new blocks or open and edit existing ones without returning to the SIMATIC Manager.

Size of user memory

The memory requirements for a block are part of the block properties (select the block from the SIMATIC Manager, then select Edit → Object Properties from the "General – Part 2" tab). You can find out how much memory your whole program needs by selecting the program (*Blocks* object) from the SIMATIC Manager and invoking the menu command Edit → Object Properties. The size of the program in load memory and in work memory, as well as the number of blocks of each type in the program, can be taken from the "Special" tab.

2.5.4 Reference Data

As a supplement to the program itself, the SIMATIC Manager shows you the reference data, which you can use as the basis for corrections or tests. These reference data include the following:

▷ Cross references

▷ Reserved locations

▷ Program structure

▷ Unused symbols

▷ Operands without symbols

To generate reference data, select the *Blocks* object and the menu command OPTIONS → REFERENCE DATA → DISPLAY. The representation of the reference data can be changed specifically for each work window with VIEW → FILTER...; you can save the settings for later editing by selecting the "Save as Standard" option. You can display and view several lists at the same time.

Cross references

The cross-reference list shows the use of the operands and blocks in the user program. It includes the absolute address, the symbol (if any), the block in which the address was used, how it was used (read or write), and language-related information. For LAD, the language-related information column shows the network in which the address was used. Click on a column header to sort the table by column contents.

The cross-reference list shows the addresses you selected with VIEW → FILTER... (for instance bit memory). When you double-click on an address, the editor opens the block displayed on that line at the location at which the address appears. STEP 7 then uses the filter saved as "Standard" every time it opens the cross-reference list.

A tip: The cross references show you whether the referenced addresses were also scanned or reset. They also show you in which blocks addresses are used (possibly more than once).

Assignments

The I/Q/M reference list shows which bits in address areas I, Q and M are assigned in the program. One byte, broken down into bits, appears on each line. Also shown is whether access is by byte, word, or doubleword. The T/C reference list shows the timers and counters used in the program. Ten timers or counters are displayed on a line.

A tip: The list shows you whether certain address areas were (improperly) assigned or where there are still addresses available.

Program structure

The program structure shows the call hierarchy of the blocks in a user program. The "starting block" for the call hierarchy is specified via filter settings. You have a choice between two different views:

The *tree structure* shows all nesting levels of the block calls. You control the display of nesting levels with the "+" and "–" boxes. The requirements for temporary local data are shown for the entire path following the starting block and/or per call path. Click the right mouse button to fade in a menu field in which you can open the block, switch to the call location, or screen additional block information.

The *parent/child structure* shows two call levels with one block call. Language-related information is also included.

A tip: Which blocks were used? Were all programmed blocks called? What are the blocks' temporary local data requirements? Is the specified local data requirement per priority class (per organization block) sufficient?

Unused symbols

This list shows all operands which have symbol table allocations but were not used in the program. The list shows the symbol, the operand, the data type, and the comment from the symbol table.

A tip: Were the operands in the list inadvertently forgotten when the program was being written? Or are they perhaps superfluous, and not really needed?

Addresses without symbols

This list shows all the operands used in the program to which no symbols were allocated. The list shows these operands and how often they were used.

A tip: Were operands used inadvertently (by accident, or because of a typing error)?

2.5.5 Rewiring

The Rewiring function allows you to replace addresses in individually compiled blocks or in the entire user program. For example, you can replace input bits I 0.0 to I 0.7 with input bits I 16.0 to I 16.7. You can replace addresses in the address areas for inputs, outputs, bit memory, timers, counters, FC functions and FB function blocks.

In the SIMATIC Manager, choose the objects in which you want to do rewiring select a single block, select a group of blocks by holding Ctrl/Strg and clicking with the mouse, or select the entire user program *(Blocks)*. Call the menu command OPTIONS → REWIRE... to screen a table in which you can enter the old and the new addresses. When you confirm with "OK", the SIMATIC Manager exchanges the addresses. A subsequently displayed info file shows you in which block changes were made, and how many.

2.6 Debugging Programs

After setting up a connection to a CPU and loading the user program, you can debug the whole program or only specific program sections, for instance individual blocks. To debug the program, you supply the variables with signals and values, using, for example, simulator modules, and evaluate the information returned by your program. If the CPU goes to stop as the result of an error, functions such as CPU Information help you locate the problem.

With the PLCSIM option package, you can simulate a CPU in the programming device, thus making it possible for you to test your program without additional hardware.

2.6.1 Connecting a PLC

The connection between the PG's and CPU's MPI interface is the mechanical requirement for an on-line connection. The connection is unique when a CPU is the only programmable module connected. If there are several CPUs in the MPI subnet, each CPU must be assigned a unique node number (MPI address). You set the MPI address when you initialize the CPU. Before linking all the CPUs to one network, connect the PG on one CPU at a time and transfer the *System Data* object from the off-line user program *Blocks* or direct with the Hardware Configuration editor using the menu command PLC → DOWNLOAD. This assigns a CPU its own special MPI address ("naming") along with the other properties. If all you want to do is change the CPU's MPI address, you can also use the *Accessible Nodes* button.

The MPI address of a CPU in the MPI network can be changed at any time by transferring a new parameter data record containing the new MPI address to the CPU. Note carefully: The new MPI address takes effect immediately. The MPI parameters are retained in the CPU even after a memory reset. The CPU can thus be addressed even after a memory reset.

A PG can always be operated on-line on a CPU, even with a module-independent program and even though no project has been set up.

- If no project has been set up, you must establish the connection to the CPU with the *Accessible Nodes* button. When you press this button, you screen a project window with the structure "*Accessible Nodes*" – "Module (MPI=n)" – "Online User Program *(Blocks)*". When you select the *Module* object, you may utilize the online functions, such as changing the operational status and checking the module status. Selecting the *Blocks* object displays the blocks in the CPU's user memory. You can then edit (modify, delete, insert) individual blocks.

- If the project window shows a module-independent program, generate the associated online project window. This window contains all on-line objects for this project. Selecting the S7 program in the on-line window makes all online functions for the connected CPU available to you. *Blocks* displays the blocks in the CPU's user memory. If the blocks in the off-line program and those in the on-line program are identical, you can edit the blocks in user memory with the information from the PG database (symbolic address, comments).

- When you switch a module-dependent program into on-line mode, you can carry out program modifications just as you would in a module-independent program. In addition, it is now possible for you to configure the SIMATIC station, that is, to set CPU parameters and address and parameterize modules.

2.6.2 CPU Information

In on-line mode, the CPU information listed below is available to you. The menu commands are screened when you have selected a module (in on-line mode and without a project) or S7 program (in the on-line project window).

- PLC → DIAGNOSE HARDWARE
Information on the status and operating mode of the accessible on-line modules in the form of the Hardware Configuration.

- PLC → MODULE INFORMATION
General information (such as version), diagnostic buffer, memory (current map of work memory and load memory, compression), cycle time (length of the last, longest, and shortest program cycle), timing system (properties of the CPU clock, clock Fynchronization, run-time meter), performance data (memory configuration, sizes of the address areas, number of available blocks, SFCs, and SFBs), communication (baud rate and communication links), stacks in STOP state (B stack, I stack, and L stack).

- PLC → OPERATING MODE
Display of the current operating mode (for instance RUN or STOP), modification of the operating mode

- PLC → CLEAR/RESET
Resetting of the CPU in STOP mode

- PLC → SET TIME AND DATE
Setting of the internal CPU clock

- PLC → CPU MESSAGES
The detection of problems in the connection to the CPU, asynchronous errors which set the CPU to STOP, and user-defined messages generated with SFC 52. You can display the messages or save them in an archive file (text files S7MSSARC.NEW or S7MSSARC.OLD) in the Windows directory) and display the archived messages with menu commands PLC → CPU MESSAGES and VIEW → NEW ARCHIVE or VIEW → OLD ARCHIVE.

- PLC → MONITOR/MODIFY VARIABLES
(Refer to Section 2.6.6, "Monitoring, Modifying and Forcing Variables")

2.6.3 Loading the User Program into the CPU

When you transfer your user program (compiled blocks and configuration data) to the CPU, it is loaded into the CPU's load memory. Physically, load memory can be RAM or Flash EPROM and either integrated in the CPU or on a memory card.

If the memory card is a Flash EPROM, you can write to it in the programming device and use it as data medium. With the card off circuit, insert it into the CPU; when it is powered up, the relevant data are transferred from the memory card to the CPU's work memory. As regards the CPU 315 in conjunction with a 5V Flash EPROM card and as regards all S7-400 CPUs, you can also write to a Flash EPROM card while it is inserted in the CPU, but only the full program.

When load memory is RAM storage, you can not only transfer the full program on-line, but you an also modify, delete or reload individual blocks. A power-down without battery backup or a memory reset erases the program in RAM.

You transfer a complete user program by switching the CPU to STOP, executing a memory reset, and transferring the complete user program. The configuration data are also transferred. If a Flash EPROM card is in the CPU when the memory reset is executed, the CPU copies its contents to work memory following the reset.

You can modify or reload individual blocks at runtime when the mode selector on the front of the CPU is in the RUN-P position. When the mode selector is at RUN, you can only read the user program. Now, however, you can remove the key, thus preventing any changes to the program.

If you only want to change the configuration data (CPU properties, the configured connections, GD communication, module parameters, and so on), you need only load the *System Data* object into the CPU (select the object and transfer it with menu command PLC → DOWNLOAD. The parameters for the CPU go into effect immediately; the CPU transfers the parameters for the remaining modules to those modules during startup. Note that the *complete* configuration is always loaded into the PLC.

2.6.4 Block Handling

Extensive programs are debugged section by section. If, for instance, you only want to debug one block, load that block into the CPU, then call it in

OB 1. If OB 1 is arranged so that the program can be debugged section by section "from front to back", you can select the blocks or program sections you want to test by jumping over the calls or program sections that are not to be executed, for example with the jump function JMP.

Special caution is advised when transferring individual blocks while the CPU is in RUN-P mode. If one block calls other blocks (which are not in the CPU's memory), you must first load the "subordinate" blocks. This also applies to data blocks. You must load the "uppermost" block last. That block is then processed immediately in the next program cycle.

Just as in the off-line user program, you can execute blocks incrementally in the on-line user program (in the CPU). But when the on-line and the off-line data management diverge, it is possible that the editor can no longer display the additional information from the off-line data manager, and this information might be lost.

The compiled blocks in the user program *Blocks* are transferred with the SIMATIC Manager. From the off-line project window, write the selected blocks to the CPU with PLC → DOWNLOAD; in the on-line project window, read out the selected blocks from the CPU with PLC → UPLOAD.

With the editor, you can edit blocks both on-line and off-line. Save the edited block in the PG off-line with FILE → SAVE; write it to the CPU with PLC → DOWNLOAD. If you want to save the open block under another number or transfer it to another project, a library, or another CPU, use the menu command FILE → SAVE AS.

When you load a new or modified block into the CPU, the CPU places the block in load memory and transfers the relevant data to work memory. If there is already a block with the same number, this "old block" is declared invalid (following a prompt for confirmation) and the new block "added on at the end" in memory. Even a deleted block is "only" declared invalid, not actually removed from memory. This results in gaps in user memory which further and further reduce the amount of memory still available. These gaps can be filled only by the *Compress* function. When you compress in RUN mode, the blocks currently being executed are not relocated; only in STOP mode can you truly achieve compression without gaps. The current memory allocation can be displayed in percent with the menu command PLC → MODULE INFORMATION, *Memory* tab. The dialog box which then appears also has a button for preventive compression.

When you load a data block from the CPU, its values are taken from work memory, for only the work memory holds the current data. You can view the values current at the moment of the read-out with VIEW → DATA. If you change the current value in the data block and write it back to the CPU, the modified value is forwarded to work memory, but the original value is still in load memory. On the non-battery-backed power-up, the CPU transfers the data blocks that are in the (Flash EPROM) load memory to work memory, thus overwriting the data block in work memory; the processed data in work memory are thus lost. In S7-300 systems, you can protect a data area against such loss by declaring it retentive.

2.6.5 Determining the Cause of a STOP

If the CPU goes to STOP because of an error, the first measure to take in order to determine the reason for the STOP is to output the diagnostic buffer. The CPU enters all messages in the diagnostic buffer, including the reason for a STOP and the errors which led to it. To output the diagnostic buffer, switch the PG to on-line, select an S7 program, and choose register card *Diagnostic Buffer* with the menu command PLC → MODULE INFORMATION. The last event (the one with the number 1) is the cause of the STOP, for instance "STOP because programming error OB not loaded". The error which led to the STOP is described in the preceding message, for example "FC not loaded". By clicking on the message number, you can screen an additional comment in the next lower display field. If the message relates to a programming error in a block, you can open and edit that block with the "Open Block" button.

If the cause of the STOP is, for example, a programming error, you can ascertain the surrounding circumstances with the *Stacks* tab. When you open *Stacks,* you will see the B stack (block stack), which shows you the call path of all non-terminated blocks up to the block containing the interrupt point. Use the "I stack" button to screen the interrupt stack, which shows you the contents of the CPU registers (accumulators, address register, data block register, status word)

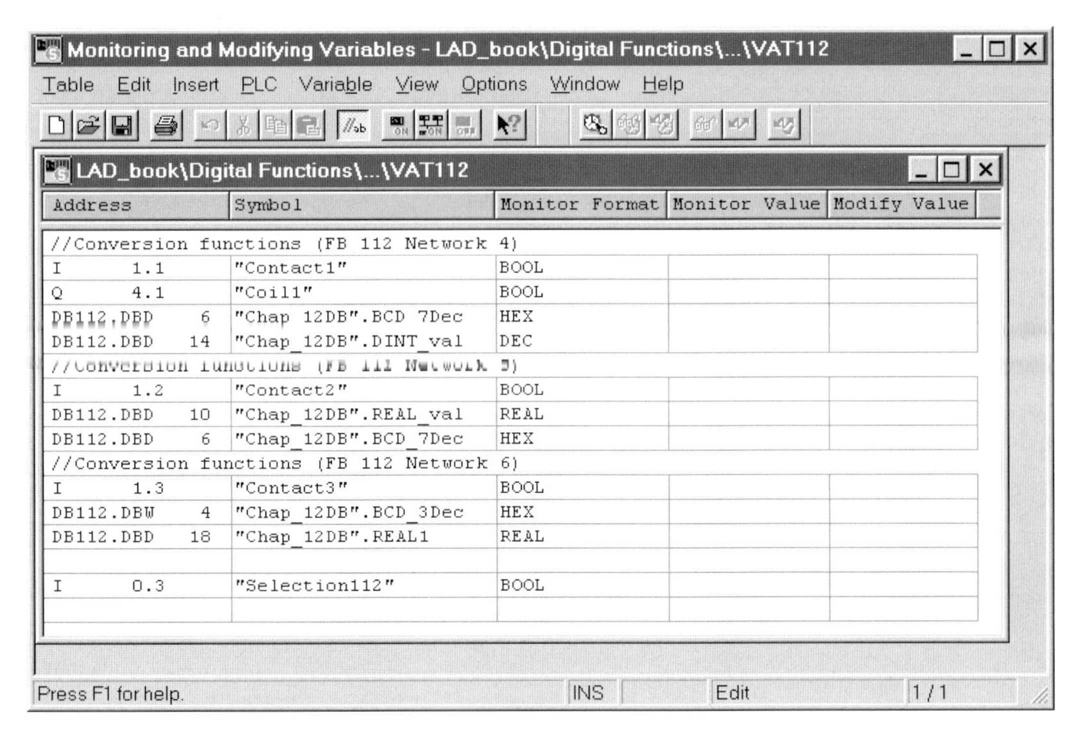

Figure 2.7 Variable Table

at the interrupt point at the instant the error occurred. The L stack (local data stack) shows the block's temporary local data, which you select in the B stack by clicking with the mouse.

2.6.6 Monitoring, Modifying and Forcing Variables

One excellent resource for debugging user programs is the monitoring, modifying and forcing of variables with VAT variable tables. These features allow you to display the signal states or values of variables of elementary data types. In RUN-P mode, you can even modify the variables, that is, change the signal state or assign new values. On appropriately designed CPUs, you can specify fixed values for certain variables which the user program cannot change ("force"). But you must exercise extreme caution; make sure that no dangerous states can result from modifying or forcing variables!

To monitor, modify or force variables, you must create a VAT variable table containing the variables and the associated data formats. You can generate up to 255 variable tables (VAT 1 to VAT 255) and assign them names in the symbol table (Figure 2.7). You can generate a VAT off-line by selecting the user program *Blocks* with INSERT → S7 BLOCK → VARIABLE TABLE, and you can generate an unnamed VAT on-line by selecting *S7 Program* with PLC → MONITOR/MODIFY VARIABLES.

- VAT entries

You can specify the variables with either absolute or symbolic addresses and choose the data type (display format) with which a variable is to be displayed and modified (with VIEW → MONITOR FORMAT or by clicking the right mouse button directly on the monitor format). Use comment lines to give specific sections of the table a header. You may also stipulate which columns are to be displayed. You can change variable or display format or add or delete lines at any time.

- Switch to on-line

To operate a variable table that was generated off-line, switch it to on-line (PLC → CONNECT TO). You must switch each VAT on-line, and you can also carry out a disconnect.

- Trigger conditions

Use VARIABLE → TRIGGER to set the trigger conditions in the variable table. Triggers are the instants at which the CPU reads the values out of or into system memory. Specify whether the read or write is to take place only once or periodically.

- Monitoring variables

Select the Monitor function with the menu command VARIABLE → MONITOR. The variables in the VAT are updated in accordance with the specified trigger conditions. Permanent monitoring allows you to follow changes in the values on the screen. The values are displayed in the data format which you set in the Monitor Format column. VARIABLE → UPDATE MONITOR VALUES updates the monitor values once only and immediately without regard to the specified trigger conditions.

- Modifying variables

Use VARIABLE → MODIFY to transfer the specified values to the CPU in dependence on the trigger conditions. Enter values only in the lines containing the variables you want to modify. You can expand the commentary for a value with "//" or with VARIABLE → MODIFY VALUE VALID; these values are not taken into account for modification. You must define the values in the data format which you set in the Monitor Format column. VARIABLE → ACTIVATE MODIFY VALUES transfers the modify values only once and immediately, without regard to the specified trigger conditions.

- Forcing variables

The menu command VARIABLE → FORCE starts a force job in the CPU. In preparation, set up a connection to a CPU with a variable table and open a window with the force values using the menu command VARIABLE → DISPLAY FORCE VALUES. Enter the respective addresses with the required force values and start the force job. The CPU fetches the force values and allows no changes to be made to them. *Note carefully: closing the variable table or interrupting the connection to the CPU does not stop forcing!* You can terminate forcing only with the menu command VARIABLE → STOP FORCING.

- Enabling peripheral outputs

VARIABLE → ENABLE PERIPHERAL OUTPUTS allows you to modify peripheral outputs while the CPU is in STOP mode. Enter the outputs to be modified (PQB, PQW, PQD) in a variable table and set up a connection to a CPU. The CPU must be in STOP mode. With VARIABLE → ACTIVATE MODIFY VALUES, transfer the modification values to the outputs. To cancel the Enable, select menu command VARIABLE → ENABLE PERIPHERAL OUTPUTS a second time or press the ESC key. Practical example for using this function: Wiring test in STOP mode and without a user program.

2.6.7 Program Status

The LAD editor provides an additional debugging option for your program with its "Program Status" function. When this function is used, the editor shows the binary signal flow and digital values within a network (current path or rung).

The block whose program you want to debug is in the CPU's user memory and is called and edited there. Open this block, for example by double-clicking on it in the SIMATIC Manager's on-line window. The editor is started and shows the program in the block.

Select the network you want to debug. Activate the Program Status function with DEBUG → MONITOR. In the block window you will see the binary signal flow, allowing you to follow any changes as they occur. Choose the mode of representation in the editor with OPTIONS → CUSTOMIZE on the "LAD/FBD" tab.

The trigger conditions are set with DEBUG → CALL ENVIRONMENT. You require this parameter when the block to be debugged is called more than once in your program. You can initiate recording of the status information either by specifying the call sequence or in dependence on the open data block. If a block is called only once, select "no condition".

The recording of the program status information requires additional execution time in the program cycle. For this reason, do not select a cycle time that is barely sufficient. You will put less of a burden on the cycle when you record only the first pass of a program loop with DEBUG → TEST ENVIRONMENT → PROCESS. If the test environment is "Laboratory", every pass through a program loop is recorded.

If the block was written in the STL language, some CPUs allow you to debug the program statement by statement in *single-step mode.* The CPU is in HOLD mode; for reasons of safety, the peripheral outputs are disabled. Using breakpoints, you can stop the program at any location and test it step by step.

3 The LAD Programming Language

This chapter discusses the elements of the LAD programming language, encompassing everything from the different priority classes (program processing methods) through the components of a LAD program (blocks and program elements) to variables (data types).

You structure your user program in the planning phase by adapting it to technological and functional conditions and circumstances; the program's structure is the determining factor in program development, debugging and startup. To ensure effective programming, it is therefore necessary to pay particular attention to the program structure.

3.1 Program Processing

The overall program for a CPU consists of the operating system and the user program.

The *operating system* is the totality of all instructions and declarations which control the system resources and the processes using these resources, and includes such things as data backup in the event of a power failure, the activation of priority classes, and so on. The operating system is a component of the CPU to which you, as user, have no write access. However, you can reload the operating system from a memory card, for instance in the event of a program update.

The *user program* is the totality of all instructions and declarations (in this case program elements) for signal processing, through which a plant (process) is affected in accordance with the defined control task.

3.1.1 Program Processing Methods

The user program may be composed of program sections which the CPU processes in dependence on certain events. Such an event might be the start of the automation system, an interrupt, or detection of a program error (Figure 3.1). The programs allocated to the events are divided into *priority classes*, which determine the program processing order (mutual interruptibility) when several events occur.

The lowest-priority program is the *main program*, which is processed cyclically by the CPU. All other events can interrupt the main program at any location, the CPU then executes the associated interrupt service routine or error handling routine and returns to the main program.

A specific *organization block (OB)* is allocated to each event. The organization blocks represent the priority classes in the user program. When an event occurs, the CPU invokes the assigned organization block. An organization block is a part of a user program which you yourself may write.

Before the CPU begins processing the main program, it executes a start-up routine. This routine can be triggered by switching on the mains power, by actuating the mode switch on the CPU's front panel, or via the programming device. Program processing following execution of the start-up routine always starts at the beginning of the main program in S7-300 systems (complete restart); in S7-400 systems, it is also possible to resume the program scan at the point at which it was interrupted (warm restart).

The main program is in organization block OB 1, which the CPU always processes. The start of the user program is identical to the first network in OB 1. After OB 1 has been processed (end of program), the CPU returns to the operating system and, after calling for the execution of various operating system functions, such as the updating of the process images, it once again calls OB 1.

Events which can intervene in the program are interrupts and errors. Interrupts can come from the process (hardware interrupts) or from the CPU (watchdog interrupts, time-of-day interrupts, etc.). As far as errors are concerned, a distinction is made between synchronous and asynchronous errors. An asynchronous error is an error which is

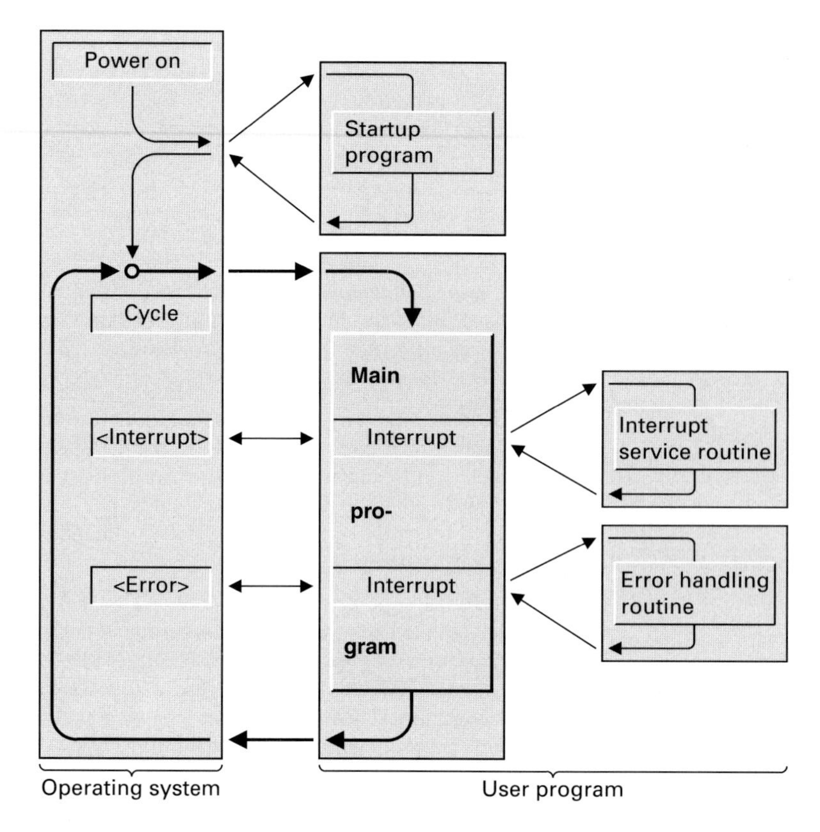

Figure 3.1
Methods of Processing the User Program

independent of the program scan, for example failure of the power to an expansion unit or an interrupt that was generated because a module was being replaced. A synchronous error is an error caused by program processing, such as referencing a non-existent address or a data type conversion error. The type and number of recorded events and the associated organization blocks are CPU-specific; not every CPU can handle all possible STEP 7 events.

3.1.2 Priority Classes

Table 3.1 lists the available SIMATIC S7 organization blocks, each with its priority. In some priority classes, you can change the assigned priority when you parameterize the CPU. The Table shows the lowest and highest possible priority classes; each CPU has a different low/high range.

Organization block OB 90 (background processing) executes alternately with organization block OB 1, and can, like OB 1, be interrupted by all other interrupts and errors.

The start-up routine may be in organization block OB 100 (complete restart) and OB 101 (warm restart), and has priority 27. Asynchronous errors occurring in the start-up routine have priority class 28. Diagnostic interrupts are regarded as asynchronous errors.

You determine which of the available priority classes you want to use when you parameterize the CPU. Unused priority classes (organization blocks) must be assigned priority 0 or L stack length 0. The relevant organization blocks must be programmed for all priority classes used; otherwise the CPU will invoke OB 85 ("Program Processing Error") or go to STOP.

3.1.3 Specifications for Program Processing

The CPU's operating system normally uses default parameters. You can change these defaults when you parameterize the CPU (in the *Hardware* object) to customize the system to suit your particular requirements. You can change the parameters at any time.

Table 3.1 SIMATIC S7 Organization Blocks

Organization block OB		Priority	
		Default	Modifiable
1	Free cycle	1	No
10 to 17	TOD interrupts	2	2 to 24
20 to 23	Time-delay interrupts	3 to 6	2 to 24
30 to 38	Watchdog interrupts	7 to 15	2 to 24
40 to 47	Hardware interrupts	16 to 23	2 to 24
60	Multiprocessor Int.	25	No
80 to 87	Asynchronous errors	26 (in start: 28)	24 to 26
90	Background processing	29 [1)]	No
100, 101	Start-up routine	27	No
121, 122	Synchronous errors	Priority of OB that caused error	

[1)] See text

Every CPU has its own specific number of parameter settings. The following list provides an overview of all STEP 7 parameters and their most important settings.

• Startup
Specifies the type of start-up (complete restart, warm restart); monitoring of Ready signals or module parameterization; maximum amount of time which may elapse before a warm restart

• Cycle/Clock Memory
Enable/disable cyclic updating of the process image; specification of the cycle monitoring time and minimum cycle time; amount of cycle time, in percent, for communication; number of the clock memory byte

• Retentive Memory
Number of retentive memory bytes, timers and counters; specification of retentive areas for data blocks

• Local Data
Local data specifications for priority classes (organization blocks)

• Interrupts
Specification of the priority for hardware interrupts, time-delay interrupts, asynchronous errors and (available soon) communication interrupts

• Time-of-Day Interrupts
Specification of the priority, specification of the start time and periodicity

• Cyclic Interrupts
Specification of the priority, the time cycle and the phase offset

• Diagnostics/Clock
Indicate the cause of a STOP; type and interval for clock Fynchronization, correction factor

• Multicomputing
Specification of the CPU number

• Integrated I/O
Activation and parameterization of the integrated I/O

On start-up, the CPU puts the user parameters into effect in place of the defaults, and they remain in force until changed.

3.2 Blocks

You can subdivide your program into as many sections as you want to in order to make it easier to read and understand. The LAD programming language supports this by providing the necessary functions. Each program section should be self-contained, and should have a technological or functional basis. These program sections are referred to as "blocks". A block is a section of a user program which is defined by its function, structure or intended purpose.

3.2.1 Block Types

LAD provides different types of blocks for different tasks:

▷ User blocks
Blocks containing user program and user data

▷ System blocks
Blocks containing system program and system data

▷ Standard blocks
Turnkey, off-the-shelf blocks, such as drivers for FM and CP modules

User blocks

In extensive and complex programs, "structuring" (subdividing) of the program into blocks is recommended, and in part necessary. You may choose among different types of blocks, depending on your application:

- Organization blocks (OBs)

These blocks serve as the interface between operating system and user program. The CPU's operating system calls the organization blocks when specific events occur, for example in the event of a hardware or time-of-day interrupt. The main program is in organization block OB 1. The other organization blocks have permanently assigned numbers based on the events they are called to handle.

- Function blocks (FBs)

These blocks are parts of the program whose calls can be programmed via block parameters. They have a variable memory which is located in a data block. This data block is permanently allocated to the function block, or, to be more precise to the function block *call*. It is even possible to assign a different data block (with the same data structure but containing different values) to each function block call. Such a permanently assigned data block is called an instance data block, and the combination of function block call and instance data block is referred to as a call instance, or "instance" for short. Function blocks can also save their variables in the instance data block of the calling function block; this is referred to as a "local instance".

- Functions (FCs)

Functions are used to program frequently recurring or complex automation functions. They can be parameterized, and return a value (called the function value) to the calling block. The function value is optional, in addition to the function value, functions may also have other output parameters. Functions do not store information, and have no assigned data block.

- Data blocks (DBs)

These blocks contain your program's data. By programming the data blocks, you determine in which form the data will be saved (in which block, in what order, and in what data type). There are two ways of using data blocks: as global data blocks and as instance data blocks. A global data block is, so to speak, a "free" data block in the user program, and is not allocated to a code block. An instance data block, however, is assigned to a function block, and stores part of that function block's local data.

The number of blocks per block type and the length of the blocks is CPU-dependent. The number of organization blocks, and their block numbers, are fixed; they are assigned by the CPU's operating system. Within the specified range, you can assign the block numbers of the other block types yourself. You also have the option of assigning every block a name (a symbol) via the symbol table, then referencing each block by the name assigned to it.

System blocks

System blocks are components of the operating system. They can contain programs (system functions (SFCs) or system function blocks (SFBs)) or data (system data blocks (SDBs)). System blocks make a number of important system functions accessible to you, such as manipulating the internal CPU clock, or various communications functions

You can call SFCs and SFBs, but you cannot modify them, nor can you program them yourself. The blocks themselves do not reserve space in user memory; only the block calls and the instance data blocks of the SFBs are in user memory.

SDBs contain information on such things as the configuration of the automation system or the parameterization of the modules. STEP 7 itself generates and manages these blocks. You, however, determine their contents, for instance when you configure the stations. As a rule, SDBs are located in load memory. You cannot access them from your user program.

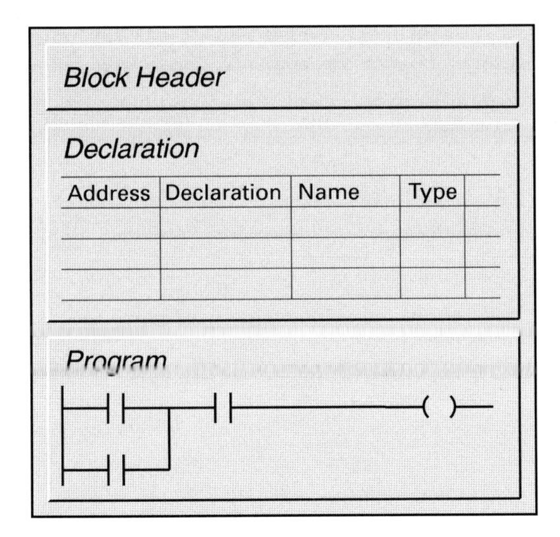

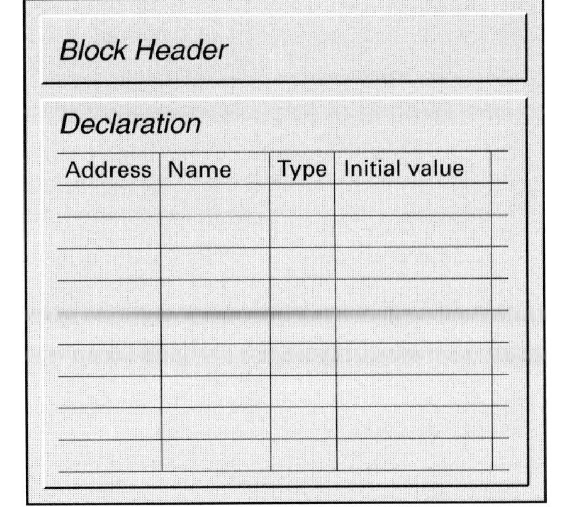

Figure 3.2 Structure of a Block

Standard blocks

In addition to the functions and function blocks you create yourself, turnkey blocks (called "standard blocks") are also available. They can either be obtained on a storage medium or are on libraries delivered as part of the STEP 7 package (for example IEC functions, or functions for the S5/S7 converter).

3.2.2 Block Structure

Essentially, code blocks comprise three parts: the block header, the declaration part, and the program part (Figure 3.2); in the case of data blocks, the last two parts are combined into a declaration part which allows for initialization and – in on-line mode – shows the current data.

3.2.3 Block Header

Every block has a block header containing the block attributes. You can view and modify the block attributes from the editor with menu command FILE → PROPERTIES (Figure 3.3).

The *Name* of the block is used to identify the block; it is not the same as the symbolic address. Different blocks may have the same name. A block *Family* allows you to assign a group of blocks a common characteristic. The block name and the block family are displayed in the comment field when you select the block in the dialog box of the program element catalog. The block's *Author* and the *Version* (two times two digits) round off the programmable block properties.

The editor keeps track of the block's modification date in two time stamps, one for the program code and one for the interface (the block parameters). Note that the modification date for the interface must be lower than (that is, must precede) the modification date of the program code in the calling block. If this is not the case, the editor reports a "time stamp conflict" upon output of the calling block. The relevant block call is then displayed in STL statements. To rectify the error, you must re-enter the block header.

Block can be created or compiled as version 1 or version 2 blocks. This serves a practical purpose only for the function blocks. If you disable the "multiple instance capability" when creating a function block, the result will be a version 1 block. You cannot call this block as local instance, nor can you call any other function block as local instance from within this block. The advantage of a version 1 function block, namely the unrestricted use of instance data in conjunction with indirect addressing, is irrelevant when programming in the LAD language.

The block header of any standard block which comes from Siemens contains the "standard block" attribute.

The attribute "DB is write-protected in the PLC" means that you can only read that data block. Output of an error message prevents the overwriting of the data in the data block. This write protection feature must not be confused with block protection. A data block with block

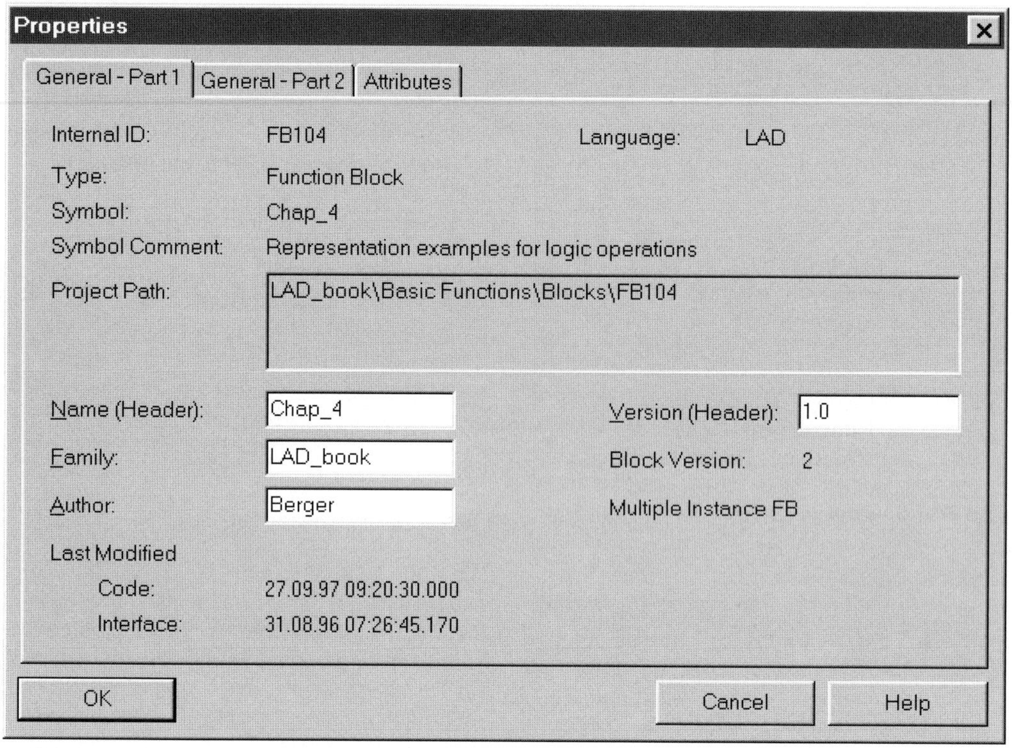

Properties

General - Part 1 | General - Part 2 | Attributes

Internal ID:	FB104	Language:	LAD
Type:	Function Block		
Symbol:	Chap_4		
Symbol Comment:	Representation examples for logic operations		
Project Path:	LAD_book\Basic Functions\Blocks\FB104		
Name (Header):	Chap_4	Version (Header):	1.0
Family:	LAD_book	Block Version:	2
Author:	Berger	Multiple Instance FB	
Last Modified			
Code:	27.09.97 09:20:30.000		
Interface:	31.08.96 07:26:45.170		

OK | Cancel | Help

Figure 3.3 Block Properties

protection can be read out and written to in the user program, but its data can no longer be viewed with a programming or operator device.

The “unlinked” attribute applies only to data blocks. Following loading into the PLC, a data block with this attribute is in load memory only; it is not “execution-relevant”. You cannot write to data blocks in load memory, and you can read them only with the system function SFC 20 BLKMOV. Note that access to load memory is extremely time-consuming. Data blocks in load memory are used, for example,

- when they contain parameter assignment data for parameterizing suitably equipped modules in the start-up routine or at runtime (with default values or machine data),

- when they contain recipes that are read only infrequently (for instance copied to a data block in work memory because of production changes).

You can set the “KNOW HOW protection” attribute (block protection) only in the STL programming language (see Chapter 24, “LAD Supplements”).

Blocks may have system attributes. System attributes control and coordinate functions between applications, for example in the SIMATIC PCS7 control system.

3.2.4 Variable Declaration

The declaration part contains the definition of the block’s local variable, that is, the variables which you use only in that block (Table 3.2). In the case of code blocks, depending on the type of block, these are the block parameters as well as the temporary and static local data; in the case of data blocks, the variables in question are the data addresses. If you do not use certain types of variables in a code block, the lines for those variables remain blank.

The declaration of a variable consists of the name, the data type, an initial value, if any, and

Table 3.2 Types of Variables in the Declaration Part

Type of variable	Declaration	Possible in block type(s)			
Input parameters	in	–	FC	FB	–
Output parameters	out	–	FC	FB	–
In/out parameters	in_out	–	FC	FB	–
Static local data	stat	–	–	FB	–
Temporary local data	temp	OB	FC	FB	–
Data addresses	struct	–	–	–	DB

a variable comment (optional). Not all variables can be assigned initial values. Temporary local data, for instance, cannot be assigned initial values. The initial values for functions and function blocks are described in detail in Chapter 19, "Block Parameters". Variables in data blocks may always be assigned initial values.

The order of the declarations in the case of code blocks is fixed (as shown in the Table), but the order within a type can be arbitrary. When declaring variables, you can save memory by bundling binary variables together in blocks of eight or 16, and BYTE variables into pairs. The editor generates a BOOL or BYTE variable at a byte boundary and variables of other data types at a word boundary (beginning at a byte with an even address).

3.2.5 Program in the Block

The block program begins with the block title and the block comment. Both precede the first network, and both are optional. The *block title* may be a brief description in keyword form. The *block comment* itself may consist of several lines, and normally describes the function of a block, how it is used, and the like.

A LAD program is broken down into networks. There is one network for each rung, or rung. LAD automatically numbers the networks, beginning with the number one. You may give each network a *network title* and a *network comment*. During editing, you can select each network directly with EDIT → GO TO → ... .

You need not terminate a block with a special element; you simply end block input. You can, however, program a final (empty) network, for instance with the title "End of block", and you will immediately see the end of the block (which is particularly advantageous in the case of very long blocks).

3.2.6 Programming Data Blocks

You have three different options for programming a data block:

- As global data block, in which case you would declare the data addresses when you program the data block
- As instance data block, in which case the data structure which you declared when you programmed the corresponding function block will be used
- As data block with user data type, in which case you must declare the data structure as user data type UDT

When programming a global data block, you can assign an *initial* value for each data address. Depending on their data type, variables normally default to either zero, the lowest possible value, or blank. An instance data block generated from a function block uses as initial value the one defined in the declaration part of the function block. If you generate a data block from a user data type (UDT), the initialization values (default values) from the UDT are taken as the initial values in the data block.

The editor displays a data block in two views. In the *declaration* view, (VIEW → DECLARATION VIEW) you define the data addresses and you see the variables as you defined them, for example a field or user data type as a single variable. In the *data* view (VIEW → DATA VIEW), the editor displays each variable and each component of a field or structure separately. You will now see an addition column titled *Actual Value*. The actual value is the value which a data address has or will have in the CPU's work memory. The standard is for the editor to take the initial value as actual value.

You can modify the actual value individually for each data address. For example: You generate several instance data blocks from one function block, but want to have slightly different initial values for some of the instance data items for each function block call (for each FB/DB pair). You could them edit each data block with VIEW → DATA VIEW and enter the values you want for this data block in the Actual Value column. You would then select menu command EDIT → INITIALIZE DATA BLOCK so that the editor will again replace all actual values with the original initial values.

When a data block is loaded into the CPU, the actual values are taken over into work memory. Each change in a value of a data address via the program corresponds to a change in the actual value. When you open a data block on-line, the actual value is the value that was in work memory when the data block was opened. If you were to change the actual value at this point using a programming device and write the data block back to the CPU, the value would be updated in work memory.

When you use a Flash EPROM memory card as load memory, the blocks on the memory card are transferred to work memory following a CPU memory reset. The data blocks are given the actual values that were originally programmed. The same thing happens when the mains power is switched on without battery backup.

3.3 Editing LAD Elements

Programming in general

The program consists of individual LAD elements arranged in series or parallel to one another. Programming of a current path, or rung, begins on the left power rail. You select the location in the rung at which you want to insert an element, then you select the program element you want with the corresponding function key (for example F2 for a normally open contact), with the corresponding button on the function bar, or from the Program Elements Catalog (with INSERT → PROGRAM ELEMENTS or VIEW → CATALOG). You terminate a rung with a coil or a box.

Most program elements must be assigned memory locations (variables). The easiest way to do this is to first arrange all program elements, then label them.

A network may contain only one rung. You can generate another network behind the current network with INSERT → NETWORK and enter the next rung in that network.

You can output the comments on the monitor with VIEW → COMMENT and then, for instance, enter the block comment and the network comment. If you want to program with symbols, enable symbolic representation with the menu command VIEW → SYMBOLIC REPRESENTATION. Before you can use a global symbol, that symbol must be in the symbol table. Even while you are programming, you can edit the symbol table with OPTIONS → SYMBOL TABLE or insert a new symbol with INSERT → SYMBOLS.

LAD elements

- Contacts

Binary addresses such as inputs are scanned using contacts. The scanned signal states are combined according to the arrangement of the contacts in a serial or parallel layout. Normally open contacts (current flows when the contact is activated) and normally closed contacts (current flows when the contact is not activated) are available. You can also scan status bits or negate the result of the logic operation (NOT contact).

- Coils

Coils are used to control binary addresses, such as outputs. A simple coil sets the binary addresses when current flows in the coil, and resets it when power no longer flows. There are coils with additional labels, such as Set and Reset coils, which serve a special function. You can also use coils to control timers and counters, call blocks without parameters, execute jumps in the program, and so on.

- Boxes

Boxes represent LAD elements with complex functions. STEP 7 provides “standard boxes” of two different types: without EN/ENO mechanism (such as memory functions, timer and counter functions, comparison boxes), and with EN/ENO (such as MOVE, arithmetic and math functions, data type conversions). When you call code blocks (FCs, FBs, SFCs and SFBs), LAD also represents the calls as boxes with EN/ENO.

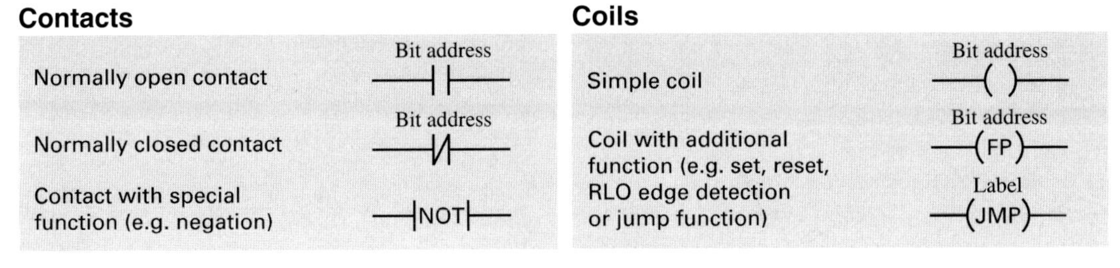

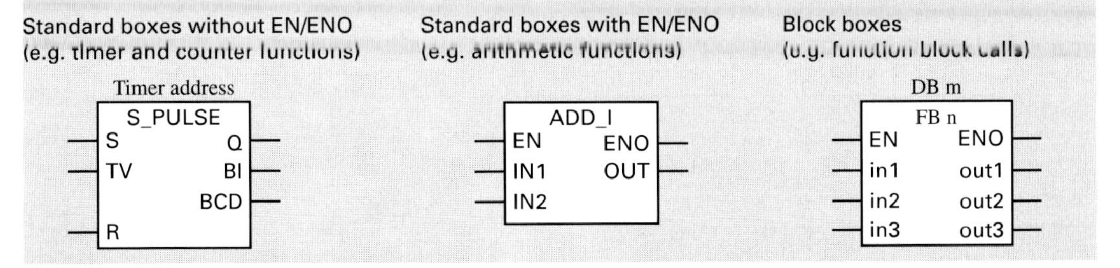

Figure 3.4 Examples of LAD Program Elements

Layout restrictions

The LAD editor sets up a network according to the "main rung" principle. This is the uppermost branch, which begins directly on the left power rail and must terminate with a coil or a box. All LAD elements can be located in this rung. In parallel branches which do not begin on the left power rail, there are sometimes restrictions in the choice of program elements.

Additional restrictions dictate that no LAD element may be "short-circuited" with an "empty" parallel branch, and that no "power" may flow through an element from right to left (a parallel branch must be closed to the branch in which it was opened). Any further rules applying to the layout of special LAD elements are discussed in the relevant chapters.

When using boxes as program elements, you can

- program only one box per network
- Terminate a current path with several simultaneously activated boxes
- arrange boxes in series by connecting the ENO output of one box with the EN input of the next box
- connect boxes in parallel via their ENO output

You can control processing of boxes connected in series jointly (also see section 15.4, "Using the Binary Result"). With the arrangement of the boxes, you evaluate the signal states of the ENO outputs; when you connect the ENO outputs to a coil, "power" flows in the coil when all boxes in a series connection have been processed without error or when one of the boxes in a parallel connection has been processed without error.

3.4 Data Types

3.4.1 General Remarks on Data Types

Data types stipulate the characteristics of data, essentially the representation of the contents of a variable, and the permissible ranges. STEP 7 provides predefined data types which you can combine into user-defined data types. The data types are available on a global basis, and can be used in every block. Depending on structure and application, the data types are classified as follows:

▷ Elementary data types

▷ Complex data types

▷ User data types

▷ Parameter types

On the diskette which accompanies this book you will find the declaration and use of variables of all data types in the "Data Types" program.

3.4.2 Elementary Data Types

Elementary data types can reserve a bit, a byte, a word or a doubleword. Table 3.3 shows the elementary data types.

Declaring elementary data types

A variable with elementary data type may be declared globally in the symbol table or locally in the declaration part of a block. The variable may be assigned an initial value in the declaration (not as block parameter for a function or as temporary variable). The initial value must have the same data type as the variable (examples see Table 3.4). A variable of elementary data type is saved in the same manner as the corresponding type of operand. All operand areas are admissible, including block parameters.

Table 3.3 Overview of Elementary Data Types

Data Type	Description		Example of Constant Notation	
BOOL	Bit	1 bit	TRUE, FALSE	
BYTE	Byte	8 bits		
	8-bit hex number		B#16#00	(min. value)
			B#16#FF	(max. value)
CHAR	One character (ASCII)	8 bits	'A'	
WORD	Word	16 bits		
	16-bit hex number		W#16#0000	(min. value)
			W#16#FFFF	(max. value)
	16-bit binary number		2#0000_0000_0000_0000	(min. value)
			2#1111_1111_1111_1111	(max. value)
	Count, 3 decades BCD		C#000	(min. value)
			C#999	(max. value)
	2 × 8-bit decimal numbers without sign		B#(0,0)	(min. value)
			B#(255,255)	(max. value)
DWORD	Double word	32 bits		
	32-bit hex number		DW#16#0000_0000	(min. value)
			DW#16#FFFF_FFFF	(max. value)
	32-bit binary number		2#00000000_00000000_00000000_00000000	(min. value)
			2#11111111_11111111_11111111_11111111	(max. value)
	4 × 8-bit decimal numbers without sign		B#(0,0,0,0)	(min. value)
			B#(255,255,255,255)	(max. value)
INT	Fixed-point number	16 bits	–32768	(min. value)
			+32767	(max. value)
DINT	Fixed-point number	32 bits	L#-2 147 483 648	(min. value)
			L#+2 147 483 647	(max. value)
REAL	Floating-point number	32 bits	+123.4567 as decimal number with comma or +1.234567E+02 as exponent (value range see text)	
S5TIME	Time value in S5 format	16 bits	S5T#0ms	(min. value)
			S5TIME#2h46m30s	(max. value)
TIME	Time value in IEC format	32 bits	TIME#-24d20h31m23s647ms	(min. value)
			TIME#24d20h31m23s647ms	(max. value)
DATE	Date	16 bits	D#1990-01-01	(min. value)
			DATE#2168-12-31	(max. value)
TIME_OF_DAY	Time of day	32 bits	TOD#00:00:00	(min. value)
			TIME_OF_DAY#23:59:59.999	(max. value)

Table 3.4 Examples of Declaration and Initial Value for Elementary Data Types

Name	Type	Initial Value	Comment
Automatic	BOOL	FALSE	Initial value is signal state "0"
Manual_off	BOOL	TRUE	Initial value is signal state "1"
Measured value	DINT	L#0	Initial value of a DINT variable
Memory	WORD	W#16#FFFF	Initial value of a WORD variable
Waiting time	S5TIME	S5T#20s	Initial value of an S5 time variable

BOOL, BYTE, WORD, DWORD

A variable of data type BOOL represents a bit value, for example I 1.0. Variables with data types BYTE, WORD and DWORD are bit strings comprising 8, 16 and 32 bits, respectively. The individual bits are not evaluated.

Special forms of these data types are the BCD numbers and the count as used in conjunction with counter functions, as well as data type CHAR, which represents an ASCII character.

BCD numbers

BCD numbers have no special identifier. Simply enter a BCD number with the data type 16# (hexadecimal) and use only digits 0 to 9.

BCD numbers occur in coded processing of time values and counts and in conjunction with conversion functions. Data type S5TIME# is used to specify a time value for starting a timer (see below), data type 16# or C# for specifying a count value. A C# count value is a BCD number between 000 and 999, whereby the sign is always 0 (positive).

As a rule, BCD numbers have no sign. In conjunction with the conversion functions, the sign of a BCD number is stored in the leftmost (highest) decade, so that there is one less decade for the number.

When a BCD number is in a 16-bit word, the sign is in the uppermost decade, whereby only bit position 15 is relevant. Signal state "0" means that the number is positive. Signal state "1" stands for a negative number. The sign has no affect on the contents of the individual decades. An equivalent assignment applies for a 32-bit word.

The available value range is 0 to ± 999 for a 16-bit BCD number and 0 to ± 9 999 999 for a 32-bit number.

CHAR

A variable with data type CHAR (character) reserves one byte. Data type CHAR represents a single character in ASCII format. Example: 'A'.

You can use any printable character in apostrophes. Some special characters require use of the notation shown in Tabel 3.5. Example: '$$' represents a dollar sign in ASCII code.

Table 3.5 Special Characters for CHAR

CHAR	Hex	Description
$$	24_{hex}	Dollar sign
$'	27_{hex}	Apostrophe
$L or $l	$0A_{hex}$	Line feed (LF)
$P or $p	$0C_{hex}$	New page (FF)
$R or $r	$0D_{hex}$	Carriage return (CR)
$T or $t	09_{hex}	Tabulator

The MOVE function allows you to use two or four ASCII characters enclosed in apostrophes as a special form of data type CHAR for writing ASCII characters in a variable.

INT

A variable with data type INT is stored as an integer (16-bit fixed-point number). Data type INT has no special identifier.

A variable with data type INT reserves one word. The signal states of bits 0 to 14 represent the digit positions of the number; the signal state of bit 15 represents the sign (S). Signal state "0" means that the number is positive, signal state "1" that it is negative. A negative number is represented as two's complement. The permissible number range is from +32 767 ($7FFF_{hex}$) to – 32 768 (8000_{hex}).

DINT

A variable with data type DINT is stored as an integer (32-bit fixed-point number). An integer is stored as a DINT variable when it exceeds 32 767 or falls below –32 768 or when the number is preceded by type identifier L#.

A variable with data type DINT reserves one doubleword. The signal states of bits 0 to 30 represent the digit positions of the number; the sign is stored in bit 31. Bit 31 is "0" for a positive and "1" for a negative number. Negative numbers are stored as two's complement. The number range is from +2 147 483 647 (7FFF $FFFF_{hex}$) to –2 147 483 648 (8000 0000_{hex}).

REAL

A variable of data type REAL represents a fraction, and is stored as a 32-bit floating-point number. An integer is stored as a REAL variable when you add a decimal point and a zero.

In exponent representation, you can precede the "e" or "E" with an integer number or fraction with seven relevant digits and a sign. The digits which follow the "e" or "E" represent the exponent to base 10. The conversion of the REAL variable into the internal representation of a floating-point number is handled by STEP 7. REAL variables are divided into numbers which can be represented with complete accuracy ("normalized" floating-point numbers) and those with limited accuracy ("denormalized" floating-point numbers). The value range of a normalized floating-point number lies between:

$-3.402\,823 \times 10^{+38}$ to $-1.175\,494 \times 10^{-38}$
± 0
$+1.175\,494 \times 10^{-38}$ to $+3.402\,823 \times 10^{+38}$

A denormalized floating-point number may be in the following range:

$-1.175\,494 \times 10^{-38}$ to $-1.401\,298 \times 10^{-45}$ and
$+1.401\,298 \times 10^{-45}$ to $+1.175\,494 \times 10^{-38}$

The S7-300 CPUs cannot calculate with denormalized floating-point numbers. The bit pattern of a denormalized number is interpreted as a zero. If a result falls within this range, it is represented as zero, and status bits OV and OS are set (overflow).

A variable of data type REAL consists internally of three components, namely the sign (bit 31), the 8-bit exponent to base 2 (bits 23 to 30), and the 23-bit mantissa (bits 0 to 22). The sign may assume the value "0" (positive) or "1" (negative). Before the exponent is stored, a constant value (bias, +127) is added to it so that it shows a value range of from 0 to 255. The mantissa represents the fractional portion of the number. The integer portion of the mantissa is not saved, as it is either always 1 (in the case of normalized floating-point numbers) or always 0 (in the case of denormalized floating-point numbers). Table 3.6 shows the internal range of a floating-point number.

S5TIME

A variable with data type S5TIME is used to set timers. It reserves one 16-bit word with 1 + 3 decades.

Table 3.6 Range of a Floating-Point Number

Sign	Exponent	Mantissa	Description
0	255	Not equal to 0	Not a valid floating-point number
0	255	0	+ infinite
0	1 to 254	Arbitrary	Positive normalized floating-point number
0	0	Not equal to 0	Positive denormalized floating-point number
0	0	0	+ zero
1	0	0	– zero
1	0	Not equal to 0	Negative denormalized floating-point number
1	1 ... 254	Arbitrary	Negative normalized floating-point number
1	255	0	– infinite
1	255	Not equal to 0	Not a valid floating-point number

Table 3.7 Examples of the Declaration of DT Variables and STRING Variables

Name	Type	Initial Value	Comment
Date1	DT	DT#1990-01-01-00:00:00	DT variable minimum value
Date2	DATE_AND_TIME	DATE_AND_TIME# 2089-12-31-23:59:59.999	DT variable maximum value
First name	STRING[10]	'Jack'	STRING variable, 4 out of 10 char. specified
Last name	STRING[7]	'Daniels'	STRING variable, all 7 char. specified
NewLine	STRING[2]	'RL'	STRING variable, special char. specified
BlankString	STRING[16]	''	STRING variable, no specification

The time is specified in hours, minutes, seconds and milliseconds. Conversion into internal representation is handled by STEP 7. Internal representation is as BCD number in the range 000 to 999. The time interval can assume the following values: 10 ms (0000), 100 ms (0001), 1 s (0010), and 10 s (0011). The duration is the product of time interval and time value.

DATE

A variable with data type DATE is stored in a word as an unsigned fixed-point number. The contents of the variable corresponds to the number of days since 1st of January 1990. Its representation shows the year, month and day, separated from one another by a hyphen.

TIME

A variable with data type TIME reserves one double word. Its representation contains the information for days (d), hours (h), minutes (m), seconds (s) and milliseconds (ms), whereby individual items of this information may be omitted. The contents of the variable are interpreted in milliseconds (ms) and stored as a signed 32-bit fixed-point number.

TIME_OF_DAY

A variable of data type TIME_OF_DAY reserves one double word. It contains the number of milliseconds since the day began (0:00 o'clock) in the form of an unsigned fixed-point number. Its representation contains the information for hours, minutes and seconds, separated by a colon. The milliseconds, which follow the seconds and are separated from them by a decimal point, may be omitted.

3.4.3 Complex Data Types

Complex data types may be used only in conjunction with variables located in data blocks or in the L stack or which are block parameters. The following complex data types are available:

- ▷ DATE_AND_TIME
  Date and time
- ▷ STRING
  Character string
- ▷ ARRAY
  Array variable (composite of variables with the same data type)
- ▷ STRUCT
  Structure variable (composite of variables with different data types)

DATE_AND_TIME

Data type DATE_AND_TIME represents an instant comprised of the date and the time. You may also use the abbreviation DT in place of DATE_AND_TIME.

In the declaration, the variable may be assigned an initial value (not as a block parameter of a function, an in/out parameter of a function block, or a temporary variable). The initial value must be of type DATE_AND_TIME or DT, and in the following format:

Keyword#Year-Month-Day-Hour:Minute:Second:Millisecond

The milliseconds may be omitted (Table 3.7).

Variables of data type DT can be used as block parameters of type DT or ANY; for example,

DT format

Byte n	Year	0 to 99
Byte n+1	Month	1 to 12
Byte n+2	Day	1 to 31
Byte n+3	Hour	0 to 23
Byte n+4	Minute	0 to 59
Byte n+5	Seconds	0 to 59
Byte n+6	ms	0 to 999
Byte n+7		Day of W.

Day of the week
From 1 = Sunday
to 7 = Saturday

STRING format

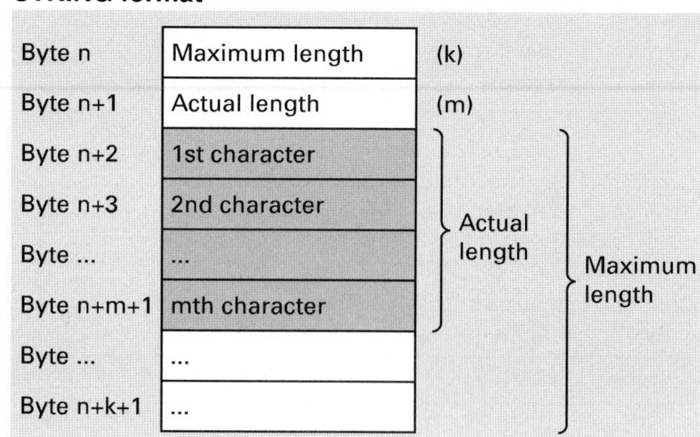

Figure 3.5 Format of DT and STRING Variables

they can be copied with system function SFC 20 BLKMOV. There are standard blocks for processing these variables ("IEC functions").

A variable of data type DATE_AND_TIME reserves eight bytes (Figure 3.5). The variable begins at a word boundary (at a byte with an even address). All specifications are in BCD format.

STRING

Data type STRING represents a character string comprising up to 254 characters. When declaring a STRING variable, follow the word STRING with the number of characters the variable can accommodate, enclosed in brackets (Table 3.7). If you make no specification, the editor will reserve 254 bytes in memory. With FC functions, the Editor does not permit length specifications or else it demands the standard length of 254.

The variable can be given an initial value in the declaration (not as a block parameter of a function, an in/out parameter for a function block, or a temporary variable). The initial value is specified in the form of ASCII characters enclosed in apostrophes or, in some cases, preceded by a dollar sign (see data type CHAR).

If the initial value is shorter than the declared maximum length, the remaining character locations are not reserved. When a variable of data type STRING is post-processed, only the currently reserved character locations are taken into consideration. It is also possible to define an "empty string" as the initial value.

Variables of data type STRING may be used as block parameters of data type STRING or ANY: for example, they can be copied with system function SFC 20 BLKMOV. There are standard blocks for processing these variables ("IEC functions").

A variable of data type STRING has a maximum length of 256 characters, with 254 bytes net data. It begins at a word boundary (at a byte with an even address).

When the variable is created, its maximum length is specified. When it is assigned an initial value, or when it is processed, its actual length (the true length of the character string = the number of valid characters) is entered. The first byte of the character string contains the maximum length, the second byte the actual length; this is followed by the characters in ASCII format (Figure 3.5).

ARRAY

Data type ARRAY represents an array comprising a fixed number of elements of the same data type. An array can accommodate up to 65,536 elements per dimension (from -32,768 to 32,767). This is, of course, purely theoretical. The actual size is determined by the maximum block size.

In the declaration, you can assign initial values to the array elements (not as a block parameter for a function, an in/out parameter for a function block, or a temporary variable). The data type of the initial values must be the same as the data type of the array. You need not assign initial values to all array elements; if the number of initial values is less than the number of array elements, only the first elements are assigned values. The number of initial values may not exceed the number of array elements. Each initial value must be separated from the one preceding it by a comma. If several array elements are to be preset to the same value, the value can be enclosed in parentheses and preceded by a repetition factor (Table 3.8).

You can create an array as a complete variable for block parameters of data type ARRAY with the same format or of data type ANY. For example, you can copy the contents of an array variable with system function SFC 20 BLKMOV. You can also specify a single array element as a block parameter when the block parameter has the same data type as the array element.

If the individual array elements are of elementary data type, you can process them with "normal" LAD functions. An array element is referenced via an array name and a index enclosed in brackets. The index is a fixed value, and cannot be changed at runtime (no variable indexing possible).

Arrays may have as many as six dimensions. In principle, the same rules apply to multi-dimensional arrays as those which are applicable for one-dimensional arrays. The areas of the dimensions are entered in the declaration in brackets and separated from one another by a comma. When addressing the elements in multi-dimensional arrays, you must always give the indices for all dimensions.

An ARRAY variable always begins at a word boundary, that is, at a byte with an even address. ARRAY variables reserve memory up to the next word boundary. Array elements of data type BOOL begin in the least-significant bit, elements of data type BYTE and CHAR in the right-hand byte. The individual elements are listed in order. In multi-dimensional arrays, the elements are stored line by line (on the basis of the dimension), beginning with the first dimension. In the case of bit and byte elements, a new dimension always begins in the next byte; in the case of elements of other data types, a new dimension always begins in the next word (with the next even byte).

STRUCT

Data type STRUCT represents a data structure with a fixed number of elements, each of which may be of a different data type.

In the declaration, the individual structure elements may be assigned initial values (not as a block parameter for a function, an in/out parameter for a function block, or a temporary variable). The initial values must have the same data type as the structure elements (Table 3.9).

You can create a structure as a complete variable for block parameters of data type STRUCT and with the same format, or of data type ANY. For example, you can copy the contents of a STRUCT variable with system function SFC 20

Table 3.8 Examples of an Array Declaration

Name	Type	Initial Value	Comment
M_Value	ARRAY[1..24]	0.4, 1.5, 11 (2.6, 3.0)	Array variable with 24 REAL elements
	REAL		
Event	ARRAY[-10..10]	21 (TOD#08:30:00)	TOD array with 21 elements
	TIME_OF_DAY		
Result	ARRAY[1..24,1..4]	96 (L#0)	Two-dimensional array with 96 elements
	DINT		
Character	ARRAY[1..2,3..4]	2 ('a'), 2 ('b')	Two-dimensional array with 4 elements
	CHAR		

Table 3.9 Example of the Declaration of a Structure

Name		Type	Initial Value	Comment
MotCtrl		STRUCT		Simple structure variable with 4 elements
	On	BOOL	FALSE	Variable MotCtrl.On of type BOOL
	Off	BOOL	TRUE	Variable MotCtrl.Off of type BOOL
	Delay	S5TIME	S5TIME#5s	Variable MotCtrl.Delay of type S5TIME
	MaxSpeed	INT	5000	Variable MotCtrl.MaxSpeed of type INT
		END_STRUCT		

BLKMOV. You can also specify an individual structure element as a block parameter when the block parameter has the same data type as the structure element.

When the individual structure elements are of elementary data type, you can process them with "normal" LAD functions. A structure element is addressed via the structure name and, separated by a decimal point, the element name.

A STRUCT variable always begins at a word boundary, that is to say at a byte with an even address; the structure elements are in memory in the order in which they were given in the declaration. STRUCT variables reserve memory up to the next word boundary.

Elements of data type BOOL begin in the least-significant bit, elements of data type BYTE and CHAR in the right-hand byte. Elements of other data types begin at a word boundary.

You can also nest structures, that is, declare a structure as an element of another structure. The maximum nesting depth is six structures. All elements can be individually referenced with "normal" FBD functions as long as they are of elementary data type. The individual names are separated from one another by a decimal point.

3.4.4 User Data Types

A user data type (UDT) is a composite of elements of arbitrary data type. It thus corresponds to a structure (Table 3.10). You program a user data type in the SIMATIC Manager by choosing the user program named *Blocks,* then selecting the menu command INSERT → S7 BLOCK → DATA TYPE, or in the editor with FILE → NEW and UDTn with n as the number of the user data type.

A user data type is global. You can address it symbolically by assigning it a name in the symbol table. The absolute address and the data type of a UDT in the symbol table is UDT*n*, where *n* = 0 to 65 535.

With user data types, you can

▷ declare local variables (block parameters, static and temporary local data)

▷ declare variables in data blocks (individual data addresses)

▷ determine the structure of whole data blocks

You address the individual elements of a UDT as you would structure elements. For example, when you declare a variable with the name *Frame Header* with the UDT shown in the Table, you can address the first element with *Frame Header.* Ident.

Table 3.10 Example of a User Data Type UDT

Name		Type	Initial Value	Comment
		STRUCT		
	Ident	WORD	W#16#F200	UDT element Ident of type WORD
	Number	INT	0	UDT element Number of type INT
	Event	TIME_OF_DAY	TOD#0:0:0.0	UDT element Event of type TOD
		END_STRUCT		

3.4.5 Parameter Types

Parameter types are data types for block parameters (Table 3.11). The length specifications in the Table refer to the memory requirements for block parameters for function blocks. You can also use TIMER and COUNTER in the symbol table as data types for timers and counters.

The use of block parameters with parameter types is discussed in Chapter 19, "Block Parameters".

Table 3.11 Overview of Parameter Types

Parameter Type	Description		Examples of Actual Addresses	
TIMER	Timer	16 bits	T 15	
COUNTER	Counter	16 bits	C 16	
BLOCK_FC	Function	16 bits	FC 17	
BLOCK_FB	Function block	16 bits	FB 18	
BLOCK_DB	Data block	16 bits	DB 19	
BLOCK_SDB	System data block	16 bits	SDB 100	
POINTER	DB pointer	48 bits	P#M10.0 P#DB20.DBX22.2 MW 20 I 1.0	(pointer) (pointer) (address) (address)
ANY	ANY pointer	80 bits	P#DB10.DBX0.0 WORD 20 or any variable	

Basic Functions

This part of the book describes the functions of the LAD programming language that represent a certain "basic functionality". These functions allow you to program a PLC as you would contactor or relay controls.

- Arrangement of the contacts in **series or parallel circuits** is determined by the bit logic combination of binary signal states. This is how you implement the Boolean AND and OR functions.
- The **memory functions** retain the result of a logic operation (RLO) so that it can, for example, be checked and further processed at another point in the program.
- You use the **transfer functions** to exchange the values of individual operands and variables or to copy whole data areas.
- The time relays in contactor controls are represented by the **timer functions** in programmable controllers. The timer functions integrated into the CPU allow you to program wait and monitoring times, for example.
- Finally, the **counter** functions represent counters that can count up and down in the range 0 to 999.

This part of the book describes the functions using the operand areas inputs, outputs and memory bits. Inputs and outputs are the link to the process or to the plant. Memory bits correspond to auxiliary contactors that store binary states. The subsequent parts of the book then describe the remaining operand areas that you can also combine according to binary logic. This includes essentially the data bits in the global data blocks as well as the temporary and the static local data bits.

Chapter 5 "Memory Functions" contains a programming example for the binary logic operations and the memory functions, and Chapter 8 "Counter Functions" has an example of timer and counter functions. In each case, the example is in an FC function without block parameters. You can find the same examples in the form of function blocks FBs with block parameters in Chapter 19 "Block Parameters".

4 Series and Parallel Circuits

Binary signal states are combined in LAD through series and parallel connection of contacts. Series connection corresponds to an AND function and parallel connection to an OR function. You use the contacts to check the signal states of the binary operands and you combine the checked signal states by connecting the contacts in series or parallel.

A rung can consist of a single contact but also of very many contacts connected together. A rung must always be terminated, usually with a coil. In the absence of any additional labeling, a coil corresponds to the assignment of a logic operation result to a binary operand. You can negate the result of the logic operation (RLO) within the rung.

You can check the following operands with a contact:

▷ Input and output bits, memory bits

▷ Timer and counter functions

▷ Global data bits

▷ Temporary local data bits

▷ Static local data bits

▷ Status bits (evaluation of calculation results)

Absolute or symbolic addressing of an operand via a contact are both possible.

The examples shown in this chapter are also contained on the diskette accompanying the book, in function block FB 104 of the "Basic Functions" program. For incremental programming, you will find the elements for binary logic operations in the Program Element Catalog (with VIEW → CATALOG [Ctrl – K] or with INSERT → PROGRAM ELEMENTS) under "Bit Logic".

4.1 NO Contact and NC Contact

There are two contact types: Normally open (NO) contact and normally closed (NC) contact. The two contact types check the binary operands, whose addresses are given above the contact, in different ways.

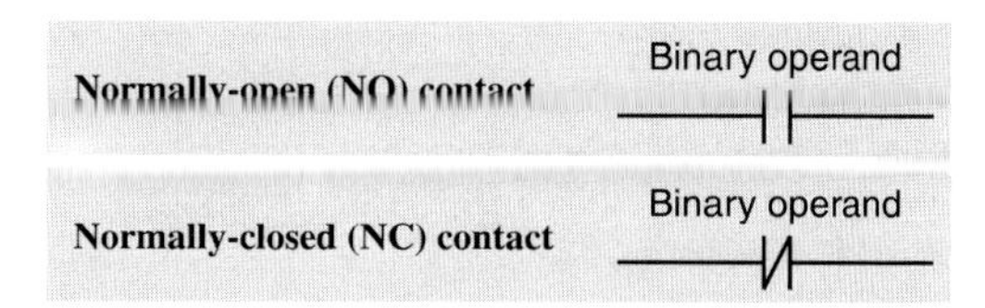

In order to explain the bit logic combinations in a ladder diagram, we will refer below as graphically as possible to "contact closed", "power flowing" and "coil energized". If "power" is flowing at a point in the ladder diagram, this means that the bit logic combination applies up to this point; the result of the logic operation (RLO) is "1". If "power" is flowing in a single coil, the coil is energized; the associated binary operand then carries signal state "1".

LAD has the NO contact and the NC contact for checking binary operands. An NO contact corresponds to a check for signal state "1": If the checked binary operand has signal state "1", the NO contact is activated, so it closes and "power flows". The example in Figure 4.1 (left side) shows sensor S1 connected to input I 1.0 and checked by an NO contact. If sensor S1 is open, input I 1.0 has signal state "0" and no power flows through the NO contact. Contactor K1, controlled by output Q 4.0 does not switch on. If sensor S1 is now activated, input I 1.0 has signal state "1". Power flows from the left power rail through the NO contact into the coil and contactor K1 connected to output Q 4.0 activates. The NO contact checks the input for signal state "1" and then closes, regardless of whether the sensor at the input is an NO or NC contact.

Power flows through an NC contact if the binary operand has the signal state "0". In the example in Figure 4.1 (right side), power flows through the NC contact if the checked sensor S2 is not closed (input I 1.1 has signal state "0"). Power also flows in the coil and energizes contactor K2 at output Q 4.1. If sensor S2 is now activated, input I 1.1 has signal state

Figure 4.1 NO Contact and NC Contact

"1" and the NC contact opens. The power flow is interrupted and contactor K2 releases. The NC contact checks the input for signal state "0" and then remains closed, regardless of whether the sensor at the input is an NO or NC contact (see also Section 4.4 "Taking Account of the Sensor Type").

4.2 Series Circuits

In series circuits, two or more contacts are connected in series. Power flows through a series circuit when all contacts are closed.

Figure 4.2 shows a typical series circuit: In network 1, the series circuit has three contacts;

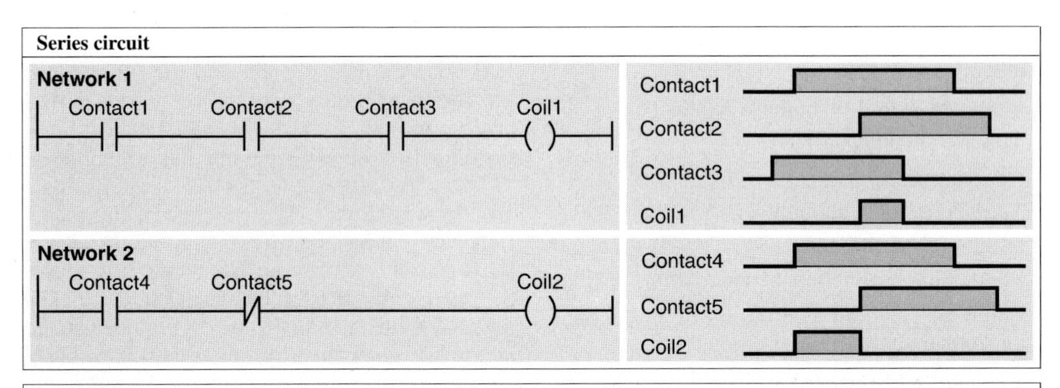

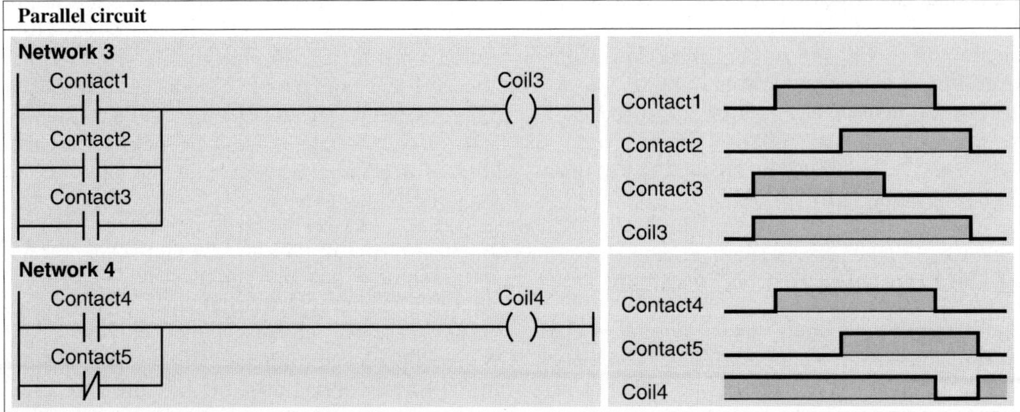

Figure 4.2 Series and Parallel Circuits

any binary operands can be checked. All contacts are NO contacts. If the associated operands all have signal state "1", (that is, if the NO contacts are activated), power flows through the rung to the coil. The operand controlled by the coil is set to signal state "1". In all other cases, no power flows and the operand *Coil1* is reset to signal state "0".

Network 2 shows a series circuit with one NC contact. Power flows through an NC contact if the associated operand has signal state "0" (that is, the NC contact is not activated). So power only flows through the series circuit in the example if the operand *Contact4* has signal state "1" and the operand *Contact5* has signal state "0".

4.3 Parallel Circuits

When two or more contacts are arranged one under the other, we refer to a parallel circuit. Power flows through a parallel circuit if one of the contacts is closed.

Figure 4.2 shows a typical parallel circuit: In network 3, the parallel circuit consists of three contacts; any binary operands can be checked. All contacts are NO contacts. If one of the operands has signal state "1", power flows through the rung to the coil. The operand controlled by the coil is set to signal state "1". If all checked operands have signal state "0", no power flows to the coil and the operand *Coil3* is reset to signal state "0".

Example 1: Both sensors are NO contacts

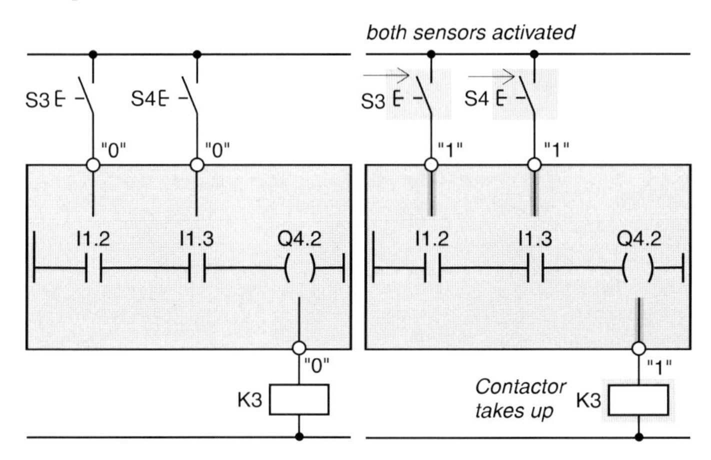

Example 2: One NO contact and one NC contact

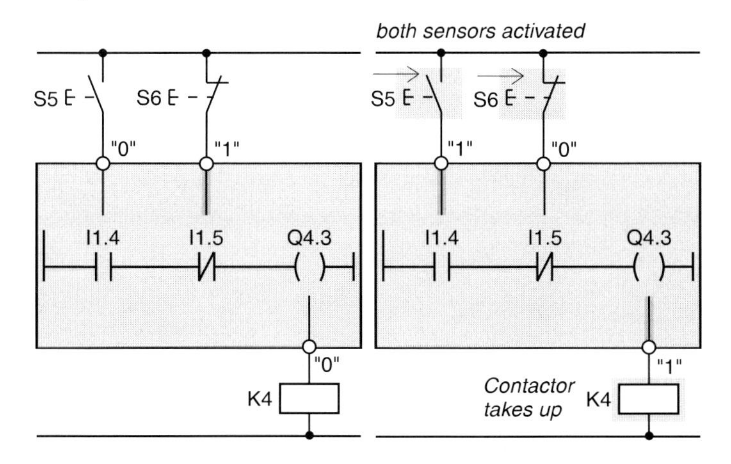

Figure 4.3
Taking Account of the Sensor Type

Network 4 shows a parallel circuit with one NC contact. Power flows through an NC contact if the associated operand has signal state "0", that is, power flows through the series circuit in the example if the operand *Contact4* has signal state "1" or the operand *Contact5* has signal state "0".

In LAD, you can also program a branch in the middle of the rung (see Figure 4.4 Network 8 as an example). You then get a parallel branch that does not begin at the left power rail. Use of LAD program elements is restricted in this parallel branch; your attention is drawn to this in the relevant chapters.

4.4 Taking Account of the Sensor Type

If you check a sensor during program creation, you must take account of whether the sensor is an NO contact or an NC contact. Depending on the sensor type, there is a different signal state at the relevant input when the sensor is activated: "1" for an NO contact and "0" for an NC contact. The CPU has no means of determining whether an input is occupied by an NO contact or by an NC contact. It can only detect signal state "1" or signal state "0".

If you design the program in such a way that you want a check result of "1" when a sensor is activated, in order to combine it further, you must check the input differently depending on the sensor type. For this purpose, the contact types NO contact and NC contact are available to you. An NO contact supplies "1" if the checked input is also "1". An NC contact supplies "1" if the checked input has "0". In this way, you can also directly check inputs that are to execute activities at signal state "0" ("zero active" inputs), and then further combine the RLO.

The example in Figure 4.3 shows programming dependent on the sensor type. In the first case,

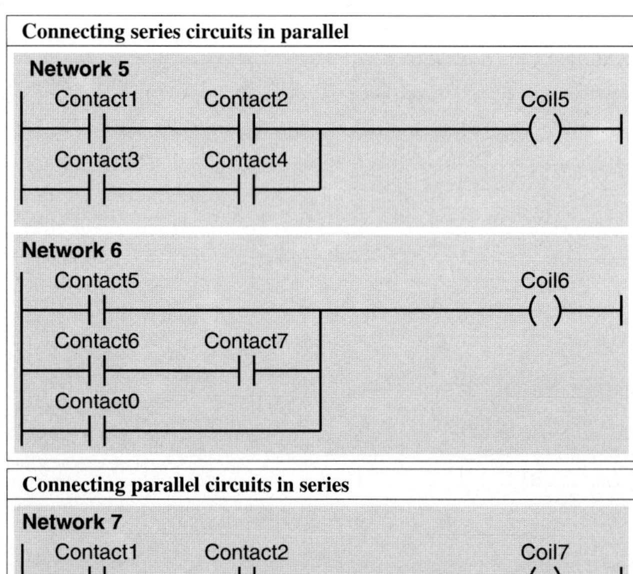

Figure 4.4
Series and parallel circuits in combination

two NO contacts are connected to the programmable controller, and in the second case one NO contact and one NC contact. In both cases, a contactor connected to an output is to take up if both sensors are activated. If an NO contact is activated, the signal state at the input is "1" and this is checked with an NO contact so that power can flow when the sensor is activated. If both NO contacts are activated, power flows through the rung to the coil and the contactor takes up.

If an NC contact is activated, the signal state at the input is "0". In order to have power flow in this case when the sensor is activated, the result must be checked with an NC contact. Therefore, in the second case, an NO contact and an NC contact must be connected in series in order to make the contactor take up when both sensors are activated.

4.5 Combinations of Binary Logic Operations

You can combine series and parallel circuits by, for example, arranging several series circuits in parallel or several parallel circuits in series, even if the series and parallel circuits themselves are complex in nature.

Connecting series circuits in parallel

Instead of contacts, you can also arrange series circuits one under the other. Figure 4.4 shows two examples. In network 5, power flows into the coil is *Contact1* and *Contact2* are closed or if *Contact3* and *Contact4* are closed. In the lower rung (network 6), power flows if *Contact5* or *Contact6* and *Contact7* or *Contact0* are closed.

Connecting parallel circuits in series

Instead of contacts, you can also arrange parallel circuits in series. Figure 4.4 shows two examples. In network 7, power flows into the coil if either *Contact1* or *Contact3* and either *Contact2* or *Contact4* are closed. To allow power to flow in the lower example (network 8), *Contact5*, *Contact0* and either *Contact6* or *Contact7* must be closed.

4.6 Negating the Result of the Logic Operation

The NOT contact negates the RLO. You can use this contact to, for example, run a series circuit negated to a coil (Figure 4.5 Network 9). Power will then only flow into the coil if there is no power in the NOT contact, that is, if either *Contact1* or *Contact2* is open (see adjacent pulse diagram).

The same applies by analogy for network 10, in which a NOT contact is inserted after a parallel circuit. Here, *Coil10* is set if neither of the contacts is closed.

You can insert NOT instead of another contact into a branch that begins at the left power rail. Inserting a NOT contact in a parallel branch that begins in the middle of a rung is not permissible.

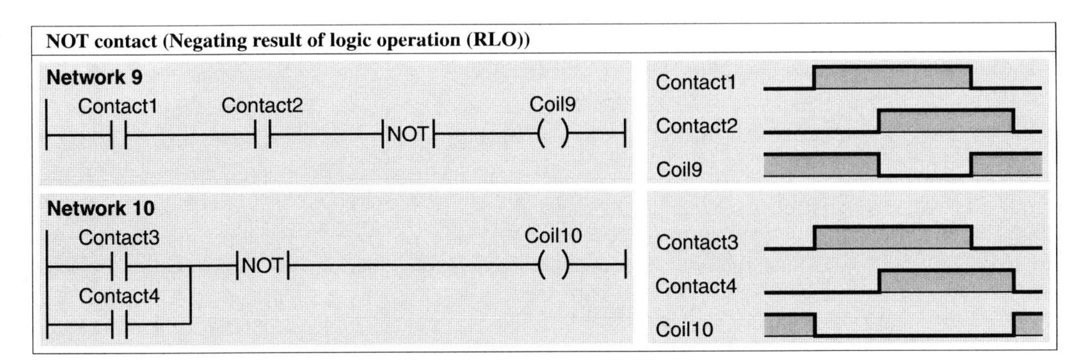

Figure 4.5 Example of the NOT Contact

5 Memory Functions

The memory functions are used in conjunction with series and parallel circuits in order to influence the signal states of the binary operands with the help of the result of logic operation (RLO) generated in the CPU.

The following are available as memory operations

▷ the single coil as an assignment of the RLO

▷ the coils S and R as individually programmed memory functions

▷ the boxes RS and SR as memory functions

▷ the midline outputs as intermediate buffer

▷ the coils P and N as edge evaluations of the power flow

▷ the boxes POS and NEG as edge evaluations of the operands

You can use the memory functions described in this chapter in conjunction with all binary operands. There are restrictions when using temporary local data bits as edge memory bits.

The examples shown in this chapter are also represented on the diskette enclosed with the book in function block FB 105 of the "Basic Functions" program. For incremental programming, you will find the program elements for the memory functions in the program element catalog (with VIEW → CATALOG [Ctrl – K] or with INSERT → PROGRAM ELEMENTS) under "Bit Logic".

5.1 Single Coil

The single coil as terminator of a rung assigns the power flow direct to the operand located at the coil. The function of the single coil depends on the Master Control Relay (MCR): If the MCR is activated, signal state "0" is assigned to the binary operand located over the coil.

If power flows into the coil, the operand is set; if there is no power, the operand is reset (Figure 5.1 Network 1). With a NOT contact before the coil, you reverse the function (Network 2).

You can also direct the power flow into several coils simultaneously by arranging the coils in parallel with the help of a branch (Network 3). All operands specified over the coils respond in the same way. Up to 16 coils can be connected in parallel.

You can arrange further contacts in series and parallel circuits after the branch and before the coil (Network 4).

See Chapter 4 "Series and Parallel Circuits" for further examples of the single coil.

5.2 Set and Reset Coil

Set and reset coils also terminate a rung. These coils only become active when power flows through them.

	Binary operand
Set coil	—(S)—
Reset coil	—(R)—

If power flows in the set coil, the operand over the coil is set to signal state "1". If power flows in the reset coil, the operand over the coil is reset to signal state "0". If there is no power in the set or reset coil, the binary operand remains unaffected (Figure 5.1 Networks 5 and 6).

The function of the set and reset coil depends on the Master Control Relay. If the MCR is activated, the binary operand over the coil is not affected.

Please note that the operand used with a set or reset coil at startup (complete restart) is usually reset. In special cases, the signal state is retained. This depends on the startup type (for ex-

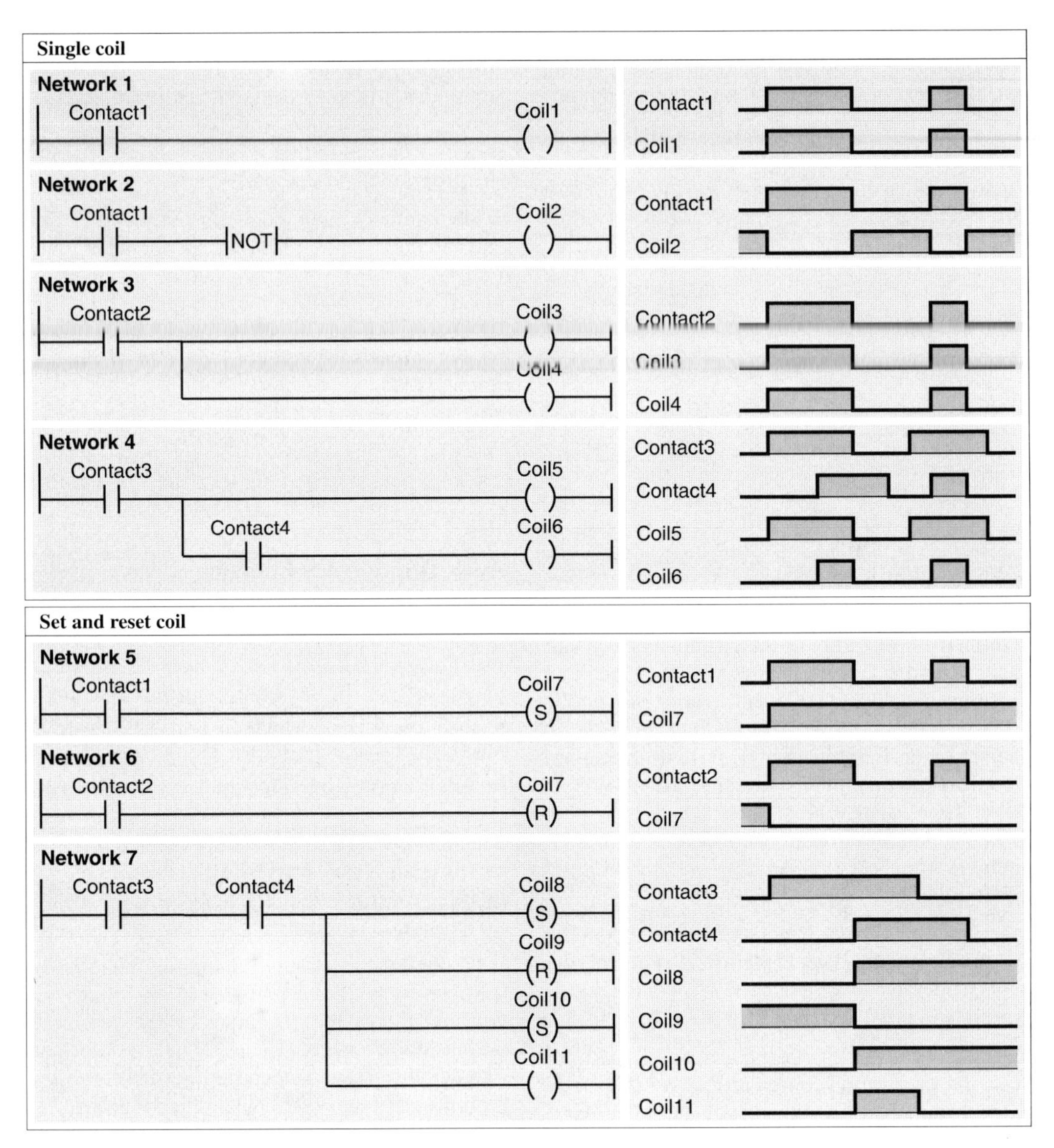

Figure 5.1 Single Coil, Set and Reset Coil

ample, restart), on the operand used (for example, static local data) and on the settings in the CPU (for example, retentive characteristic).

You can arrange several set and reset coils in any combination and together with single coils in the same rung (Network 7). To achieve clarity in your programming, it is advisable to group the set and reset coils affecting an operand together in pairs, and to use them only once in each case. You should also avoid additionally controlling these operands with a single coil.

As with the single coil, you can also arrange contacts after the branch and before a set and reset coil.

5.3 Memory Box

The functions of the set and reset coil are summarized in the box of a memory function. The common binary operand is located over the box. Input S of the box corresponds here to the set coil, input R to the reset coil. The signal state of the binary operand assigned to the memory function is at output Q of the memory function.

There are two versions of the memory function: As SR box (reset priority) and as RS box (set priority). Apart from the labeling, the boxes also differ from each other in the arrangement of the S and R input.

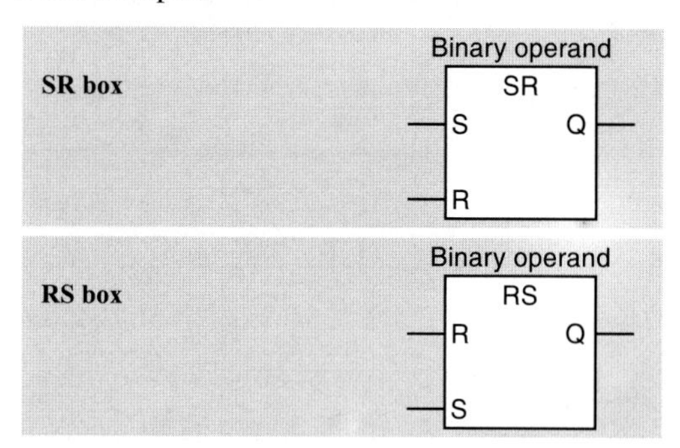

A memory function is set (or more precisely: the binary operand over the memory box is set) if the set input has signal state "1" and the reset input has signal state "0". A memory function is reset if there is "1" at the reset input and "0" at the set input. Signal state "0" at both inputs has no effect on the memory function. If both inputs have signal state "1" simultaneously, the two memory functions respond differently: the SR memory function is reset and the RS memory function is set.

The function of the memory box depends on the Master Control Relay: If the MCR is active, the binary operand of a memory box is no longer affected.

Please note that the operand used with memory function at startup (complete restart) is usually reset. In special cases, the signal state of a memory box is retained. This depends on the startup type (for example, restart), on the operand used (for example, static local data) and on the settings in the CPU (for example, retentive characteristic).

5.3.1 SR Memory Function

With the SR memory box, the reset input has priority. Reset priority means that the memory function is or remains reset if power flows "simultaneously" in the set input and the reset input. The reset input has priority over the set input (Figure 5.2 Network 8).

In accordance with sequential execution of the instructions, the CPU sets the memory operand with the set input processed first but then resets it again when processing the reset input. The memory operand remains reset while the rest of the program is processed.

If the memory operand is an output, this brief setting only takes place in the process-image output, and the (external) output on the associated output module remains unaffected. The CPU does not transfer the process-image output to the output modules until the end of the program cycle.

The memory function with reset priority is the "normal" application of the memory function since the reset state (signal state "0") is normally the safer or less hazardous state.

5.3.2 RS Memory Function

With the RS memory box, the set input has priority. Set priority means that the memory function is or remains set if power flows "simultaneously" in the set input and the reset input. The set input then has priority over the reset input (Figure 5.2 Network 9).

In accordance with sequential execution of the instructions, the CPU resets the memory operand with the reset input first processed, but then sets it again when processing the set input. The memory operand remains set while the rest of the program is processed.

If the memory operand is an output, this brief resetting only takes place in the process-image output, and the (external) output on the associated output module is not affected. The CPU does not transfer the process-image output to the output modules until the end of the program cycle.

Set priority is the exception in the application of the memory function. It is used, for example, in the implementation of a fault message buffer if the still current fault message at the set input is to continue to set the memory function despite an acknowledgment at the reset input.

5.3.3 Memory Function within a Rung

You can also place a memory box within a rung. Contacts can be connected in series and in parallel both at the inputs and at the output (Figure 5.2 Network 10). It is also possible to leave the second input of a memory box unswitched. You can

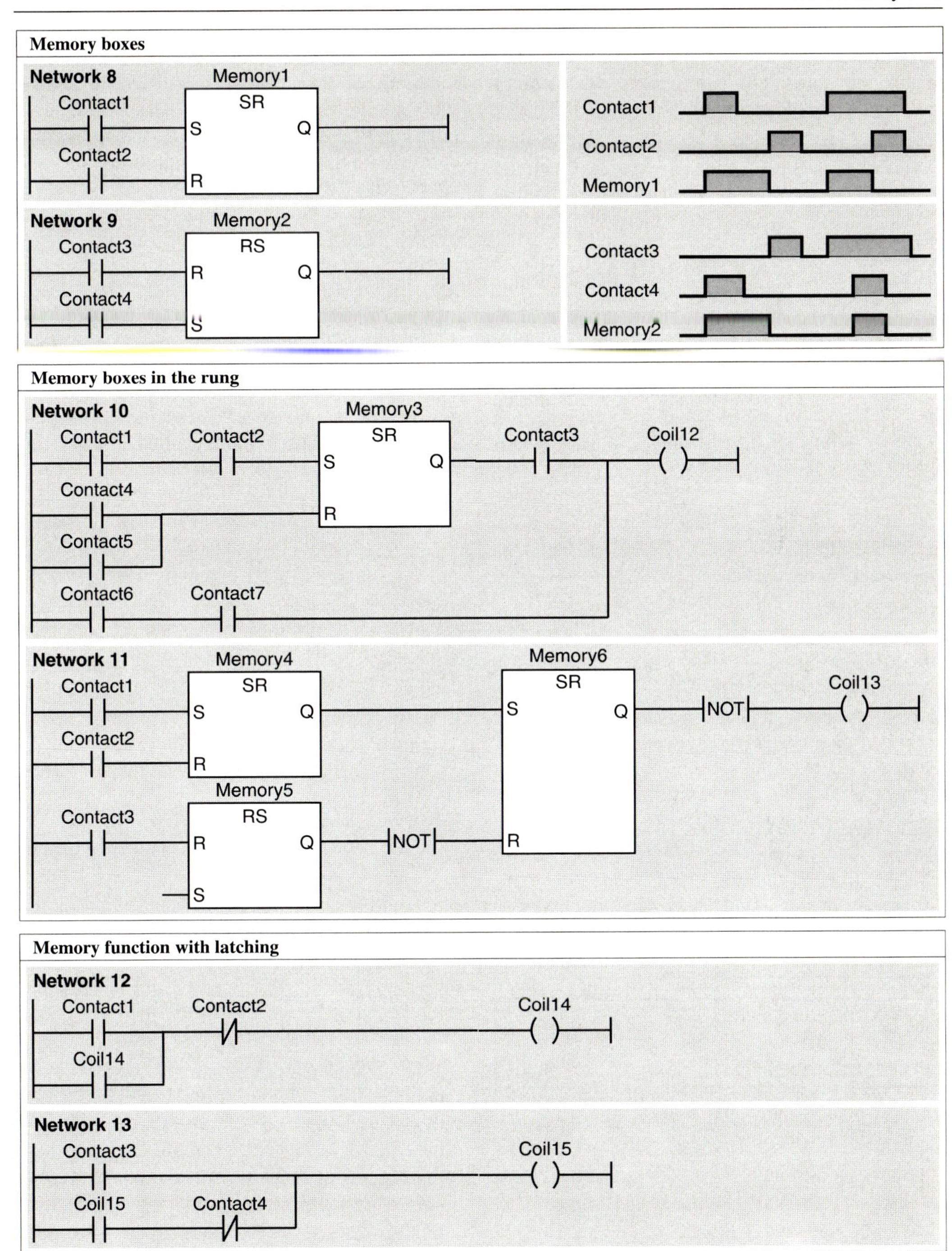

Figure 5.2 Memory Functions

also connect several memory boxes together within one rung. You can arrange the memory boxes in series or in parallel (Network 11).

You can only arrange a memory function in a branch that starts at the left power rail. Placing a memory function after a branch is not permissible.

5.4 Memory Function with Latching

In a relay logic diagram, the memory function is usually implemented by latching the output to be controlled. This method can also be used when programming in ladder logic. However, it has the disadvantage, when compared with the memory box, that the memory function is not immediately recognizable.

Networks 12 and 13 in Figure 5.2 show both types of memory function, set priority and reset priority, using latching. The principle of latching is a simple one. The binary operand controlled with the coil is checked and this check (the "contact of the coil") is connected in parallel to the set condition. If *Contact1* closes, *Coil14* energizes and closes the contact parallel to *Contact1*. If *Contact1* now opens again, *Coil14* remains energized. *Coil14* de-energizes, if *Contact2* opens. If signal state "1" is present both at *Contact1* and at *Contact2*, power does not flow into the coil (reset priority). This situation looks different in the lower network: If signal state "1" is present both at *Contact3* and at *Contact4*, power flows into the coil (set priority).

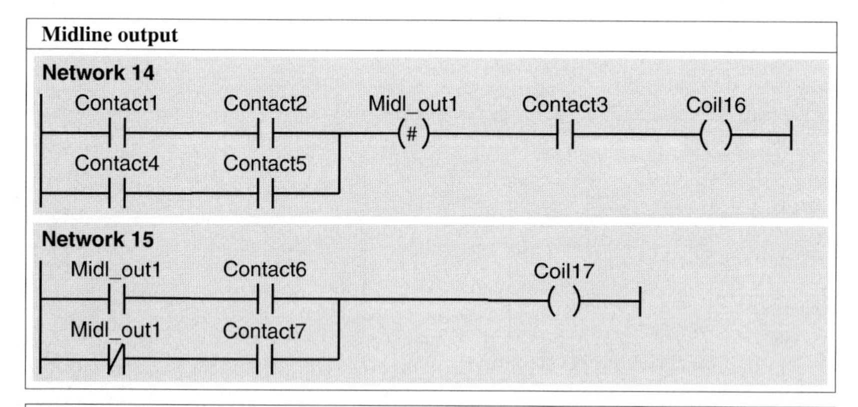

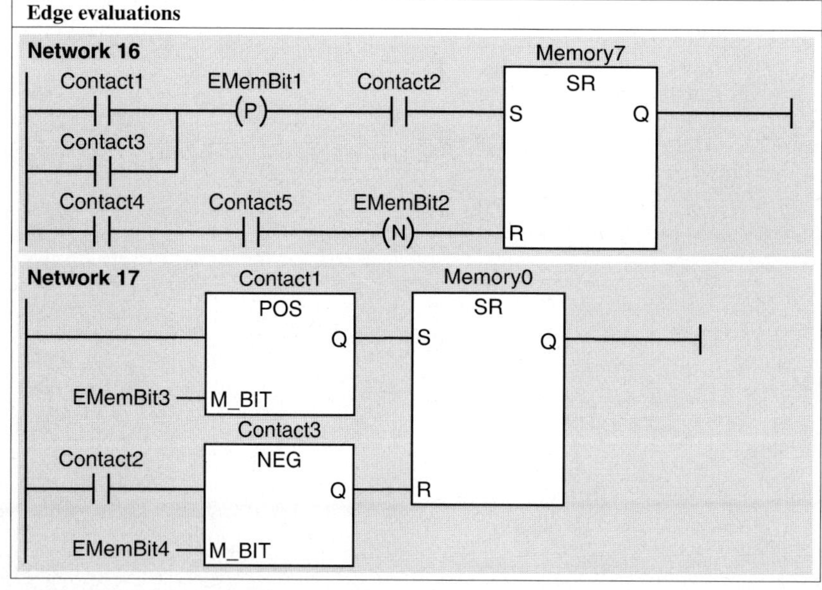

Figure 5.3
Midline Outputs and Edge Evaluations

5.5 Midline Outputs

A midline output is a single coil within a rung. The RLO valid up to this point (the power that flows in the rung at this point), is stored in the binary operand over the midline output. The midline output itself has no effect on the power flow.

The function of the midline output is dependent on the Master Control Relay: If the MCR is activated, signal state "0" is assigned to the binary operand at the midline output. There is then no more power flow after the midline output (the RLO is then "0").

You can check the binary operand over the midline output at another point in the program with NO and NC contacts. Several midline outputs can be programmed in one rung.

You can only arrange a midline output in a branch that starts at the left power rail. However, it must not be located direct at the power rail. Placing it after a branch is not permissible. A midline output must not terminate a rung; the single coil is available for this purpose.

Figure 5.3 shows an example of how an intermediate result is stored in a midline output. The RLO from the circuit formed by *Contact1*, *Contact2*, *Contact4* and *Contact5* is stored in midline output *Midl_out1*. If the logic operation is fulfilled (power flows in the midline output) and if *Contact3* is closed, *Coil16* is energized. The RLO stored is used in two ways in the next network. On the one hand, a check is made to see if the logic operation was fulfilled and the bit logic combination made with *Contact6*, and on the other hand, a check is made to see if the logic operation was not fulfilled and a bit logic combination made with *Contact7*.

The following binary operands are suitable for intermediate storage of binary results:

▷ You can use temporary local data bits if you only require the intermediate result within the block. All code blocks have temporary local data.

▷ Static local data bits are available only within a function block; they store the signal state until the next control, even beyond processing in the current block.

▷ Memory bits are available globally in a fixed CPU-specific quantity; for clarity of programming, try to avoid multiple use of memory bits (the same memory for different tasks).

▷ Data bits in global data blocks are also available in the entire program, but before they are used they require the relevant data block to be opened (even if this happens implicitly with the complete addressing).

Note: You can replace the "scratchpad memory" used in STEP 5 with the temporary local data available in every block.

5.6 Edge Evaluation

The LAD programming language provides four different elements for edge evaluation:

5.6.1 Principle of Operation of an Edge Evaluation

With an edge evaluation, you detect the change in a signal state, a signal edge. A positive (rising) edge exists when the signal changes from "0" to "1". The opposite is referred to as a negative (falling) edge.

In the relay logic diagram, the equivalent to an edge evaluation is the pulse contact element. If this pulse contact element emits a pulse when the relay is switched on, this corresponds to the rising edge. A pulse from the pulse contact element on switching off corresponds to a falling edge.

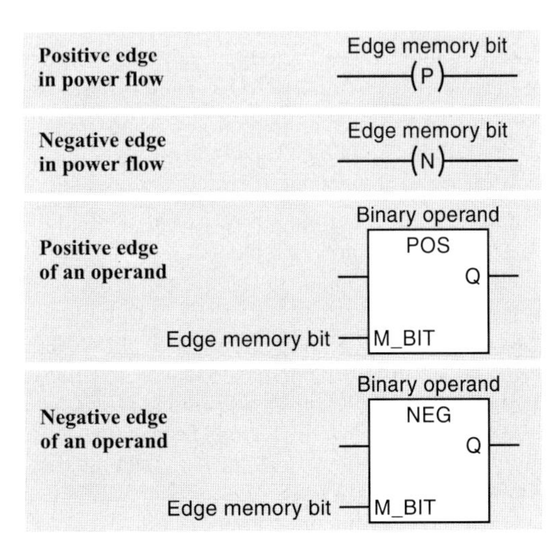

Detection of a signal edge (change in a signal state) is implemented in the program. The CPU compares the current RLO (for example, the result of an input check) with a stored RLO. If the two signal states are different, a signal edge is present.

The stored RLO is located in an "edge memory bit" (it does not have to be a memory bit). It must be an operand whose signal state is available again when edge evaluation is next processed (in the next program cycle) and which you do not intend to use further in the program. Memory bits, data bits in global data blocks and static local data bits in function blocks are all suitable as operands.

This edge memory bit stores the "old" RLO, with which the CPU last processed the edge evaluation. If a signal edge is now present, that is, if the current RLO is different to the signal state of the edge memory bit, the CPU corrects the signal state of the edge memory bit by assigning to it the now current RLO. When the edge evaluation is next processed (usually in the next program cycle), the signal state of the edge memory bit is the same as the current RLO (if this has not changed in the meantime) and the CPU no longer detects an edge.

A detected edge is indicated by the RLO after edge evaluation. If the CPU detects a signal edge, it sets the RLO to "1" after edge evaluation (power then flows). If there is no signal edge, the RLO equals "0".

Signal state "1" after an edge evaluation therefore means "edge detected". Signal state "1" is present only briefly, usually only for the length of one execution cycle. Since the CPU does not detect an edge when the edge evaluation is next processed (if the "input RLO" of the edge evaluation does not change) it sets the RLO back to "0" after edge evaluation.

You can process the RLO direct after an edge evaluation, for example, store it with a set coil, combine it with subsequent contacts or store it in a binary operand ("pulse memory bit"). A pulse memory bit is used if the RLO of the edge evaluation is also to be processed at another point in the program; it is effectively the intermediate memory for a detected edge (the pulse contact element in the relay logic diagram). Memory bits, data bits in global data blocks, temporary and static local data bits are all suitable operands for the pulse memory bit.

Please note the behavior of edge evaluation when switching on the CPU. If no edge is to be evaluated, the RLO prior to edge evaluation and the signal state of the edge memory bit must be identical on switching on. In certain circumstances, the edge memory bit must be reset at startup (depending on the desired behavior and on the operands used).

5.6.2 Edge Evaluation in the Power Flow

An edge evaluation in the power flow is indicated by a coil that contains a P (for positive, rising edge) or an N (for negative, falling edge). Above the coil is an edge memory bit, a binary operand, in which the "old" RLO from the preceding processing of the edge evaluation is stored. An edge evaluation like this detects a change in the power flow from "power flowing" to "power not flowing" and vice versa.

The example in Figure 5.3 shows a positive and a negative edge evaluation in Network 16. If the parallel circuit consisting of *Contact1* and *Contact3* is fulfilled, the edge evaluation emits a brief pulse with *EMemBit1*. If *Contact2* is closed at this instant, *Memory7* is set. *Memory7* is reset again by a pulse from *EMemBit2* if the series circuit consisting of *Contact4* and *Contact5* interrupts the power flow.

You must only arrange an edge evaluation with coil in a branch that starts at the left power rail. Placing it direct at the left power rail and after a branch is not permissible.

5.6.3 Edge Evaluation of an Operand

LAD represents the edge evaluation of an operand using a box. Above the box is the operand whose signal state change is to be evaluated. The edge memory bit that stores the "old" signal state from the preceding processing is located at input M_BIT.

With the unlabeled input and the output Q, the edge evaluation is "inserted" in the rung instead of a contact. If power flows into the unlabeled input, output Q emits a pulse at an edge; if no power flows in this input, output Q is also always reset. You can arrange this edge evaluation in the place of any contact even in a parallel branch that does not begin at the left power rail.

Figure 5.3 shows the use of an edge evaluation of an operand in Network 17. The edge evaluation in the upper branch emits a pulse if the operand *Contact1* changes its signal state from "0" to "1" (positive edge). This pulse sets *Memory0*. The edge evaluation is always enabled by the direct connection of the unlabeled input to the left power rail. The lower edge evaluation is enabled by *Contact2*. If it is enabled with "1" at this input, it emits a pulse if the binary operand *Contact3* changes its signal state from "1" to "0" (negative edge).

5.7 Binary Scaler

A binary scaler has one input and one output. If the signal at the input of the binary scaler changes its state, for example, from "0" to "1", the output also changes its signal state (Figure 5.4). This (new) signal state is retained until the next, in our example positive, signal state change. Only then does the signal state of the output change again. This means that half the input frequency appears at the output of the binary scaler.

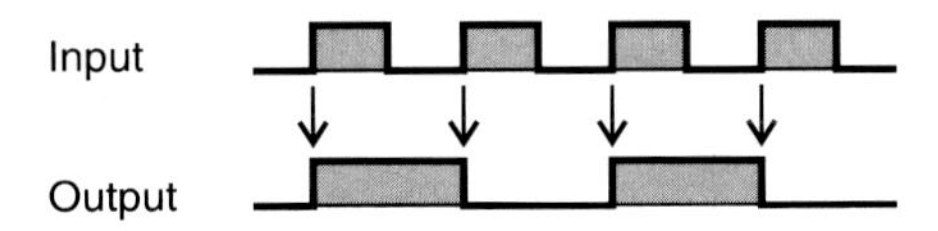

Figure 5.4 Pulse Diagram of a Binary Scaler

There are different methods of solving this task, two of which are introduce below.

The first solution uses memory functions (Figure 5.5, Networks 18 and 19). If the operand *Input_1* has signal state "1", the operand *Output_1* is set (the operand *Memory_1* is still reset). If the signal state at *Input_1* changes to "0", *Memory_1* is also set (*Output_1* now has "1"). If *Input_1* has "1" next time, *Output_1* is reset again (*Memory_1* is now "1"). If *Input_1* now has "0" again, *Memory_1* is reset (since *Output_1* is now also reset). Now the "basic state" has been reached again after two input pulses and one output pulse.

The second solution uses the latching function usual to relay logic diagrams (Networks 20 and 21). The principle of operation is the same as for the first solution although the reset condition is "zero-active" as is usual in the case of latching.

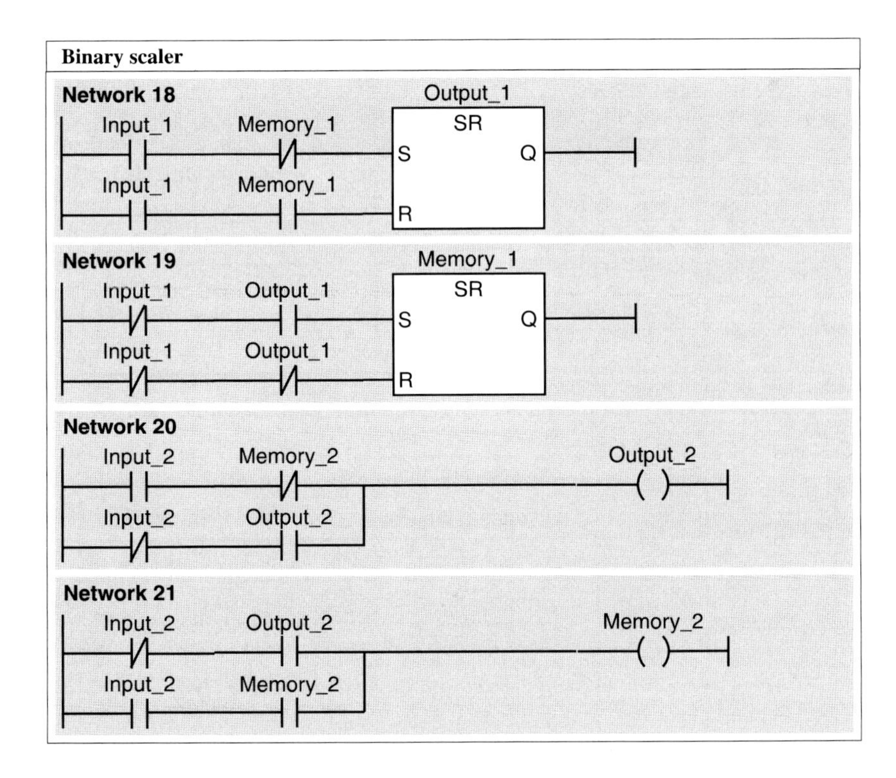

Figure 5.5 Binary Scaler Examples

5.8 Conveyor Belt Control Example

Using the example of a functionally extremely simple conveyor belt control, the principle of operation of the binary logic operations and memory functions is illustrated in conjunction with inputs, outputs and memory bits.

Functional Description

Parts are to be transported on a conveyor belt; one crate or pallet per belt. The essential functions are as follows:

• when the belt is empty, the controller requests more parts with the signal "readyload" (ready to load)

• with the signal "Start", the belt starts up and transports the parts

• at the end of the conveyor belt, an "end-of-belt" sensor (a light barrier, for example) detects the parts, at which point the belt motor switches off and triggers the signal " ready_rem" (ready to remove)

• with the signal "continue", the parts are transported further until the "end-belt" (end-of-belt) sensor no longer detects them.

The example is programmed with inputs, outputs and memory bits. It can be located at any point in any block. In this case, a function without functional value has been selected as the block.

At this point, the example is programmed with memory boxes. In Chapter 19 "Block Parameters", the same example is given with latching functions. The program is located there in a function block with block parameters that can be called several times (for several conveyor belts).

Signals, symbols

A few additional signals supplement the functionality of the conveyor belt control:

▷ Basic_st Sets the controller to the basic state

▷ Man_on Switches on the belt, regardless of conditions

Table 5.1 Symbol Table for the "Conveyor Belt Control" Example

Symbol	Address	Data Type	Comment
Belt_controller	FC 11	FC 11	Control of the conveyor belt
Basic_st	I 0.0	BOOL	Set controllers to the basic state
Man_on	I 0.1	BOOL	Switch on conveyor belt motor
/Stop	I 0.2	BOOL	Stop conveyor belt motor (zero-active)
Start	I 0.3	BOOL	Start conveyor belt
Continue	I 0.4	BOOL	Acknowledgment that parts removed
Lbarr1	I 1.0	BOOL	(Light barrier) sensor signal "End of belt" for conveyor belt 1
/Mfault1	I 2.0	BOOL	Motor protection switch conveyor belt 1, zero-active
Readyload	Q 4.0	BOOL	Load new parts onto belt (ready to load)
Ready_rem	Q 4.1	BOOL	Remove parts from belt (ready to remove)
Belt_mot1	Q 5.0	BOOL	Switch on belt motor for conveyor belt 1
Load	M 2.0	BOOL	Load parts command
Remove	M 2.1	BOOL	Remove parts command
EM_Rem_N	M 2.2	BOOL	Edge memory bit for negative edge of "remove"
EM_Rem_P	M 2.3	BOOL	Edge memory bit for positive edge of "remove"
EM_Loa_N	M 2.4	BOOL	Edge memory bit for negative edge of "load"
EM_Loa_P	M 2.5	BOOL	Edge memory bit for positive edge of "load"

▷ /Stop — Stops the conveyor as long as the signal "0" is present (an NC contact as sensor, "zero active")

▷ End_belt — The parts have reached the end of the belt

▷ /Mfault — Fault signal of the belt motor (e.g. motor protection switch); designed as "zero active" signal so that, for example, a wirebreak also results in a fault signal

We want symbolic addressing, that is, the operands receive names with which we then program. Before entering the program we create a symbol table (Table 5.1) that contains the inputs, outputs, memory bits and blocks.

Program

The example is located in a function block that you call in organization block OB 1 (selected from the Program Elements Catalog "FC Blocks") for processing in a CPU.

When programming, the global symbols can also be used without quotation marks provided they do not contain any special characters. If a symbol does contain a special character (for example an Umlaut or a space), it must be placed within quotation marks. In the compiled block, the Editor indicates all global symbols with quotation marks.

Figure 5.6 shows the program of the conveyor belt control. On the diskette supplied with the book, you can find this program in function block FC 11 under "Conveyor Example".

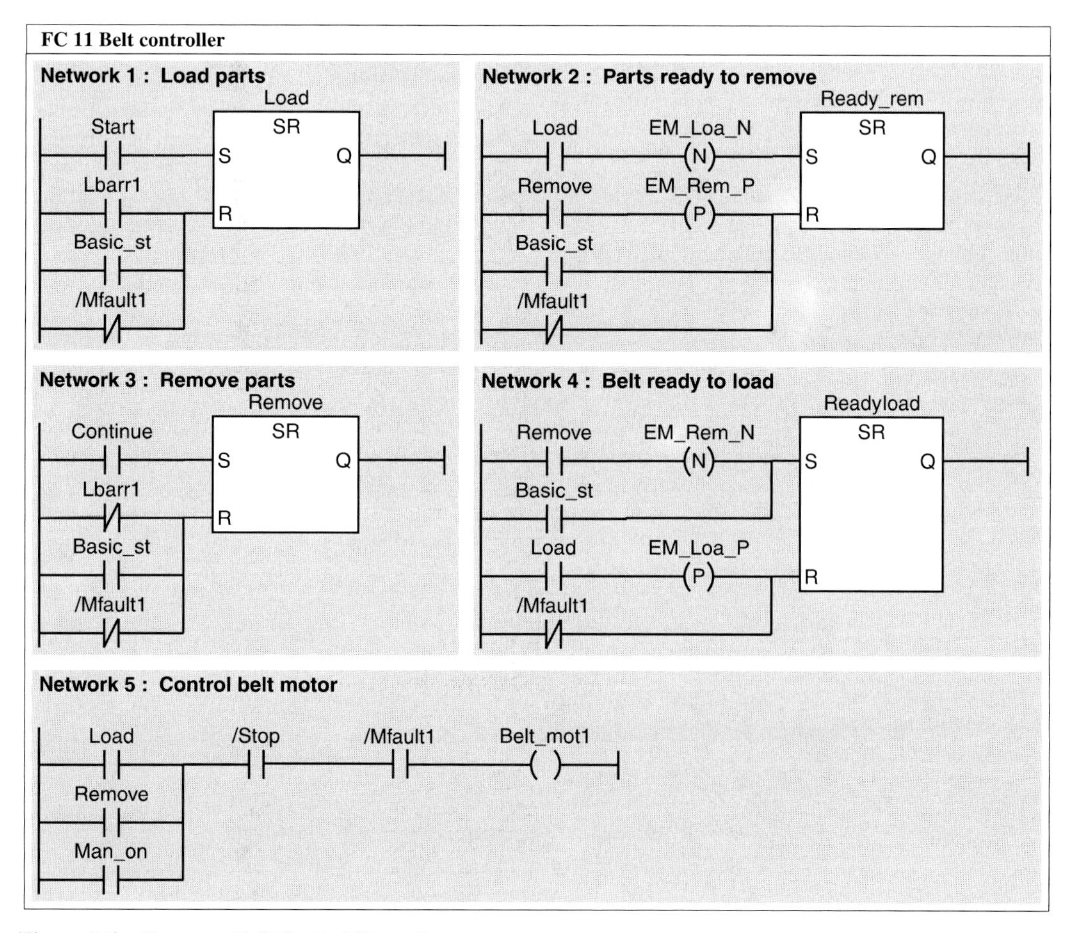

Figure 5.6 Conveyor Belt Control Example

6 Move Functions

The LAD programming language provides the following move functions

▷ MOVE box
Copy operands and variables with elementary data types

▷ SFC 20 BLKMOV
Copy data area

▷ SFC 21 FILL
Fill data area

6.1 General

You use the move functions to copy information between the system memory, the work memory and the user data area of the modules (Figure 6.1).

Information is transferred via a CPU-internal register that functions as intermediate storage. This register is called accumulator 1. Moving information from memory to accumulator 1 is referred to as "loading" and moving from accumulator 1 to memory is called "transferring". The MOVE box contains both transmission paths. It moves information pending at the input IN to accumulator 1 (load) and immediately following this from accumulator 1 to the operand at the output OUT (transfer).

6.2 MOVE Box

6.2.1 Processing the MOVE Box

Representation

In addition to the enable input EN and the enable output ENO, the MOVE box has an input IN and an output OUT. At the input IN and the output OUT, you can apply all digital operands and digital variables that are of elementary data types (except BOOL). The variables at input IN and output OUT can have different data types.

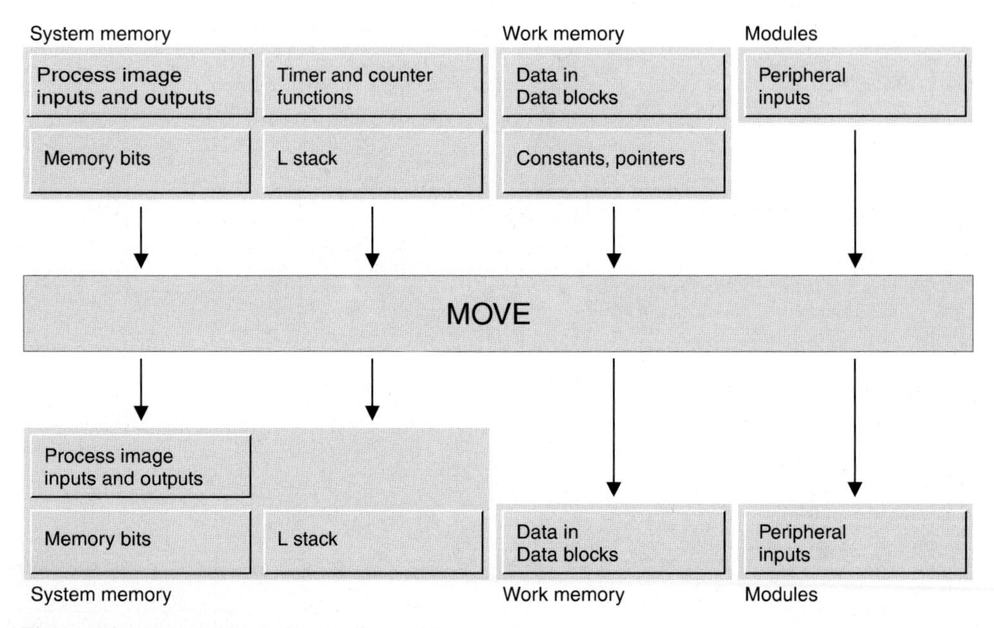

Figure 6.1 Memory Areas for Loading and Transferring

See Section 3.4 “Data Types” for the bit assignments of the data formats.

Different operand widths

The operand widths (byte, word, doubleword) at the input and the output of the MOVE box may vary. If the operand at the input is “less” than at the output, it is moved to the output operand right-justified and the location on the left is filled with zero. If the input operand is “greater than” the output operand, only that part of the input operand on the right that fits into the output operand is moved.

Figure 6.2 explains this. A byte or a word at the input is loaded right-justified into accumulator 1 and the remainder is filled with zero. A byte or a word at output OUT is removed right-justified from accumulator 1.

Function

The MOVE box moves the information of the operand at input IN to the operand at output OUT. The MOVE box only moves information when the enable input has “1” or is unswitched and when the master control relay is switched off. If EN = “1” and the MCR is switched on, zero is written to output OUT. With “0” at the enable input, the operand at output OUT is unaffected. MOVE does not signal an error (Figure 6.3).

<table>
<tr><td colspan="2">IF EN == "1"
THEN</td><td>ELSE</td></tr>
<tr><td colspan="2">ENO := "1"</td><td rowspan="3">ENO := "0"</td></tr>
<tr><td>IF MCR active
THEN</td><td>ELSE</td></tr>
<tr><td>**OUT := 0**</td><td>**OUT := IN**</td></tr>
</table>

Figure 6.3 Function of the MOVE Box

Example

The contents of input word IW 0 is moved to memory word MW 60.

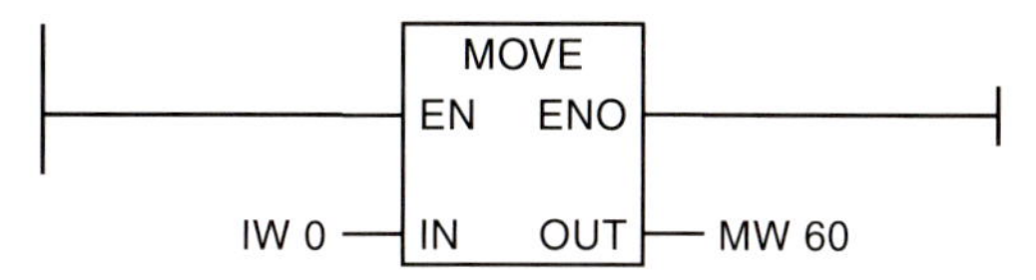

MOVE box in a rung

You can arrange contacts in series and parallel before the input EN and after the output ENO.

The MOVE box itself must only be placed in a branch that leads direct to the left power rail. This branch can also have contacts before the input EN and it need not be the top branch. With the direct connection to the left power rail you can con-

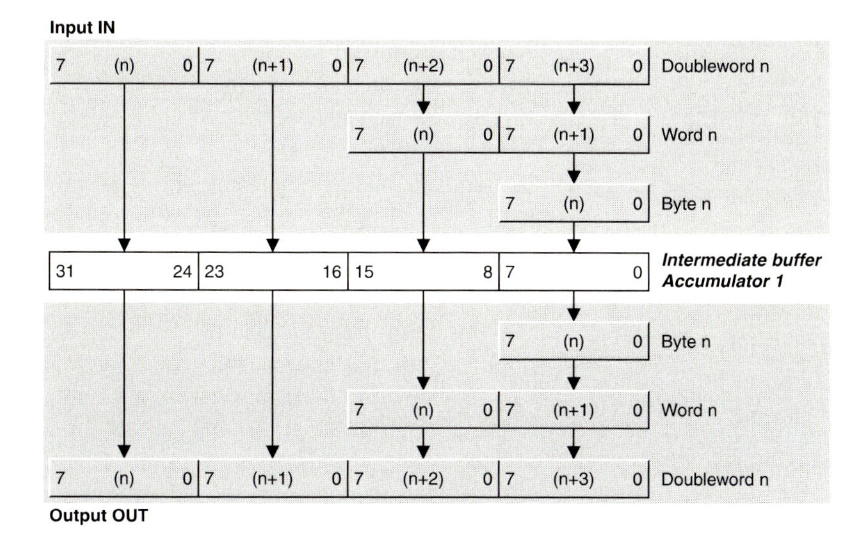

Figure 6.2 Moving Different Operand Widths

nect MOVE boxes in parallel. When connecting boxes in parallel, you require a coil to terminate the rung. If you have not provided error evaluation, assign a “dummy” operand to the coil, for example, with a temporary local data bit.

You can connect MOVE boxes in series. In doing so, the ENO output of the preceding box leads to the EN input of the following box.

If you arrange several MOVE boxes in one rung (parallel at the left power rail and then continuing in series), the boxes in the uppermost branch are processed first, from left to right, and then the boxes in the parallel branch from left to right, etc.

You can find examples of the move functions in the diskette supplied with the book (FB 106 in “Basic Functions” program).

For incremental programming, you will find the MOVE box in the Program Elements Catalog (with VIEW → CATALOG[Ctrl – K] or with INSERT → PROGRAM ELEMENTS) under “Move”.

6.2.2 Moving Operands

In addition to the operands mentioned in this chapter, you can also move timer and counter values with the MOVE box (see Chapter 7 “Timer Functions” and 8 “Counter Functions”). Section 18.2 “Block Functions for Data Blocks” deals with using data operands.

Moving inputs

IB n — Moving an input byte
IW n — Moving an input word
ID n — Moving an input doubleword

On the S7-400, loading from the process image input or transferring to the process-image input is permissible only for the input bytes that are also available as input module. On the S7-300, you can access the entire available process image.

Moving outputs

QB n — Moving an output byte
QW n — Moving an output work
QD n — Moving an output doubleword

On the S7-400, loading from the process-image output or transferring to the process-image output is permissible only for the output bytes that are also available as output module. On the S7-300, you can access the entire available process image.

Moving from the I/O

PIB n — Loading a peripheral input byte
PIW n — Loading a peripheral input word
PID n — Loading a peripheral input doubleword
PQB n — Transferring to a peripheral output byte
PQW n — Transferring to a peripheral output word
PQD n — Transferring to a peripheral output doubleword

When moving in the I/O area, you access different operands depending on the direction of the move. You specify I/O inputs PI at the IN input of the MOVE box, and I/O outputs at the OUT output.

When moving from the I/O to the memory (loading), the input modules are accessed as peripheral inputs PI. Only the available modules must be accessed. Please note that direct loading from the I/O modules can move a different value than loading from the inputs of the module with the same address: While the signal state of the inputs corresponds to the value at the start of the program cycle (when the CPU updated the process image) the value loaded direct from the I/O modules is the current value.

Transferring to the I/O uses the operand I/O outputs PQ. Only those addresses can be accessed that are also occupied by output modules. Transferring to I/O modules that have a process-image output, simultaneously corrects this process-image, so that there is no difference between identically addressed outputs and peripheral outputs.

Moving memory bits

MB n — Moving a memory byte
MW n — Moving a memory word
MD n — Moving a memory doubleword

Moving from and to the memory area is always permissible since the memory bits are available in full in the CPU. Please note here the differences in memory area size between CPUs.

Moving temporary local data

LB n — Moving a local data byte
LW n — Moving a local data word
LD n — Moving a local data doubleword

Moving from and to the L stack is always permissible. Please note here the information in Section 1.5.3 “Temporary Local Data”.

6.2.3 Moving Constants

You can specify constant values only at the input IN of the MOVE box.

Moving constants with elementary data types

You can movea fixed value to an operand immediately. You can specify this constant in different representation formats for enhanced clarity. In Section 3.4.2 "Elementary Data Types", you will find an overview of the possible representations. All constants that can be moved using the MOVE box belong to the elementary data types. Examples:

B#16#F1 Moving a 2-digit hexadecimal number
–1000 Moving an INT number
5.0 Moving a REAL number
S5T#2s Moving an S5 timer
TOD#8:30 Moving a time-of-day

Moving pointers

Pointers are a special form of constant used for calculating addresses in standard blocks. You can use the MOVE box to store these pointers in operands.

P#1.0 Moving an area-internal pointer
P#M2.1 Moving an area-crossing pointer

6.3 System Functions for Data Transfer

The following system functions are available for data transfer

▷ SFC 20 BLKMOV
Copy data area

▷ SFC 21 FILL
Fill data area

These system functions have two parameters each with data type ANY (Table 6.1). In principle, you can assign any operand, any variable or any absolute addressed area to these parameters. If you use a variable with complex data type, it can only be a "complete" variable; components of a variable (for example, individual field or structure components) are not permissible. Use the ANY pointer for specifying an absolute addressed area.

6.3.1 ANY Pointer

You require the ANY pointer when you want to specify an absolute addressed operand area at a block parameter of the type ANY. The general representation of the ANY pointer is as follows:

P#[DataBlock.]Operand Type Quantity

Examples:

P#M16.0 BYTE 8
Area with 8 bytes from MB 16

P#DB11.DBX30.0 INT 12
Area with 12 words in DB 11 from DBB 30

P#I18.0 WORD 1
Input word IW 18

P#I1.0 BOOL 1
Input I 1.0

Please note that the operand address in the ANY pointer must always be a bit address.

It makes sense to specify a constant pointer when you want to access a data area for which you have not declared variables. In principle, you can assign variables or operands to an ANY

Table 6.1 SFC 20 and 21 Parameters

SFC	Parameter	Declaration	Data Type	Assignment, Description
20	SRCBLK	INPUT	ANY	Source area to be copied from
	RET_VAL	OUTPUT	INT	Error information
	DSTBLK	OUTPUT	ANY	Destination area to be copied to
21	BVAL	INPUT	ANY	Source area to be copied
	RET_VAL	OUTPUT	INT	Error information
	BLK	OUTPUT	ANY	Destination area to which the source area is to be copied (multiple copies also)

parameter. For example, the representation “P#I1.0 BOOL 1” is identical with “I 1.0” or the relevant symbolic address.

6.3.2 Copy Data Area

The system function SFC 20 BLKMOV copies the contents of a source area (parameter SRCBLK) in increasing address direction (incrementing) to a destination area (parameter DSTBLK). At the parameters, you can assign any variables from the operand areas inputs I, outputs Q, memory bits M, local data L (except data type ANY) and data blocks (variables from global data blocks and from instance data blocks), as well as absolute addressed data areas specifying an ANY pointer.

You cannot use SFC 20 to copy timer and counter functions or information from and to the modules (I/O operand area) and system data blocks (SDBs).

In the case of inputs and outputs, the specified area is copied regardless of the input or output modules actually assigned. You can also specify a variable or an area from a data block as the source area in the load memory (data block programmed with the keyword UNLINKED).

Source and destination area must not overlap. If the source and destination area are of different lengths the transfer is made only to the length of the smaller area.

Example (Figure 6.4 Network 4): Starting from memory byte MB 64, 16 bytes are to be copied to data block DB 124 starting from data byte DBB 0.

6.3.3 Fill Data Area

System function SFC 21 FILL copies a specified value (source area) to a memory area (destination area) as often as required to fully overwrite the destination area. Transfer takes place in increasing address direction (incrementing). At the parameters, you can assign any variables from the operand areas inputs I, outputs Q, memory bits M, local data L (except data type ANY) and data blocks (variables from global data blocks and from instance data blocks), as well as absolute addressed data areas specifying an ANY pointer.

You cannot use SFC 21 to copy timer and counter functions or information from and to the modules (I/O operand area) and system data blocks (SDBs).

In the case of inputs and outputs, the specified area is copied regardless of the input or output modules actually assigned.

The source and destination area must not overlap. The destination is always completely overwritten even if the source area is greater than the destination area or if the length of the destination area is not an integer multiple of the length of the source area.

Example (Figure 6.4 Network 5): The contents of memory byte MB 80 is to be copied 16 times to data block DB 124 from DBB 16.

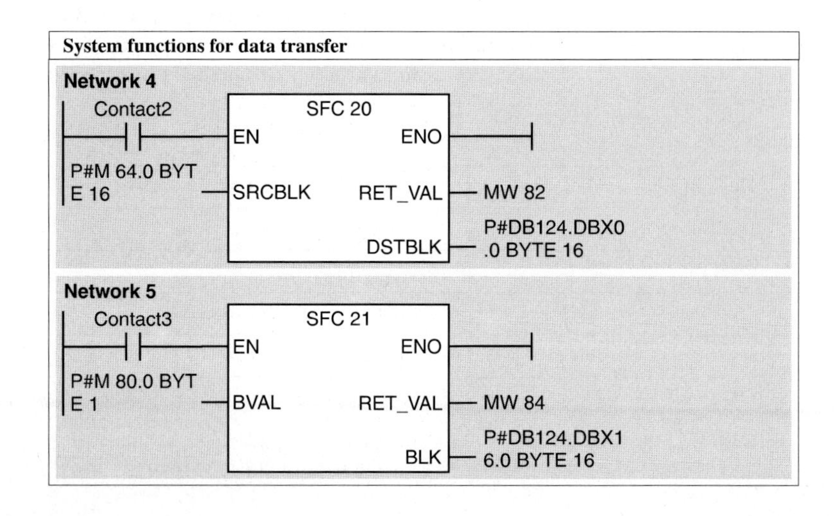

Figure 6.4 Examples of SFC 20 BLKMOV and SFC 21 FILL

7 Timer Functions

The timer functions allow software implementation of time sequences such as waiting and monitoring times, measurements of a time interval or the generation of pulses. The timer functions are located in the system memory of the CPU and the number of timer functions is CPU-specific. When parameterizing some CPUs, you can use the "Cycle/Clock Memory" thumb index to specify how many timer functions from T 0 the operating system is to updated. On the S7-300, the timer functions are only updated if the main program in OB 1 is being processed.

The following timer types are available for a timer function:

▷ Pulse generation

▷ Extended pulse

▷ On delay

▷ Retentive on delay

▷ Off delay

You can program a timer function complete as a box or with individual program elements. When starting a timer function, you specify the timer type and the duration; you can also reset a timer function. A timer function is checked via the timer status ("Timer running") or via the current time value that you can see from the timer function coded either in binary or BCD.

7.1 Programming a Timer Function

7.1.1 General Representation of a Timer Function

You can perform the following operations with a timer function:

▷ Start a timer specifying the time value

▷ Reset a timer

▷ Check the (binary) timer status

▷ Check the (digital) time value in binary coding

▷ Check the (digital) timer duration in BCD code

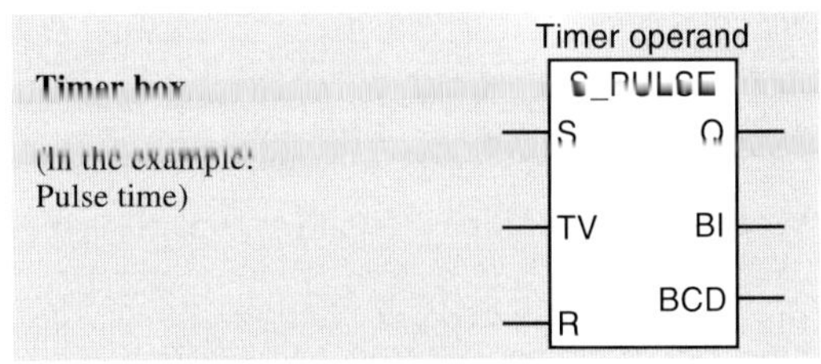

Name	Data type	Description
S	BOOL	Start input
TV	S5TIME	Time duration specification
R	BOOL	Reset input
Q	BOOL	Timer status
BI	WORD	Current binary-coded timer value
BCD	WORD	Current BCD-coded timer value

Figure 7.1 Timer Function in Box Representation

The timer function box contains the coherent representation of all timer operations in the form of function inputs and outputs (Figure 7.1). Above the box is the timer function address in absolute or symbolic form. In the box, as a header, is the timer type (S_PULSE means "Start as pulse"). Assignment of the S and TV inputs is mandatory and assignment of the other inputs and outputs is optional.

You can also program a timer function with individual program elements (Figure 7.2). The timer is then started via a coil. The timer type is in the coil (SP = start as pulse), and below the coil is the duration, in S5TIME data format, with which the timer function is started. To reset a timer function, use the reset coil, and to check the timer status, use an NO or NC contact. You can then store the current time value in binary coded form in a word-width operand using the MOVE box.

7.1.2 Starting a Timer Function

A timer function starts (the time begins) when the RLO changes before the start input or before the start coil. A signal state change like this is always required to start a timer function. With

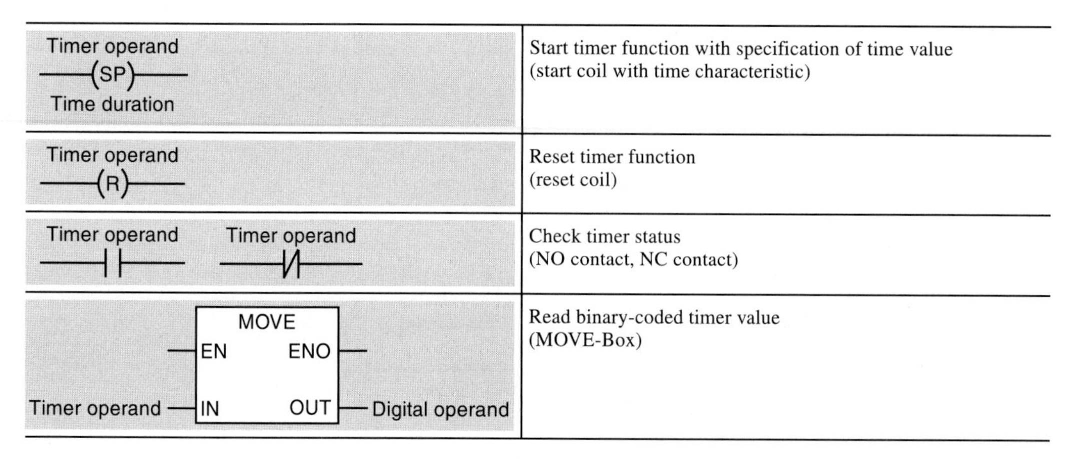

Figure 7.2 Individual Elements in a Timer Function

an off delay, the RLO must change from "1" to "0", and in all other cases the time sequence starts on a change from "0" to "1".

You can start each timer function with one of the five possible timer types (Figure 7.3). There is, however, no point in assigning several timer types to a timer function.

7.1.3 Specifying the Time Duration

The timer function adopts the value below the coil or the value at the input TV as the time duration when starting. You can specify the time duration as constant, a word-width operand or as variable with data type S5TIME.

- Specifying the time duration as a constant

S5TIME#10s   Duration 10 s
S5T#1m10ms   Duration 1 min + 10 ms

The duration is specified in hours, minutes, seconds and milliseconds. The range is from S5TIME#10ms to S5TIME#2h46m30s (corresponding to 9990 s). Intermediate values are rounded down to base 10 ms. You can use S5TIME# or S5T# as the identifier of the constant representation.

- Specifying the duration as operand or variable

MW 20   Word-width operand with duration
"Duration1"   Variable with data type S5TIME

The value in the word-width operand must correspond to the data type S5TIME (see under "Structure of the Time Duration").

Structure of the Time Duration

The time duration consists internally of the time value and the time base: Duration = Time value × Time base. The duration is the time during which

with	Start timer function as	Start signal
SP S_PULSE	pulse	t
SE S_PEXT	extended pulse	t
SD S_ODT	on delay	t
SS S_ODTS	retentive on delay	t
SF S_OFFDT	off delay	t

a timer function is active (timer "running"). The time value represents the number of time intervals for which the timer runs. The time base specifies the time interval with which the operating system of the CPU changes the time value (Figure 7.4).

You can also build up the duration direct in a word operand. The smaller the time base you select here, the more accurate is the actual duration processed. For example, if you want to implement a duration of one second, three specifications are possible:

Duration = 2001_{hex}	Time base 1 s
Duration = 1010_{hex}	Time base 100 ms
Duration = 0100_{hex}	Time base 10 ms

The last specification in the example is the preferred one in this case.

When starting a timer function, the CPU adopts the programmed time value. The operating system updates the timer functions in a fixed time base and independently of user program execution, that is, with active timers the operating system counts the time value down in the interval of the time base. When the value zero is reached, the time has elapsed. The CPU then sets the timer status (signal state "0" or "1" depending on the timer type) and drops all further activities until the timer function is started again. If you specify a duration of zero (0 ms or W#16#0000) when starting a timer function, the timer function remains active until the CPU processes the timer function and determines that the time has elapsed.

The timer function is updated asynchronously to program execution. This can result in the timer status having a different value at the start of the cycle to its value at the end of the cycle. If you use the timer operations only at one point in the program and in the suggested order (see further below), errrored functions will not occur as a result of asynchronous timer updating.

7.1.4 Resetting a Timer Function

A timer function is reset when power flows in the reset input or in the reset coil (RLO = "1" is present). As long as the timer is reset, a timer check with an NO contact will supply "0" and with an NC contact will supply "1". Resetting the timer function sets the time value and the time base to zero. Input R at the timer box need not be switched.

7.1.5 Checking a Timer Function

Checking the timer status

The timer status is found at output Q of the timer box. You can also check the timer status with an NO contact (corresponds to output Q) or with an NC contact. Depending on the type of the timer function, the check with an NO contact or with output Q indicates different variants in the time sequence (see the description of the timer type below). As with inputs, for example, the check with an NC contact results in the exact opposite check result as that produced by a check with an NO contact. Output Q need not be switched at the timer box.

Checking the time value

Outputs BI and BCD provide the time value in the timer function either binary coded (BI) or BCD-coded (BCD). It is the current value available at the time of the check (with a running timer function, the time value is counted from the set value down towards zero). The value is stored in the operand specified (transfer as with a MOVE box). You do not need to switch these outputs at the timer box.

• Direct checking of a time value
The time value is available binary coded and can be fetched in this form from the timer function. The time base is lost in so doing and is replaced with "0". The value corresponds to a positive number in data format INT. Please note:

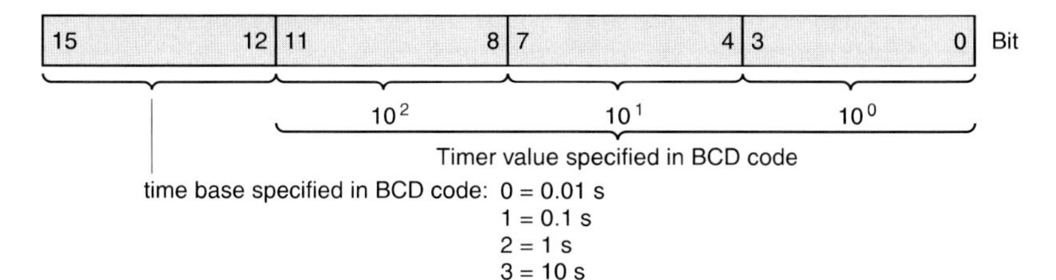

Figure 7.4
Bit Assignment of the Time Duration

The *time value* is checked and not the *duration*! You can also program direct checking of a time value with the MOVE box.

- Coded checking of a time value

You can also fetch the time value available in binary form "coded" from the timer function. In this case, the time base is also available in BCD coded form in addition to the available BCD-coded time value. The BCD-coded value is structured in the same way as for the specification of a time value (see above).

7.1.6 Order of the Timer Operations

When programming a timer function, you do not need to use all the operations available for the timer function. It is enough to use the operations that are necessary for the desired function. Under normal circumstances, these are the operations to start the timer function specifying the duration, and to check the timer status.

So that the timer function behaves as described in this chapter, it is advisable to observe the following order when programming with individual program elements:

▷ Start the timer

▷ Reset the timer

▷ Check the time value or the duration

▷ Check the timer status

Omit the unrequired individual elements when programming. If, in the order shown, the timer function is started and reset "simultaneously", the timer will start but will be immediately reset. The subsequent check of the timer function will fail to detect that the timer started.

7.1.7 Timer Box in a Rung

You can connect contacts in series and in parallel before the start input and the reset input as well as after the output Q.

The timer box itself must only be located in a branch that is connected direct to the left power rail. This branch can also have contacts before the start input and it need not be the uppermost branch.

You can find further examples of the representation and arrangement of timer functions in the diskette supplied with the book (FB 107 in the "Basic Functions" program).

With incremental programming, you will find the timer functions in the Program Elements Catalog (with VIEW → CATALOG [Ctrl – K] or INSERT → PROGRAM ELEMENTS) under "Timers".

7.2 Pulse Timer

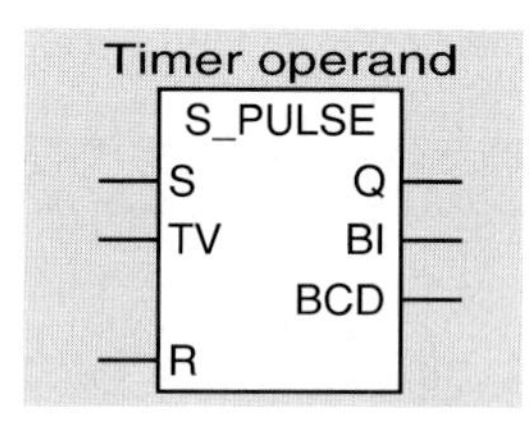

The diagram in Figure 7.5 describes the characteristics of a timer function after starting as a pulse and on resetting. The description applies if you observe the order shown in Section 7.1.6 when programming with individual elements.

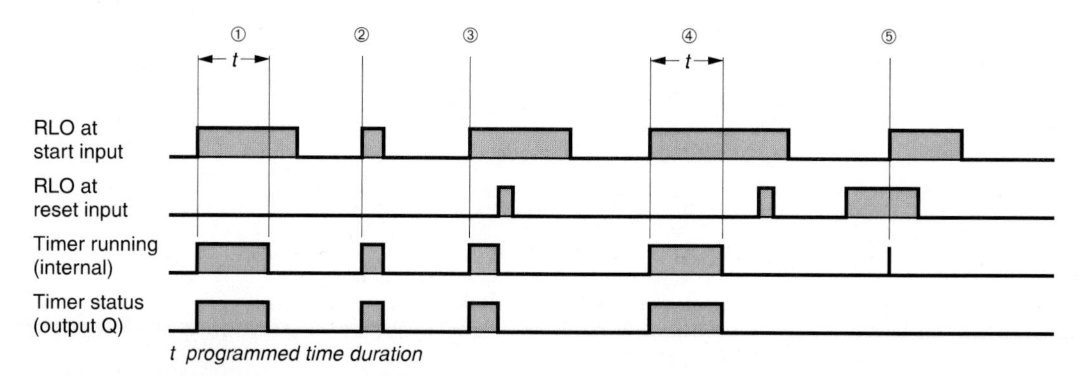

Figure 7.5 Timer Characteristics when Starting and Resetting as Pulse

Starting a pulse timer

① If the RLO at the start input of the timer function changes from "0" to "1", the timer function starts. It runs for the programmed duration as long as the RLO at the start input remains "1". Output Q supplies signal state "1" as long as the timer runs.

② If the RLO at the start input of the timer function changes to "0", before the time has elapsed, the timer function stops. Output Q then supplies signal state "0". The time value indicates the remaining duration by which the time sequence was prematurely interrupted.

Resetting a pulse timer

③ RLO "1" at the reset input of the timer function with a running time resets the timer function. Signal state "0" is then present at output Q. The time value and the time base are also set to zero. If the RLO at the reset input changes from the "1" to "0" while RLO "1" is present at the start input, the timer function remains unaffected.

④ When a timer is not running, RLO "1" at the reset input has no effect.

⑤ If the RLO start input changes from "0" to "1" when the reset signal is present, the timer function starts but is reset immediately (indicated by a line in the diagram). If the prescribed order of the timer operations is observed, the brief starting does not affect the timer function check.

7.3 Extended Pulse Timer

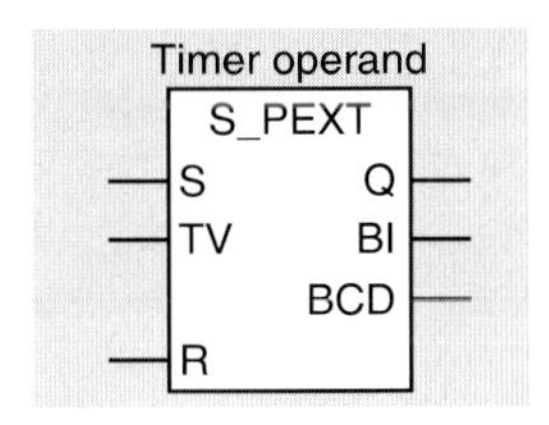

The diagram in Figure 7.6 describes the characteristics of the timer function after starting as an extended pulse and on resetting. The description applies if you observe the order shown in Section 7.1.6 when programming with individual elements.

Starting with extended pulse

❶❷ If the RLO at the start input of the timer function changes from "0" to "1", the timer function starts. It runs for the programmed duration regardless of the further development of the RLO at the Start input. Output Q supplies signal state "1", as long as the timer is running.

❸ If the RLO at the start input changes from "0" to "1" while the timer is running, the timer function starts again with the currently specified time value (the timer function is "retriggered"). It can be restarted any number of times without elapsing first.

Resetting with extended pulse

❹❺ RLO "1" at the reset input of the timer function while the timer is running resets

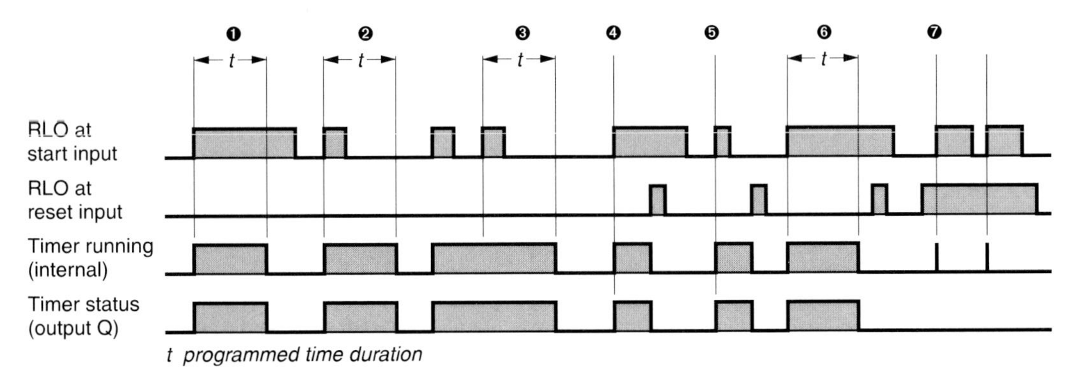

Figure 7.6 Timer Characteristics when Starting and Resetting as Extended Pulse

the timer function. Output Q then supplies signal state "0". The time value and the time base are also set to zero.

❻ When the timer is not running, RLO "1" at the reset input has no effect.

❼ If the RLO at the start input changes from "0" to "1" when the reset signal is present, the timer function starts, but is immediately reset (indicated by a line in the diagram). If the prescribed order of the timer operations is observed, the brief starting does not affect the timer function check.

7.4 On-Delay Timer

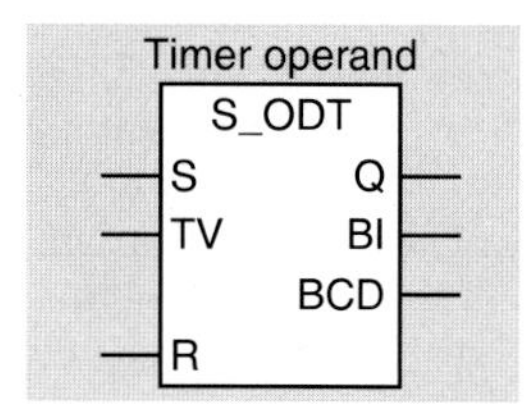

The diagram in Figure 7.7 describes the characteristics of the timer function after starting as an on-delay and on resetting. The description applies if you observe the order shown in Section 7.1.6 when programming with individual elements.

Starting as on-delay

① If the RLO at the start input of the timer function changes from "0" to "1", the timer function starts. It runs for the programmed duration. Output Q supplies signal state "1" if the timer has elapsed without errors and the start input is still controlled with RLO "1" (on-delay).

② If the RLO at the start input changes from "1" to "0" when the timer is running, the timer function stops. Output Q always supplies signal state "0" in such cases. The time value indicates the remaining duration by which the time sequence was prematurely interrupted.

Resetting as on-delay

③ RLO "1" at the reset input of the timer function while the timer is running resets the timer. Output Q then supplies signal state "0", even if the timer is not running and RLO "1" is still present at the start input. Time value and time base are also set to zero. If the RLO at the reset input changes from "1" to "0" while RLO "1" is still present at the start input, the timer function remains unaffected.

④ RLO "1" at the reset input when the timer is no longer running also resets the timer function. As a result of this, output Q supplies signal state "0".

⑤ If the RLO at the start input changes from "0" to "1" while the reset signal is present, the timer function starts, but is immediately reset (indicated by a line in the diagram). If you observe the prescribed order of the timer operations, the checking of the timer function will not be affected by this brief starting.

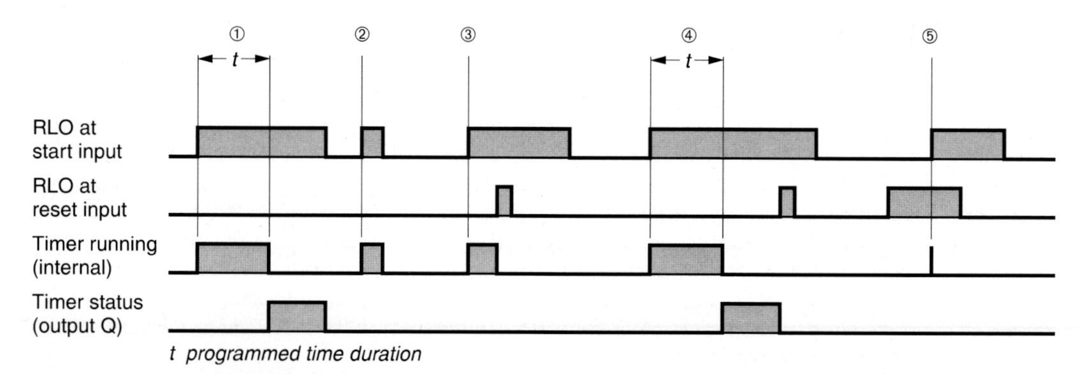

Figure 7.7 Time Characteristics when Starting and Resetting as On-Delay

7.5 Retentive On-Delay Timer

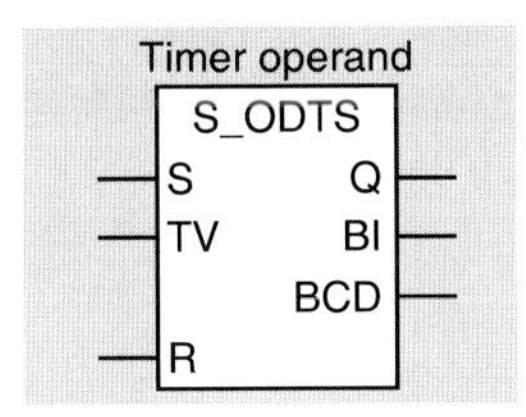

The diagram in Figure 7.8 describes the characteristics of the timer function after starting as a retentive on-delay and on resetting. The description applies if you observe the order shown in Section 7.1.6 when programming with individual elements.

Starting as a retentive on-delay

❶❷ If the RLO at the start input of the timer function changes from "0" to "1", the timer function starts. It runs for the programmed duration regardless of the further development of the RLO at the start input. When the time has elapsed, output Q supplies signal state "1" regardless of the RLO at the start input. The result of the check does not return to "0" until the timer function has been reset, regardless of the RLO at the start input.

❸ If the RLO at the start input changes from "0" to "1" while the timer is running, the timer function starts again with the currently specified time value (the timer function is "retriggered"). It can be started as often as desired without first having to elapse.

Resetting as retentive on-delay

❹❺ RLO "1" at the reset input of the timer function resets the timer function regardless of the RLO at the start input. Output Q then supplies signal state "0". Time value and time base are set to zero.

❻ If the RLO at the start input changes from "0" to "1" while the reset signal is present, the timer function starts, but is immediately reset (indicated by a line in the diagram). If you observe the prescribed order of the timer operations, the checking of the timer function will not be affected by this brief starting.

7.6 Off-Delay Timer

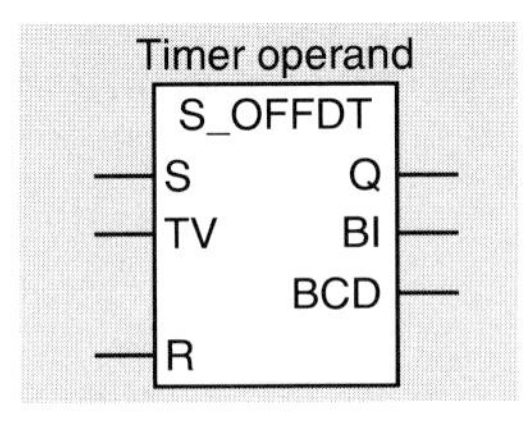

The diagram in Figure 7.9 describes the characteristics of the timer function after starting as off-delay and on resetting. The description applies if you observe the order shown in Section 7.1.6 when programming with individual elements.

Starting as off-delay

①③ If the RLO at the start input of the timer function changes from "1" to "0" the timer

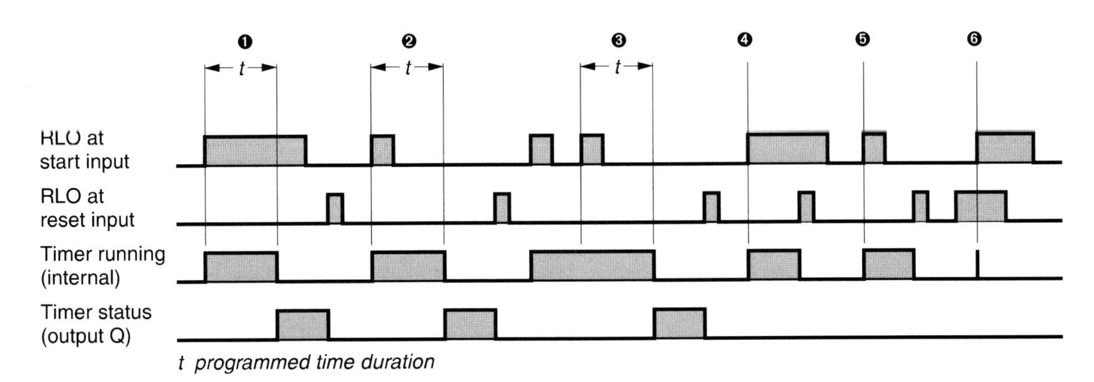

Figure 7.8 Timer Characteristics when Starting and Resetting as a Retentive On-Delay

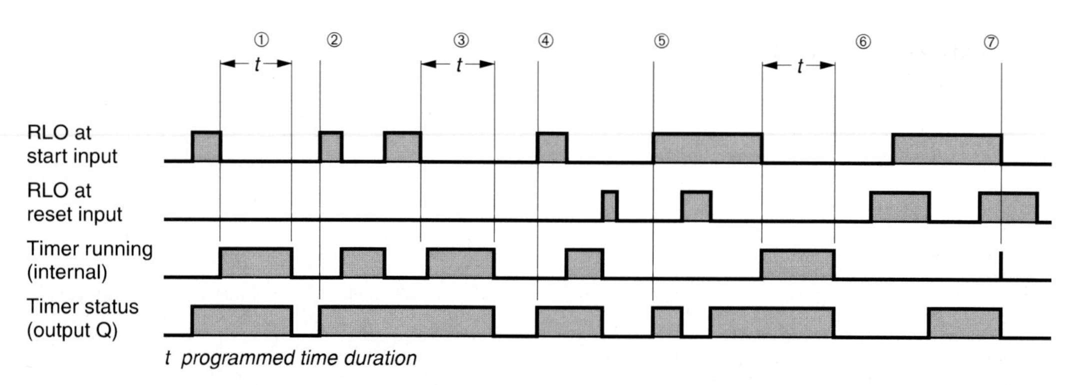

Figure 7.9 Time Characteristics when Starting and Resetting as Off-Delay

function starts. It runs for the programmed duration. Output Q supplies signal state “1” if the RLO at the start input is “1” or if the timer is running (off-delay).

② If the RLO at the start input changes from “0” to “1” while the timer is running, the timer is reset. The timer does not start again until there is a falling edge at the start input.

Resetting as off-delay

④ RLO “1” at the reset input of the timer function while the timer is running resets the timer function. Output Q then supplies signal state “0”. Time value and time base are also set to zero.

⑤⑥ RLO “1” at the start input and at the reset input resets output Q of the timer function. If the RLO at the reset input now changes back to “0”, output Q again has signal state “1”.

⑦ If the RLO at the start input changes from “1” to “0” while the reset signal is present, the timer function starts but is immediately reset again (indicated by a line in the diagram). Output Q then immediately supplies signal state “0”.

8 Counter Functions

With the counter functions, you can execute counting tasks direct via the CPU. The counter functions can count both up and down; the counter range extends over three decades (000 to 999). The counter functions are located in the system memory of the CPU; the number of counter functions is CPU-specific.

You can program a counter function complete as a box or with individual program elements. You can set the count value to a specific starting value or delete it, and you can count up and down. The counter function is checked via the counter status (counter value zero or not zero) or via the current count value that you can obtain from the counter function either binary-coded or BCD-coded.

8.1 Programming a Counter Function

You can perform the following operations with a counter function:

- ▷ Set counter specifying the count value
- ▷ Count up
- ▷ Count down
- ▷ Reset counter
- ▷ Check (binary) count status
- ▷ Check (digital) count value binary-coded
- ▷ Check (digital) count value BCD-coded

Representation of a counter function as box

The box of a counter function contains the coherent representation of all counter operations in the form of function inputs and outputs (Figure 8.1). Above the box is the address of the counter function in absolute or symbolic form. In the box, as a header, is the counter type (S_CUD means "up/down counter"). The assignment of the first input (CU in the example) is mandatory; assignment of the other inputs and outputs is optional.

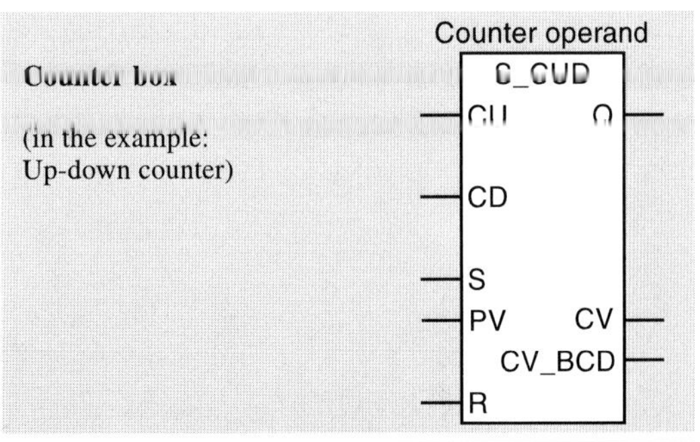

Name	Data typ	Description
CU	BOOL	Up counter
CD	BOOL	Down counter
S	BOOL	Set counter
PV	WORD	Counter value specification
R	BOOL	Reset counter
Q	BOOL	Counter status (count value > 0)
CV	WORD	Current binary-coded count value
CV_BCD	WORD	Current BCD-coded count value

Figure 8.1 Counter Function in Box Representation

The boxes for the counter functions are available in three versions: As up-down counter (S_CUD), as up only counter (S_CU) and as down only counter (S_CD). The differences in functionality are explained below.

Representation of a counter function with individual elements

You can also program a counter function with individual elements (Figure 8.2). Setting and counting is then done via coils. The set coil contains the count operation (SC = Set counter), below the coil, in WORD data format, is the count value with which the counter function is set. In the coils for counting, CU stands for up counting and CD for down counting. To reset a counter function, use the reset coil and to check the counter status, use an NO or NC contact. You can then transfer the current count value in binary-coded form using the MOVE box.

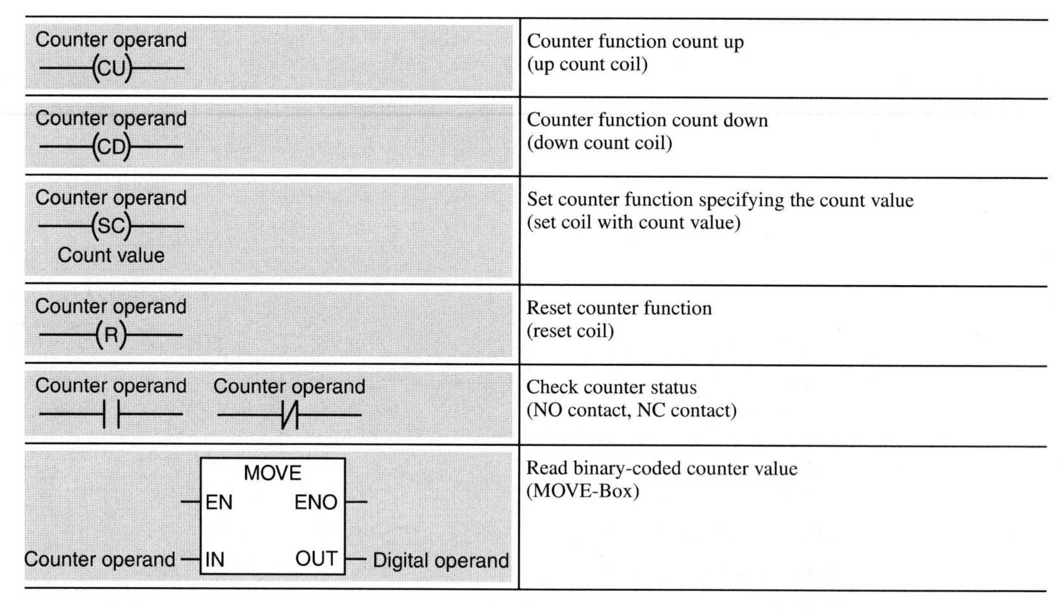

Figure 8.2 Individual Elements of a Counter Function

Count operations sequence

When programming a counter function, you do not need to use all the operations available for the counter function. The operations required for the desired function are enough. For example, for a down counter, only Set to starting value, Count down and Counter status check are necessary.

So that a counter function behaves as described in this chapter, it is advisable to observe the following order when programming with individual program elements:

▷ Count (up or down in any order)

▷ Set counter

▷ Reset counter

▷ Check count value

▷ Check counter status

Omit the individual elements not required for programming. If, in the order shown, the counter function is counted, set and reset "simultaneously", the count value will first be changed accordingly but then immediately deleted again. The subsequent check of the counter function will therefore show the count value at zero and a counter status of "0".

If, in the order shown, the counter function is counted and set "simultaneously", the count value will first be changed accordingly but then set to the programmed value that it will retain for the remaining program execution.

The order of the operations for up and down counting is not significant.

Counter box in a rung

You can arrange contacts in series and parallel before the counter inputs, before the start input and the reset input, and after the Q output.

The counter itself must only be placed in a branch that is connected direct to the left power rail. This branch can also have contacts before the inputs and need not be the uppermost branch.

You can find further examples of the representation and arrangement of the counter functions in the diskette supplied with the book (FB 108 in the "Basic Functions" program).

With incremental programming, you will find the counter functions in the Program Element Catalog (with VIEW → CATALOG [Ctrl – K] or INSERT → PROGRAM ELEMENTS) under "Counters".

8.2 Setting and Resetting Counters

Setting counters

A counter is set when the RLO before the set input S or before the set coil changes from "0" to "1". A positive edge is always required to set a counter.

"Set counter" means the counter function is set to a starting value. The value range is from 0 to 999.

Specifying the count value

The count function adopts the value at the input PV or the value below the set coil as the count value for setting. You can specify the count value as a constant, as a word-width operand or as a variable with data type WORD.

- Specifying the count value as a constant

C#100	Count value 100
W#16#0100	Count value 100

The size of the count value is three decades in the range 000 to 999. Only positive values in BCD code are permissible; the counter function cannot process negative values. As the identifier for the constant representation, you can use C# or W#16# (in conjunction with decimal digits only).

- Specifying the count value as an operand or a variable

MW 56	Word-width operand with count value
"Count_value1"	Variable with data type WORD

Reset counter

A counter function is reset when power flows to the reset input or to the reset coil (RLO "1" is present). As long as RLO "1" is present, a check on the counter function using an NO contact will result in "0" and a check with an NC contact will result in "1". Resetting the counter function sets the count value to "zero". The input R of the counter box need not be switched.

8.3 Counting

The counting frequency of the counter function is determined by the execution time of your program! To be able to count, the CPU must detect a signal state change of the input pulse, that is, an input pulse (or a space) must be present for at least one program cycle. The longer the program execution time, therefore, the lower the counting frequency.

Up counting

A counter function is counted up when the RLO changes from "0" to "1" before the up count input CU, or at the up count coil. A positive edge is always required for counting up.

When counting up, each positive edge increments the count value by one unit until it reaches the upper limit of 999. Each subsequent positive edge for counting up then has no further effect. No remainder occurs.

Counting down

A counter function is counted down if the RLO before the down count input CD, or at the down count coil, changes from "0" to "1". A positive edge is always required for counting down.

Each positive edge when counting down decrements the count value by one unit until it reaches the lower limit of 0. Each subsequent positive edge for counting down then has no further effect. Counting does not take place with a negative count value.

Different counter boxes

The Editor provides three different counter boxes:

▷ S_CUD	Up/down counter
▷ S_CU	Up counter
▷ S_CD	Down counter

These counter boxes differ only in the type and number of the counter inputs. While S_CUD has the inputs for both count directions, S_CU has only the up count input and S_CD only the down count input.

You must always switch the first input of a counter box. If, for example, you do not switch the second input (the down count input CD) on S_CUD, this box will then take on the same characteristics as S_CU.

8.4 Checking a Counter Function

Checking the counter status

The counter status is at output Q of the counter box. You an also check the counter status with an NO contact (corresponding to output Q) or with an NC contact.

Output Q has signal state "1" (power flows from the output), if the current count value is greater than zero. Output Q has "0", if the current count value equals zero. Output Q does not need to be switched at the counter box.

Checking the count value

The CV and CV_BCD outputs provide the count value present in the counter function in binary-coded form (CV) or BCD-coded form (CV_BCD). It is the count value currently available at the time of checking. The value is stored in the operand specified (transferred as with a MOVE box). You need not switch these outputs at the counter box.

• Direct checking of the count value
The count value is available in binary-coded form and can be fetched in this form from the counter function. The value corresponds to a positive number in data form INT. You can also program direct checking of a count value with the MOVE box.

• Coded checking of the count value
You can also fetch the count value available in binary form "coded" from the counter function. The BCD-coded value is structured in the same way as for specifying the count value (see above).

8.5 Parts Counter Example

The example illustrates the handling of timer and counter functions. It is programmed with inputs, outputs and memory bits so that it can be programmed at any point in any block. At this point, a function without block parameters is used; the timer and counter functions are represented with complete boxes. You will find the same example programmed as a function block with block parameters and with individual elements in Chapter 19 "Block Parameters".

Function description

Parts are transported on a conveyor belt. A light barrier detects and counts the parts. After a set number, the counter function sends the signal "Finished". The counter is equipped with a monitoring circuit. If the signal state of the light barrier does not change within a specified time, the monitor sends a signal.

The "Set" input specifies the starting value (the number to be counted) to the counter. A positive edge at the light barrier decrements the counter by one unit. When the value zero is reached, the counter sends the signal "Finished". It is a requirement that the parts are individually arranged (at intervals) on the conveyor.

The "Set" input also sets the "Active" signal. The controller only monitors a signal state change of the light barrier in the active state. When counting finishes and the last counted item has exited the light barrier, "Active" is switched off.

In the active state, a positive edge of the light barrier starts the timer with the time value "Duration1" ("Dura1") as a retentive pulse. If the start input of the timer is processed in the next cycle with "0", it nevertheless continues to run. The timer is "retriggered" with a renewed positive edge, in other words, it starts again. The next positive edge for renewed starting of the timer function if generated when the light barrier signals a negative edge. Then the timer starts with the time value "Duration2" ("Dura2"). If the light barrier is now covered for longer than the time value"-Dura1" or free for longer than the time value " Dura2", the timer elapses and signals "Fault". The first time it is activated, the timer is started with the time value "Dura2".

Signals, symbols

The "Set" signal activates the counter and the monitor. The light barrier controls the counter, the "active" state, the time value selection and the starting (retriggering) of the monitoring time via positive and negative edges.

Evaluation of the positive and negative edge of the light barrier is required often and temporary local data are suitable here as "scratchpad memory". Temporary local data are block-local variables; they are declared in blocks (and not in the symbol table). In the example, the pulse mem-

Table 8.1 Symbol Table for the Parts Counter Example

Symbol	Address	Data Type	Comment
Counter control	FC 12	FC 12	Counter and monitoring control for parts
Acknowl	I 0.6	BOOL	Acknowledge fault
Set	I 0.7	BOOL	Set counter, activate monitor
Lbarr1	I 1.0	BOOL	Detection of parts (light barrier)
Finished	Q 4.2	BOOL	Number of items reached
Fault	Q 4.3	BOOL	Monitor responded
Active	M 3.0	BOOL	Monitor activated
EM_LB_P	M 3.1	BOOL	Edge memory bit light barrier positive edge
EM_LB_N	M 3.2	BOOL	Edge memory bit light barrier negative edge
EM_Ac_P	M 3.3	BOOL	Edge memory bit "Monitor active" positive edge
EM_ST_P	M 3.4	BOOL	Edge memory bit "Set" positive edge
Quantity	MW 4	WORD	Quantity of parts
Dura1	MW 6	S5TIME	Monitoring time for light barrier covered
Dura2	MW 8	S5TIME	Monitoring time for light barrier not covered
Count	C 1	COUNTER	Counter for parts
Monitor	T 1	TIMER	Timer function for monitor

ory bits of the edge evaluation are stored in temporary local data. (The edge memory bits also require their signal state in the next cycle; they must therefore not be local data.)

We want symbolic addressing, that is, the operands receive names with which we then program. Before entering the program, we create a symbol table (Table 8.1) that contains the inputs, outputs, memory bits, timers, counters and blocks.

Program

The program is located in a function that you call in the CPU in organization block OB 1 (selected from the Program Elements Catalog under "FC Blocks").

During programming, the global symbols can also be used without quotation marks provided they do not contain any special characters. If a symbol contains a special character (for example, an Umlaut or a space),it must be placed within quotation marks. In the compiled block, the Editor indicates all global symbols with quotation marks.

Figure 8.3 shows the program of the transported parts counter. You can find this program under "Conveyor Example" in the function FC 12 on the diskette supplied with the book.

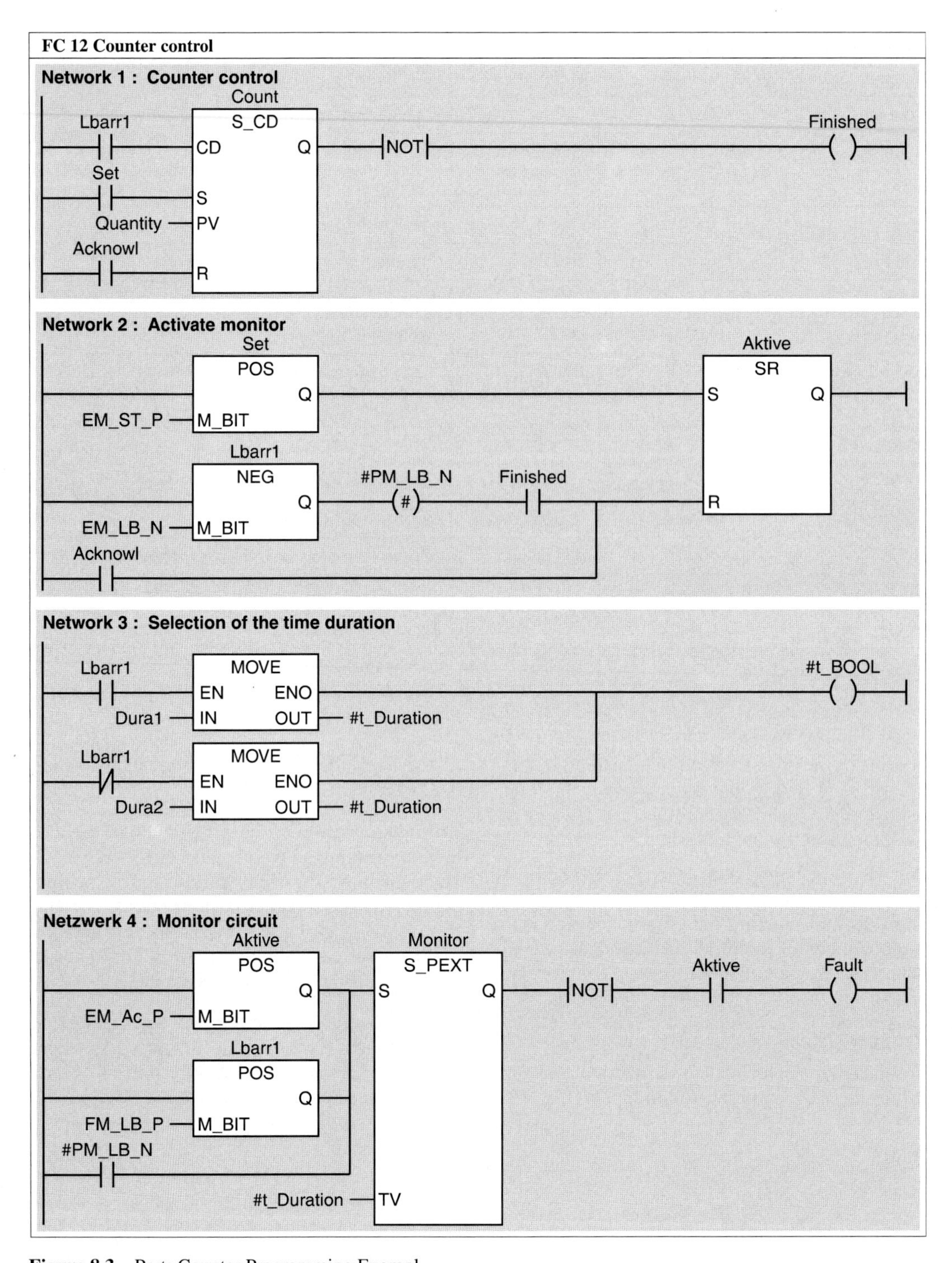

Figure 8.3 Parts Counter Programming Exampl

Digital Functions

The digital functions process digital values predominantly with the data types INT, DINT and REAL and thus extend the functionality of the PLC.

- The **comparison functions** form a binary result from the comparison of two values. They take account of the data types INT, DINT and REAL.
- You use the **arithmetic functions** to make calculations in your program. All the basic arithmetic functions in data types INT, DINT and REAL are available.
- The **mathematical functions** extend the calculation possibilities beyond the basic arithmetic functions to include, for example, trigonometric functions.
- Before and after performing calculations, you adapt the digital values to the desired data type using the **conversion functions**.
- The **shift functions** allow justification of the contents of a variable by shifting to the right or the left.
- With **word logic**, you mask digital values by targeting individual bits and setting them to "1" or "0".

The digital logic operations work mainly with values stored in data blocks. These can be global data blocks or instance data blocks if static local data are used. Section 18.2 "Block Functions for Data Blocks" shows how to handle data blocks and gives the methods of addressing data operands.

09 Comparison Functions
Comparison for equal to, not equal to, greater than, greater than or equal to, less than and less than or equal to; comparison function in a rung

10 Arithmetic Functions
Basic arithmetic functions with data types INT, DINT and REAL; arithmetic box in a rung

11 Mathematical Functions
Trigonometric functions; inverse trigonometric functions; squaring, square-root extraction, exponentiation and logarithms

12 Conversion Functions
Conversion from INT/DINT to BCD and vice versa; conversion from DINT to REAL and vice versa with different rounding types; one's complement, negation and absolute value generation

13 Shift Functions
Shifting to left and right, by word and doubleword, shifting with correct sign; rotating to left and right

14 Word Logic
AND, OR, exclusive OR; word and doubleword combinations

9 Comparison Functions

The comparison functions compare two digital variables of data types INT, DINT and REAL for equal to, not equal to, greater than, greater than or equal to, less than and less than or equal to. The comparison result is then available as a binary value (Table 9.1).

9.1 Processing a Comparison Function

Representation

The box of a comparison function has two inputs IN1 and IN2 and an unlabeled binary output, in addition to the (unlabeled) binary input. The "heading" in the box represents the comparison operation (CMP, compare) and the type of comparison executed (for example, CMP ==I stands for the comparison of two INT numbers for equal to).

Comparison box
(in the example: Comparison for equal to according to INT)

CMP ==I
IN1
IN2

You can arrange a comparator in a rung in place of a contact. The unlabeled input and the unlabeled output establish the connection to the (binary) contacts. A successful comparison is the same as a closed contact ("power" flows across the comparator). If the comparison is not successful, the contact is open. The output of the comparator must always be switched.

The values to be compared are at inputs IN1 and IN2 and the comparison result is at the output. When the comparison is successful, the comparator output has signal state "1", otherwise it has "0". The data types of the inputs differ depending on the comparison function. For example, with the comparison function CMP >R (comparison of REAL numbers for greater than), the inputs have data type REAL. The variables applied must be of the same data type as the inputs. If you use operands with absolute addresses, the operand widths must be adapted to the data types, for example, you can use a word-width operand for data type INT.

You can find the bit assignments of the data formats in Section 3.4 "Data Types".

Function

The comparison box corresponds to a contact that is closed when the comparison is successful. That is, the power that flows into the unlabeled input is switched through to the output if the comparison is successful.

Table 9.1 Overview of the Comparison Functions

Comparison Function	Comparison According to Data Type		
	INT	DINT	REAL
Comparison for equal to	CMP ==I	CMP ==D	CMP ==R
Comparison for not equal to	CMP <>I	CMP <>D	CMP <>R
Comparison for greater than	CMP >I	CMP >D	CMP >R
Comparison for greater than or equal to	CMP >=I	CMP >=D	CMP >=R
Comparison for less than	CMP <I	CMP <D	CMP <R
Comparison for less than or equal to	CMP <=I	CMP <=D	CMP <=R

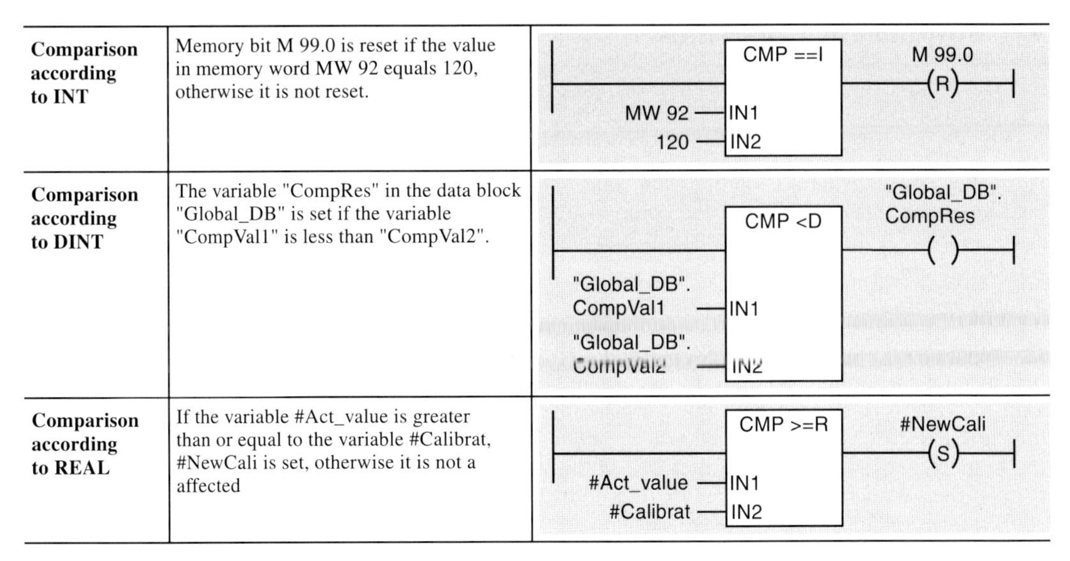

Comparison according to INT	Memory bit M 99.0 is reset if the value in memory word MW 92 equals 120, otherwise it is not reset.
Comparison according to DINT	The variable "CompRes" in the data block "Global_DB" is set if the variable "CompVal1" is less than "CompVal2".
Comparison according to REAL	If the variable #Act_value is greater than or equal to the variable #Calibrat, #NewCali is set, otherwise it is not a affected

Figure 9.1 Comparison Function Examples

The comparison of REAL numbers is not successful if one or both REAL numbers are invalid. In addition, the status bits OV and OS are set in this case. You can find out how the comparison functions set the remaining status bits in Chapter 15 "Status Bits".

Examples

Figure 9.1 shows an example of each of the different data types. A comparison function executes the comparison according to the characteristic specified even if no data types have been declared when using operands with absolute addresses.

Comparison function in a rung

You can use the comparison function in a rung in place of a contact.

You can connect contacts before and after the comparison function in series and in parallel. The comparison boxes themselves can also be connected in series or in parallel. In the case of comparison functions connected in series, both comparisons must be successful for power to flow in the rung. In the case of comparison functions connected in parallel, one successful comparison condition is sufficient for power to flow in the parallel circuit.

The diskette supplied with the book contains further examples of representation and arrangement of comparison functions (FB 109 in the "Digital Functions" program).

In the case of incremental programming, you will find the comparison functions in the Program Elements Catalog (with VIEW → CATALOG [Ctrl – K] or INSERT → PROGRAM ELEMENTS) under "Compare".

9.2 Description of the Comparison Functions

Comparison for equal to

The "comparison for equal to" interprets the contents of the input variables in accordance with the data type specified in the comparison function and checks to see if both values are equal. The comparison is successful (the comparison function allows power to flow), if both variables have the same value.

If, in the case of a REAL comparison, one or both input variables are invalid, the comparison is not successful. In addition, the status bits OV and OS are set.

Comparison for not equal to

The "comparison for not equal to" interprets the contents of the input variables in accordance with the data type specified in the comparison function and checks to see if the two values differ. The comparison is successful (the comparison function allows power to flow), if the two variables have different values.

If, in the case of a REAL comparison, one or both input variables are invalid, the comparison is not successful. In addition, the status bits OV and OS are set.

Comparison for greater than

The "comparison for greater than" interprets the contents of the input variables in accordance with the data type specified in the comparison function and checks to see if the value at input IN1 is greater than the value at input IN2. If this is case, the comparison is successful (the comparison function allows power to flow).

If, in the case of a REAL comparison, one or both input variables are invalid, the comparison is not successful. In addition, the status bits OV and OS are set.

Comparison for greater than or equal to

The "comparison for greater than or equal to" interprets the contents of the input variables in accordance with the data type specified in the comparison function and checks to see if the value at input IN1 is greater than or equal to the value at input IN2. If this is the case, the comparison is successful (the comparison function allows power to flow).

If, in the case of a REAL comparison, one or both input variables are invalid, the comparison is not successful. In addition, the status bits OV and OS are set.

Comparison for less than

The "comparison for less than" interprets the contents of the input variables in accordance with the data type specified in the comparison function and checks to see if the value at input IN1 is less than the value at input IN2. If this is the case, the comparison is successful (the comparison function allows power to flow).

If, in the case of a REAL comparison, one or both input variables are invalid, the comparison is not successful. In addition, the status bits OV and OS are set.

Comparison for less than or equal to

The "comparison for less than or equal to" interprets the contents of the input variables in accordance with the data type specified in the comparison function and checks to see if the value at input IN1 is less than or equal to the value at input IN2. If this is the case, the comparison is successful (the comparison function allows power to flow).

If, in the case of a REAL comparison, one or both input variables are invalid, the comparison is not successful. In addition, the status bits OV and OS are set.

10 Arithmetic Functions

The arithmetic functions combine two values in accordance with the basic calculations of adding, subtraction, multiplication and division. You can use the arithmetic functions on variables of the data types INT, DINT and REAL (Table 10.1).

10.1 Processing an Arithmetic Function

Representation

The box of an arithmetic function has two inputs IN1 and IN2 and one output OUT in addition to the enable input EN and the enable output ENO. The "header" in the box represents the type of arithmetic function executed (for example, ADD_I stands for the addition of INT numbers).

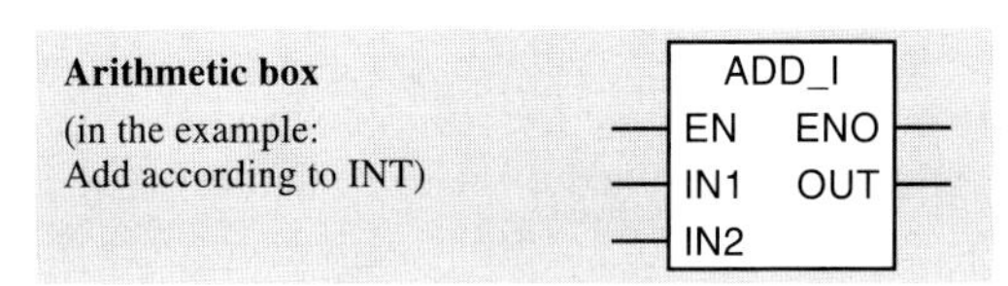

The values to be combined are at the inputs IN1 and IN2 and the result of the calculation is at output OUT. The inputs and the output have different data types depending on the arithmetic function. For example, in the case of the arithmetic function ADD_R (addition of REAL numbers), the inputs and the output are of the data type REAL. The variables applied must be of the same data type as the inputs or the output. If you use operands with absolute addresses, the operand width must be adapted to the data types; for example, you can use a word-width operand for data type INT.

You can find the bit assignments of the data formats in Section 3.4 "Data Types".

Function

The arithmetic function is executed if "1" is present at the enable input (power flows into the input EN). If an error occurs during the calculation, the enable input is set to "0", otherwise, it is set to "1". If execution of the function is not enabled (EN = "0"), the calculation does not take place and ENO is also "0" (Figure 10.1).

If the Master Control Relay MCR is activated, the output OUT is set to zero if the arithmetic function is processed (EN = "1"). The MCR does not affect the ENO output.

The following errors can occur with an arithmetic function:

▷ Overflow of the number range in INT and DINT calculations

▷ Underflow and overflow in REAL calculations

▷ No valid REAL number in a REAL calculation

Table 10.1 Overview of the Arithmetic Functions

Arithmetic Function	with Data Type INT	DINT	REAL
Addition	ADD_I	ADD_DI	ADD_R
Subtraction	SUB_I	SUB_DI	SUB_R
Multiplication	MUL_I	MUL_DI	MUL_R
Division with quotient as result	DIV_I	DIV_DI	DIV_R
Division with remainder as result	–	MOD_DI	–

<table>
<tr><td colspan="2">IF EN == "1"
THEN</td><td>ELSE</td></tr>
<tr><td colspan="2">OUT := IN1 Cfct IN2</td><td rowspan="3">ENO := "0"</td></tr>
<tr><td>IF error occurred
THEN</td><td>ELSE</td></tr>
<tr><td>ENO := "0"</td><td>ENO := "1"</td></tr>
</table>

with Cfct as calculation function

Figure 10.1 Function of an Arithmetic Box

See Chapter 15 "Status Bits" to find out how the arithmetic functions set the status bits.

Examples

Figure 10.2 shows an example of each of the different data types. An arithmetic function executes the calculation in accordance with the characteristic specified, even if no data types have been declared when using operands with absolute addresses.

Arithmetic function in a rung

You can connect contacts in series and in parallel before the EN input and after the ENO output.

The arithmetic box itself must only be placed in a branch that leads direct to the left-hand power rail. This branch can also have contacts before the EN input and it need not be the uppermost branch. The direct connection to the left-hand power rail means that you can connect arithmetic boxes in parallel. With the parallel connection of boxes, you require a coil to terminate the rung. If you have not provided any error evaluation, assign a "dummy" operand to the coil, for example, a temporary local data bit.

You can connect arithmetic boxes in series. If the ENO output of the preceding box leads to the EN input of the subsequent box, the subsequent box is only processed if the preceding box has been completed without errors. If you want to use the result of the preceding box as the input value for a subsequent box, variables from the temporary local data are a convenient intermediate buffer.

If you arrange several arithmetic boxes in one rung (parallel to the left-hand power rail and then further in series), the boxes in the uppermost branch are the first to be processed from left to right, followed by the boxes in the second branch from left to right, and so on.

The diskette supplied with the book contains further examples of the representation and arrangement of the arithmetic functions (FB 110 in the "Digital Functions" program).

With incremental programming, you will find the arithmetic functions in the Program Elements Catalog (with VIEW → CATALOG [Ctrl – K] or INSERT → PROGRAM ELEMENTS) under

Calculating according to INT	The value in memory word MW 100 is divided by 250; the integer result is stored in memory word MW 102.	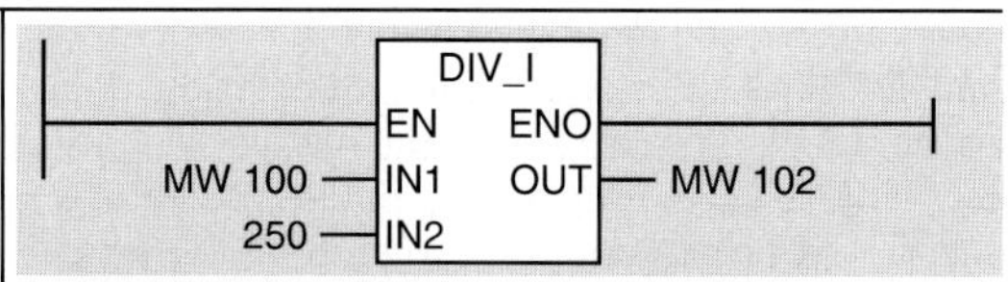 DIV_I EN ENO MW 100 — IN1 OUT — MW 102 250 — IN2
Calculating according to DINT	The values in the variables "CalcVal1" and "CalcVal2" are added and stored in the variable "CalcRes". All variables are contained in the "Global_DB" data block.	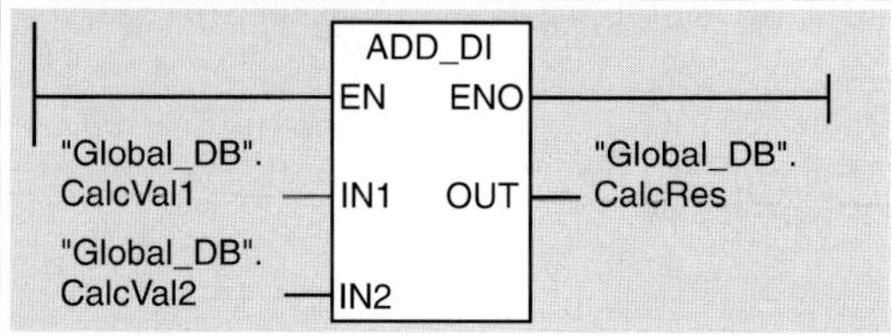 ADD_DI EN ENO "Global_DB". CalcVal1 — IN1 OUT — "Global_DB". CalcRes "Global_DB". CalcVal2 — IN2
Calculating according to REAL	The variable #Act_value is multiplied by the variable #Factor; the product is transferred to variable #Indicator.	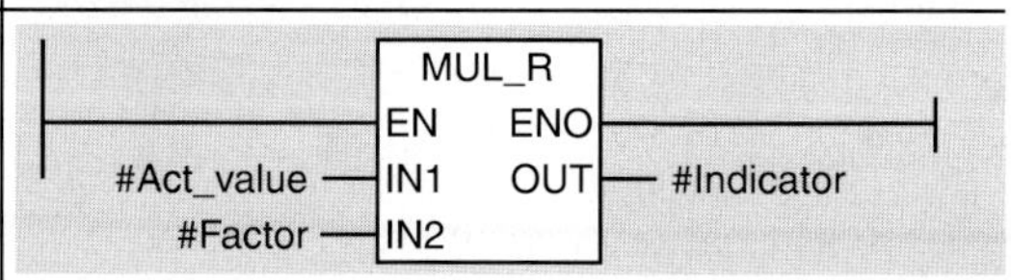 MUL_R EN ENO #Act_value — IN1 OUT — #Indicator #Factor — IN2

Figure 10.2 Examples of Arithmetic Functions

"Integer Math" (INT and DINT calculations) and under "Floating-Point Math" (REAL calculations).

10.2 Calculating with Data Type INT

INT addition

The function ADD_I interprets the values at inputs IN1 and IN2 as numbers of data type INT. It adds both numbers and stores the sum at output OUT.

After execution of the calculation, status bits CC0 and CC1 indicate whether the sum is negative, zero or positive. Status bits OV and OS indicate if the permissible number range has been exceeded.

INT subtraction

The function SUB_ I interprets the values at inputs IN1 and IN2 as numbers of data type INT. It subtracts the value at IN2 from the value at IN1 and stores the difference at output OUT.

After execution of the calculation, status bits CC0 and CC1 indicate whether the difference is negative, zero or positive. Status bits OV and OS indicate if the permissible number range has been exceeded.

INT multiplication

The function MUL_I interprets the values at inputs IN1 and IN2 as numbers of data type INT. It multiplies both numbers and stores the product at output OUT.

After execution of the calculation, status bits CC0 and CC1 indicate whether the product is negative, zero or positive. Status bits OV and OS indicate if the permissible INT number range has been exceeded.

INT-Division

The function DIV_ I interprets the values at inputs IN1 and IN2 as numbers of data type INT. It divides the value at input IN1 (dividend) by the value at input IN2 (divisor) and supplies the quotient at output OUT. It is the integer result of the division. The quotient is zero if the dividend is equal to zero and the divisor is not equal to zero or if the amount of the dividend is less than the amount of the divisor. The quotient is negative if the divisor is negative.

After execution of the calculation, status bits CC0 and CC1 indicate whether the quotient is negative, zero or positive. Status bits OV and OS indicate if the permissible number range has been exceeded. Division by zero produces zero as quotient and sets the status bits CC0, CC1, OV and OS to "1".

10.3 Calculating with Data Type DINT

DINT Addition

The function ADD_DI interprets the values at inputs IN1 and IN2 as numbers of data type DINT. It adds both numbers and stores the sum at output OUT.

After execution of the calculation, status bits CC0 and CC1 indicate whether the sum is negative, zero or positive. Status bits OV and OS indicate if the permissible number range has been exceeded.

DINT subtraction

The function SUB_DI interprets the values at inputs IN1 and IN2 as numbers of data type DINT. It subtracts the value at input IN2 from the value at input IN1 and stores the difference at output OUT.

After execution of the calculation, status bits CC0 and CC1 indicate whether the difference is negative, zero or positive. Status bits OV and OS indicate if the permissible number range has been exceeded.

DINT multiplication

The function MUL_ DI interprets the values at inputs IN1 and IN2 as numbers of data type DINT. It multiplies both numbers and stores the product at output OUT.

After execution of the calculation, status bits CC0 and CC1 indicate whether the product is negative, zero or positive. Status bits OV and OS indicate if the permissible number range has been exceeded.

DINT division with quotients as result

The function DIV_ DI interprets the values at inputs IN1 and IN2 as numbers of data type DINT. It divides the value at input IN1 (dividend) by the value at input IN2 (divisor) and stores the quotient at output OUT. It is the integer result of the division. The quotient is zero if the dividend is equal to zero and the divisor is not equal to zero or if the amount of the dividend is less than the amount of the divisor. The quotient is negative if the divisor is negative.

After execution of the calculation, status bits CC0 and CC1 indicate whether the quotient is negative, zero or positive. Status bits OV and OS indicate if the permissible number range has been exceeded. Division by zero produces zero as quotient and sets the status bits CC0, CC1, OV and OS to "1".

DINT division with remainder as result

The function MOD_DI interprets the values at inputs IN1 and IN2 as numbers of data type DINT. It divides the value at input IN1 (dividend) by the value at input IN2 (divisor) and stores the remainder of the division at output OUT. The remainder is what is left over from the division; it does not correspond to the decimal places. With a negative dividend, the remainder is also negative.

After execution of the calculation, status bits CC0 and CC1 indicate whether the remainder is negative, zero or positive. Status bits OV and OS indicate if the permissible number range has been exceeded. Division by zero produces zero as the remainder and set status bits CC0, CC1, OV and OS to "1".

10.4 Calculating with Data Type REAL

REAL numbers are represented internally as floating-point numbers with two number ranges: One range with full accuracy ("normalized" floating-point numbers) and one range with limited accuracy ("denormalized" floating-point numbers, see also Section 3.4 "Data Types"). S7-400 CPUs calculate in both areas, S7-300 CPUs only in the area with full accuracy. If a calculation result occurs in the area with limited accuracy with an S7-300 CPU, zero is supplied as the result and a number range violation is signaled.

REAL addition

The function ADD_R interprets the values at inputs IN1 and IN2 as numbers of data types REAL. It adds both numbers and stores the sum at output OUT.

After execution of the calculation, status bits CC0 and CC1 indicate whether the sum is negative, zero or positive. Status bits OV and OS indicate if the permissible number range has been exceeded.

In the case of an impermissible calculation (one of the input values is an invalid REAL number, or you try to add $+\infty$ and $-\infty$), ADD_R supplies an invalid value at output OUT and sets the status bits CC0, CC1, OV and OS to "1".

REAL subtraction

The function SUB_R interprets the values at inputs IN1 and IN2 as numbers of data types REAL. It subtracts the number at input IN2 from the number at input IN1 and stores the difference at output OUT.

After execution of the calculation, status bits CC0 and CC1 indicate whether the difference is negative, zero or positive. Status bits OV and OS indicate if the permissible number range has been exceeded.

In the case of an impermissible calculation (one of the input values is an invalid REAL number, or you try to subtract $+\infty$ from $+\infty$), SUB_R supplies an invalid value at output OUT and sets the status bits CC0, CC1, OV and OS to "1".

REAL multiplication

The function MUL_R interprets the values at inputs IN1 and IN2 as numbers of data types REAL. It multiplies both numbers and stores the product at output OUT.

After execution of the calculation, status bits CC0 and CC1 indicate whether the product is negative, zero or positive. Status bits OV and OS indicate if the permissible number range has been exceeded.

In the case of an impermissible calculation (one of the input values is an invalid REAL number, or you try to multiply ∞ and 0), MUL_R supplies an invalid value at output OUT and sets the status bits CC0, CC1, OV and OS to "1".

REAL division

The function DIV_R interprets the values at inputs IN1 and IN2 as numbers of data types REAL. It divides the number at input IN1 (dividend) by the number at input IN2 (divisor) and stores the quotient at output OUT.

After execution of the calculation, status bits CC0 and CC1 indicate whether the quotient is negative, zero or positive. Status bits OV and OS indicate if the permissible number range has been exceeded.

In the case of an impermissible calculation (one of the input values is an invalid REAL number, or you try to divide ∞ by ∞ or 0 by 0), DIV_R supplies an invalid value at output OUT and sets the status bits CC0, CC1, OV and OS to “1”.

11 Math Functions

The following math functions are available in LAD:

- Sine, cosine, tangent
- Arc sine, arc cosine, arc tangent
- Establishing the square and/or square root
- Exponential function on base e, natural logarithm

All math functions process REAL numbers.

11.1 Processing a Math Function

Representation

The box of a math function has an input IN and an output OUT in addition to the enable input EN and the enable output ENO. The "header" in the box represents the math function executed (for example, SIN stands for sine).

The input value is at input IN and the result of the math function is at output OUT. Input and output are of data type REAL. If you use operands with absolute addresses, these operands must be double-width.

See Section 3.4 "Data Types" for the bit assignments of the REAL data format.

Function

The math function is executed if "1" is present at the enable input (if power flows into input EN). If an error occurs in the calculation, the enable input is set to "0", otherwise it is set to "1". If execution of the function is not enabled (EN = "0"), the calculation does not take place and ENO is also "0" (Figure 11.1).

If the Master Control Relay MCR is active, output OUT is set to zero if the math function is processed (EN = "1"). The MCR does not influence the ENO.

The following errors can occur in a math function:

▷ Number range underflow and overflow

▷ No valid REAL number as input value

Chapter 15 "Status Bits" explains how the math functions set the status bits.

Examples

Figure 11.2 shows three examples of math functions. A math function performs the calculation in accordance with REAL even if no data types have been declared when using operands with absolute addresses.

Math function in a rung

You can connect contacts in series and in parallel before the input EN and after the output ENO.

The math box itself must only be placed in a branch that leads direct to the left-hand power rail. This branch can also have contacts before the input EN and it need not be the uppermost branch. With the direct connection to the left-hand power rail, you can thus connect math boxes in parallel. When connecting boxes in

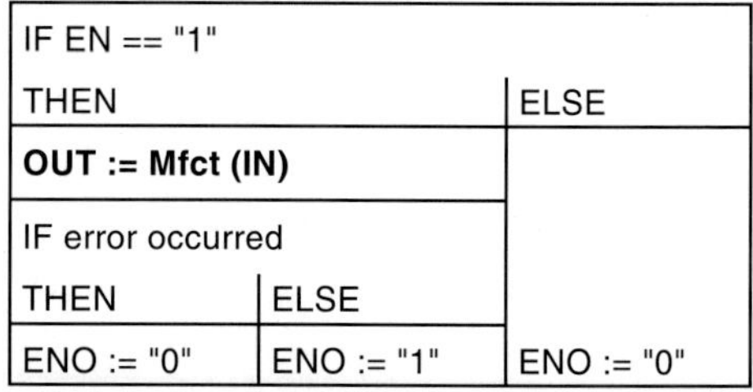

Figure 11.1 Function of a Math Box

Sine	The value in memory doubleword MD 110 contains an angle in the radian measure. The sine is generated from this and stored in memory doubleword MD 104.	
Square root	The square root of the variable "MathVal1" is generated and stored in the variable "MathRoot".	
Exponent	The variable #Result contains the power from e and #Exponent.	

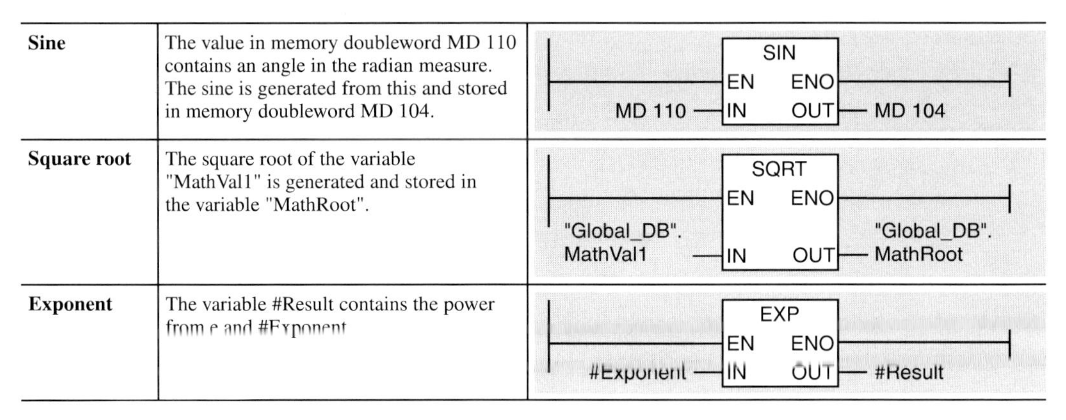

Figure 11.2 Examples of Math Functions

parallel, you require a coil to terminate the rung. If you have not provided error evaluation, assign a "dummy" operand to the coil, for example, a temporary local data bit.

You can connect math boxes in series. If the ENO output of the preceding box leads to the EN input of the subsequent box, the subsequent box is only processed if the preceding box has been completed without errors. If you want to use the result of the preceding box as the input value for a subsequent box, variables from the temporary local data are a convenient intermediate buffer.

If you arrange several math boxes in one rung (parallel to the left-hand power rail and then further in series), the boxes in the uppermost branch are the first to be processed from left to right, followed by the boxes in the second branch from left to right, and so on.

The diskette supplied with the book contains further examples of the representation and arrangement of the math functions (FB 111 in the "Digital Functions" program).

With incremental programming, you will find the math functions in the Program Elements Catalog (with VIEW → CATALOG [Ctrl – K] or INSERT → PROGRAM ELEMENTS) under "Floating-Point Math".

11.2 Trigonometric Functions

The trigonometric functions

- SIN Sine,
- COS Cosine and
- TAN Tangent

assume an angle in the radian measure as a REAL number at the input.

Two units are conventionally used for giving the size of an angle, degrees from 0° to 360° and the radian measure from 0 to 2π (where π = +3.141593e+00). Both can be converted proportionally. For example, the radian measure for a 90° angle is $\pi/2$, or +1.570796e+00. With values greater than 2π (+6.283185e+00), 2π or a multiple of 2π is subtracted until the input value for the trigonometric function is less than 2π.

Example (Figure 11.3 Network 4): Calculating the idle power $P_s = U \cdot I \cdot \text{sine}(\varphi)$

11.3 Arc Functions

The arc functions (inverse trigonometric functions)

- ASIN Arc sine,
- ACOS Arc cosine and
- ATAN Arc tangent

are the inverse functions of the corresponding trigonometric functions. They assume a REAL number in a specific number range at the input IN and return an angle in the radian measure (Table 11.1).

If the permissible value range is exceeded at the input IN, the arc function returns an invalid REAL number and ENO = "0" and sets the status bits CC0, CC1, OV and OS to "1".

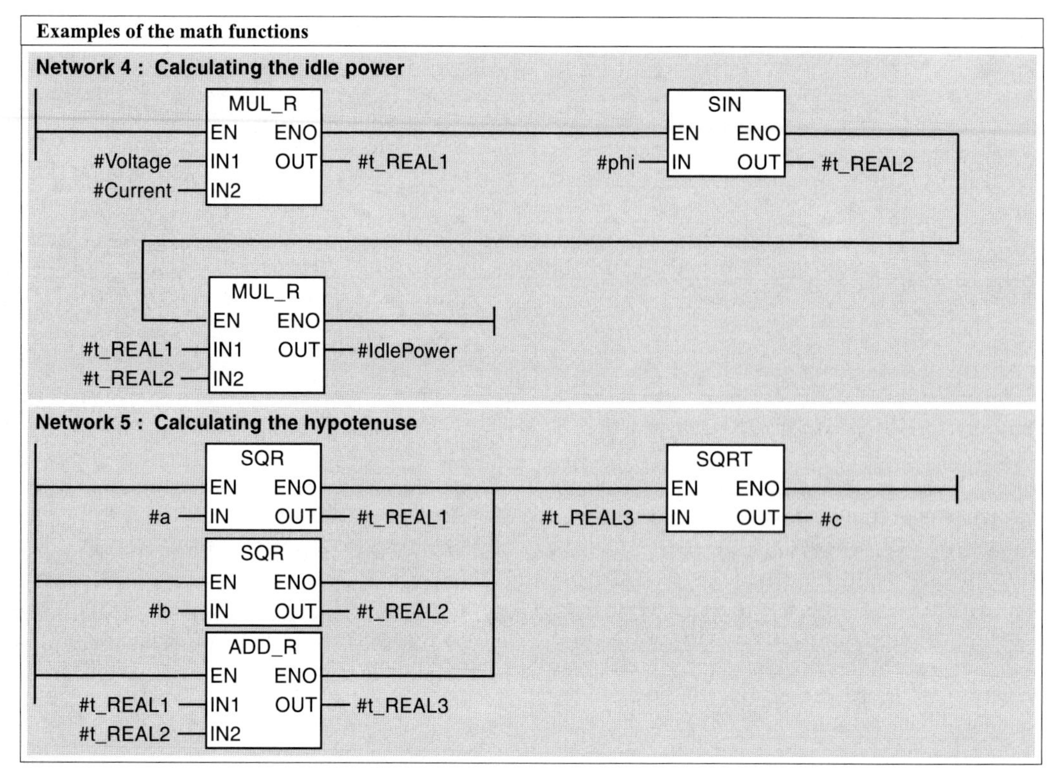

Figure 11.3 Examples of Math Functions

11.4 Other Math Functions

The following math functions are also available

- ▷ SQR
  Establish the square of a number,
- ▷ SQRT
  Establish the square root of a number,
- ▷ EXP
  Establish the exponential function on base e and
- ▷ LN
  Calculate the natural logarithm (logarithm on base e).

Table 11.1 Value Ranges of the Arc Functions

Function	Permissible Value Range	Returned Value
ASIN	–1 to +1	$-\pi/2$ to $+\pi/2$
ACOS	–1 to +1	0 to π
ATAN	Entire range	$-\pi/2$ to $+\pi/2$

Establishing the square

The function SQR squares the value at input IN and stores the result at output OUT.

Example: See “Establishing the square root”.

Establishing the square root

The function SQRT finds the square root of the value at input IN and stores the result at output OUT. If the value at input IN is less than zero, SQRT sets the status bits CC0, CC1, OV and OS to “1” and returns an invalid REAL number. If the value at input IN is –0 (minus zero), –0 is returned.

Example: $c = \sqrt{a^2 + b^2}$

Figure 11.3 Network 5: First, the squares of the variables a and b are found and then added. Finally, the square root is established from the sum. Temporary local data are used as intermediate memory.

(If you have declared *b* or *c* as a local variable, you must precede it with #, so that the Editor recognizes it as a local variable; if *b* or *c* is a global variable, it must be placed within quotation marks.)

Establishing the exponential value on base e

The function EXP calculates the exponential value from base e ($=2.718282e+00$) and the values at input IN (e^{IN}) and stores the result at output OUT.

You can calculate any exponential value using the formula $a^b = e^{b \cdot \ln a}$.

Finding the natural logarithm

The LN finds the natural logarithm on base e ($=2.718282e+00$) from the number at input IN and stores it at output OUT. If the value at input IN is less than or equal to zero, LN sets the status bits CC0, CC1, OV and OS to "1" and returns an invalid REAL number.

The natural logarithm is the inverse function of the exponential function: If $y = e^x$ then $x = \ln y$.

To find any logarithm, use the formula

$$\log_b a = \frac{\log_n a}{\log_n b}$$

where *b* or *n* is any base. If you make $n = e$, you can find a logarithm to any base using the natural logarithm:

$$\log_b a = \frac{\ln a}{\ln b}$$

in the special case for base 10, the formula is as follows:

$$\lg a = \frac{\ln a}{\ln 10} = 0.4342945 \cdot \ln a$$

12 Conversion Functions

The conversion functions convert the data type of a variable. Figure 12.1 gives an overview of the data type conversions described in this chapter.

12.1 Processing a Conversion Function

Representation

The box of a conversion function has an input IN and an output OUT in addition to the enable input EN and the enable output ENO. The "header" in the box represents the conversion function executed (for example, I_BCD stands for the conversion of INT to BCD).

The value to be converted is at input IN and the result of the conversion is at output OUT. The input and output are of different data types depending on the conversion function. For example, with the conversion function DI_R (DINT to REAL), the input is of data type DINT and the output is of data type REAL. The variables applied must be of the same data type as the input or the output. If you use operands with absolute addresses, the operand widths must be adapted to the data types, for example, you can use a word-width operand for data type INT.

You can find the bit assignments of the data formats in Chapter 3.4 "Data Types".

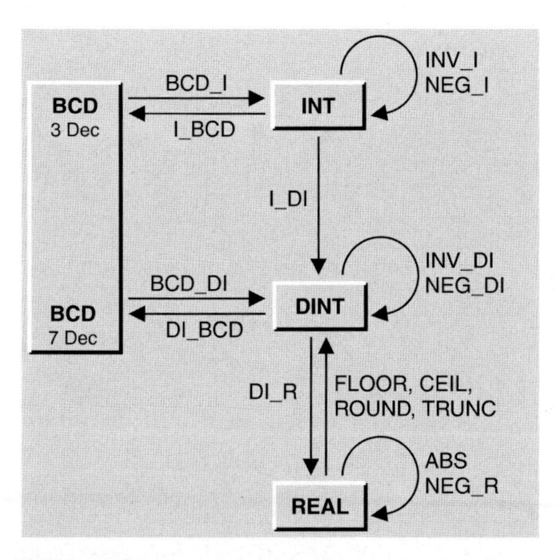

Figure 12.1 Overview of the Conversion Functions

Function

The conversion function is executed if "1" is present at the enable input (if current flows into input EN). If an error occurs during conversion, the enable input ENO is set to "0", otherwise, it is set to "1". If execution of the function is not enabled (EN = "0"), the conversion does not take place and ENO is also "0" (Figure 12.2).

If the Master Control Relay MCR is active, output OUT is set to zero if the conversion function is processed (EN = "1"). The MCR does not influence the ENO output.

Not all conversion functions signal an error. An error only occurs if the permissible number range is exceeded (I_BCD, DI_BCD) or an invalid REAL number is specified (FLOOR, CEIL, ROUND, TRUNC).

<table>
<tr><td colspan="3">IF EN == "1"</td></tr>
<tr><td colspan="2">THEN</td><td>ELSE</td></tr>
<tr><td colspan="2">OUT := Confct (IN)</td><td rowspan="4"></td></tr>
<tr><td colspan="2">IF error occurred</td></tr>
<tr><td>THEN</td><td>ELSE</td></tr>
<tr><td>ENO := "0"</td><td>ENO := "1"</td><td>ENO := "0"</td></tr>
</table>

with Confct as conversion function

Figure 12.2 Function of a Conversion Box

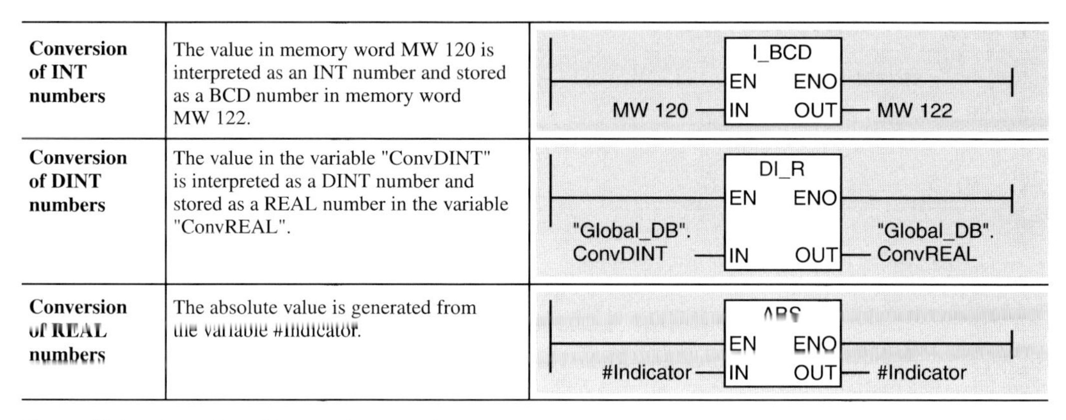

Conversion of INT numbers	The value in memory word MW 120 is interpreted as an INT number and stored as a BCD number in memory word MW 122.
Conversion of DINT numbers	The value in the variable "ConvDINT" is interpreted as a DINT number and stored as a REAL number in the variable "ConvREAL".
Conversion of REAL numbers	The absolute value is generated from the variable #Indicator.

Figure 12.3 Examples of Conversion Functions

If, in the case of the BCD conversion functions BCD_I and BCD_DI the input value contains a pseudo tetrad, program execution is interrupted and the error organization block OB 121 (programming errors) is called.

Chapter 15 "Status Bits" explains how the conversion functions set the status bits.

Figure 12.3 shows one example for each of the different data types. A conversion function converts according to the specified characteristic even if no data types have been declared when using operands with absolute addresses.

Conversion function in a rung

You can connect contacts in series and in parallel before the input EN and after the output ENO.

The conversion box itself must only be placed in a branch that leads direct to the left-hand power rail. This branch can also have contacts before the input EN and it need not be the uppermost branch. With the direct connection to the left-hand power rail, you can thus connect conversion boxes in parallel. When connecting boxes in parallel, you require a coil to terminate the rung. If you have not provided error evaluation, assign a "dummy" operand to the coil, for example, a temporary local data bit.

You can connect conversion boxes in series. If the ENO output of the preceding box leads to the EN input of the subsequent box, the subsequent box is only processed if the preceding box has been completed without errors. If you want to use the result of the preceding box as the input value for a subsequent box, variables from the temporary local data are a convenient intermediate buffer.

If you arrange several conversion boxes in one rung (parallel to the left-hand power rail and then further in series), if the boxes in the uppermost branch are the first to be processed first from left to right, followed by the boxes in the second branch from left to right, and so on.

The diskette supplied with the book contains further examples of the representation and arrangement of the conversion functions (FB 112 in the "Digital Functions" program).

With incremental programming, you will find the conversion functions in the Program Elements Catalog (with VIEW → CATALOG [Ctrl – K] or INSERT → PROGRAM ELEMENTS) under "Convert".

12.2 Conversion of INT and DINT Numbers

Table 12.1 shows the conversion functions provided by LAD for converting INT and DINT numbers. Variables of the specified data type or absolute-addressed operands with the relevant width must be applied to the inputs and outputs of the boxes (for example, a word-width operand with data type INT).

Conversion from INT to DINT

The function I_DI interprets the value at input IN as a number of data type INT and transfers it to the right-hand word of the output OUT. The

Table 12.1 Conversion of INT and DINT Numbers

Data Type Conversion	Box	Data Type for Parameter	
		IN	OUT
INT to DINT	I_DI	INT	DINT
INT to BCD	I_BCD	INT	WORD
DINT to BCD	DI_BCD	DINT	DWORD
DINT to REAL	DI_R	DINT	REAL

signal state of bit 15 (the sign) of the input is transferred to the left-hand word of output OUT into bits 16 to 31.

The conversion from INT to DINT does not signal errors.

Conversion from INT to BCD

The function I_BCD interprets the value at input IN as a number of data type INT and converts it to a BCD-coded number with 3 decades at output OUT. The 3 right-justified decades represent the amount of the decimal number. The sign is in bits 12 to 15. If all bits are set to signal state "0", the sign is positive; if all bits are at signal state "1" this signifies a negative sign.

If the INT number is too large for conversion to a BCD number (>999), the function I_BCD sets the status bits OV and OS. The conversion then does not take place.

Conversion from DINT to BCD

The DI_BCD interprets the value at input IN as a number of data type DINT and converts it to a BCD-coded number with 7 decades at output OUT. The 7 right-justified decades represent the amount of the decimal number. The sign is in bits 28 to 31. If all bits are set to signal state "0", the sign is positive; if all bits are at signal state "1" this signifies a negative sign.

If the DINT number is too large for conversion to a BCD number (>9 999 999), the status bits OV and OS are set. The conversion then does not take place.

Conversion from DINT to REAL

The function DI_R interprets the value at input IN as a number of data type DINT and converts it to a REAL number at output OUT.

Since a number in DINT format has a higher accuracy than a number in REAL format, rounding may take place during the conversion. It is then rounded up to the next whole number (in accordance with the function ROUND).

The function DI_R does not signal errors.

12.3 Conversion of BCD Numbers

Table 12.2 shows the functions for converting BCD numbers provided by LAD. Variables of the specified data type or absolute-addressed operands with the relevant width must be applied to the inputs and outputs of the boxes (for example, a word-width operand with data type INT).

Conversion from BCD to INT

The function BCD_I interprets the value at input IN as a BCD-coded number with 3 decades and converts it to an INT number at output OUT. The 3 right-justified decades represent the amount of the decimal number. The sign is in

Table 12.2 Conversion of BCD Numbers

Data Type Conversion	Box	Data Type for Parameter	
		IN	OUT
BCD to INT	BCD_I	WORD	INT
BCD to DINT	BCD_DI	DWORD	DINT

Table 12.3 Conversion of REAL Numbers to DINT Numbers

Data Type Conversion with Rounding	Box	Data Type for Parameter	
		IN	OUT
to the next higher whole number	CEIL	REAL	DINT
to the next lower whole number	FLOOR	REAL	DINT
to the next whole number	ROUND	REAL	DINT
without rounding	TRUNC	REAL	DINT

bits 12 to 15. Signal state "0" of these bits signifies "positive", and signal state "1" signifies "negative". Only the signal state of bit 15 is taken into account in the conversion.

If there is a pseudo tetrad (number value 10 to 15 or A to F in hexadecimal representation) in the BCD-coded number, the CPU signals a programming error and calls organization block OB 121 (synchronization errors). If this block is not available, the CPU goes to Stop.

The function BCD_I does not set status bits.

Conversion from BCD to DINT

The function BCD_DI interprets the value at input IN as a BCD-coded number with 7 decades and converts it to an INT number at output OUT. The 7 right-justified decades represent the amount of the decimal number. The sign is in bits 28 to 31. Signal state "0" of these bits signifies "positive", and signal state "1" signifies "negative". Only the signal state of bit 31 is taken into account in the conversion.

If there is a pseudo tetrad (number value 10 to 15 or A to F in hexadecimal representation) in the BCD-coded number, the CPU signals a programming error and calls organization block OB 121 (synchronization errors). If this block is not available, the CPU goes to Stop.

The function BCD_I does not set status bits.

12.4 Conversion of REAL Numbers

There are several functions for converting a number in REAL format to DINT format (conversion of a fractured value to a whole number) (Table 12.3). They differ in the way they perform rounding. Variables of the specified data type or absolute-addressed double-word operands must be applied to the inputs and outputs of the boxes.

Rounding to the next higher whole number

The function CEIL interprets the value at input IN as a number in REAL format and converts it to a number in DINT format at output OUT. CEIL returns a whole number greater than or equal to the number to be converted.

If the value at input IN is higher or lower than range permissible for a number in DINT format, or if it does not correspond to a number in REAL format, CEIL sets status bits OV and OS. Conversion then does not take place.

Rounding to the next lower whole number

The function FLOOR interprets the value at input IN as a number in REAL format and converts it to a number in DINT format at output OUT. FLOOR returns a whole number less than or equal to the number to be converted.

If the value at input IN is higher or lower than the range permissible for a number in DINT format, or if it does not correspond to a number in REAL format, FLOOR sets status bits OV and OS. Conversion then does not take place.

Rounding to the next whole number

The function ROUND interprets the value at input IN as a number in REAL format and converts it to a number in DINT format at output OUT. ROUND returns the next whole number. If the result lies exactly between an odd and an even number, the even number is selected.

If the value at input IN is higher or lower than the range permissible for a number in DINT format, or if it does not correspond to a number in REAL format, ROUND sets status bits OV and OS. Conversion then does not take place.

Without rounding

The function TRUNC interprets the value at input IN as a number in REAL format and converts it to a number in DINT format at output OUT. TRUNC returns the whole-number component of the number to be converted; the fracture component is "truncated".

If the value at input IN is higher or lower than the range permissible for a number in DINT format, or if it does not correspond to a number in REAL format, TRUNC sets status bits OV and OS. Conversion then does not take place.

Summary of conversion from REAL to DINT

Table 12.4 shows the different effects of the functions for converting from REAL to DINT. For the example, the range –1 to +1 has been selected.

12.5 Other Conversion Functions

Other conversion functions available are one's complement generation, negation functions, absolute value generation of a REAL number (Table 12.5). Variables of the specified data type or absolute-addressed operands with the relevant width must be applied to the inputs and outputs of the boxes (for example, a doubleword operand for data type DINT).

One's complement INT

The function INV_I negates the value at input IN bit for bit and writes it to output OUT. INV_I replaces the zeros with ones and vice versa.

The function INV_I does not signal errors.

One's complement DINT

The function INV_DI negates the value at input IN bit for bit and writes it to output OUT. INV_DI replaces the zeros with ones and vice versa.

The function INV_DI does not signal errors.

Negation INT

The function NEG_I interprets the value at input IN as an INT number, changes the sign through two's complement generation and writes the changed value to output OUT. NEG_I is identical to multiplication by –1.

Table 12.4 Rounding Modi for the Conversion of REAL Numbers

Input value		Result			
REAL	DW#16#	ROUND	CEIL	FLOOR	TRUNC
1.0000001	3F80 0001	1	2	1	1
1.00000000	3F80 0000	1	1	1	1
0.99999995	3F7F FFFF	1	1	0	0
0.50000005	3F00 0001	1	1	0	0
0.50000000	3F00 0000	0	1	0	0
0.49999996	3EFF FFFF	0	1	0	0
5.877476E-39	0080 0000	0	1	0	0
0.0	0000 0000	0	0	0	0
–5.877476E-39	8080 0000	0	0	–1	0
–0.49999996	BEFF FFFF	0	0	–1	0
–0.50000000	BF00 0000	0	0	–1	0
–0.50000005	BF00 0001	–1	0	–1	0
–0.99999995	BF7F FFFF	–1	0	–1	0
–1.00000000	BF80 0000	–1	–1	–1	–1
–1.0000001	BF80 0001	–1	–1	–2	–1

Table 12.5 Other Conversion Functions

Conversion	Box	Data Type for Parameter	
		IN	OUT
One's complement INT	INV_I	INT	INT
One's complement DINT	INV_DI	DINT	DINT
Negation of an INT number	NEG_I	INT	INT
Negation of a DINT number	NEG_DI	DINT	DINT
Negation of a REAL number	NEG_R	REAL	REAL
Absolute value generation of a REAL number	ABS	REAL	REAL

The function NEG_I sets status bits CC0, CC1, OV and OS.

Negation DINT

The function NEG_DI interprets the value at input IN as an DINT number, changes the sign through two's complement generation and writes the changed value to output OUT. NEG_DI is identical to multiplication by –1.

The function NEG_DI sets status bits CC0, CC1, OV and OS.

Negation REAL

The function NEG_R interprets the value at input IN as an REAL number, multiplies this number by –1 and writes it to output OUT. NEG_R changes the sign of the mantissa even on an invalid REAL number.

The function NEG_R does not signal errors.

Absolute value generation REAL

The function ABS interprets the value at input IN as an REAL number, generates the absolute value from this number and writes it to output OUT. ABS sets the sign of the mantissa to "0" even on an invalid REAL number.

The function ABS does not signal errors.

13 Shift Functions

The shift functions shift the contents of a variable bitwise to the left or to the right. The bits shifted out are either lost or are adopted again on the other side of the variable (Table 13.1).

13.1 Processing a Shift Function

Representation

The box of a shift function has an input IN, an input N and an output OUT in addition to the enable input EN and the enable output ENO. The "header" in the box represents the conversion function executed (for example, SHL_W stands for shifting a word variable to the left).

Shift box
(in the example: wordwise shifting to the left)

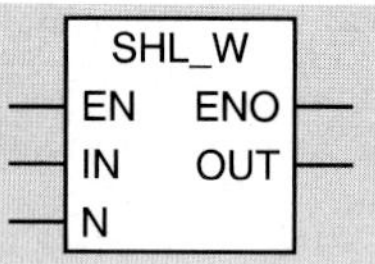

SHL_W
EN ENO
IN OUT
N

The value to be shifted is at input IN, the shift number is at input N and the result at output OUT. The input and the output have different data types depending on the shift function. For example, the input and output in the case of the shift function SHR_DW (shift a doubleword variable to the right) are of data type DWORD. The variables applied must be of the same data type as the input or output. If you use operands with absolute addresses, the operand widths must be adapted to the data types, for example, you can use a word-width operand for data type INT. Input N has data type WORD for every shift function.

See Chapter 3.4 "Data Types" for the bit assignments of the data formats.

Function

The shift function is executed if "1" is present at the enable input (when power flows into input EN). ENO then has "1". If execution of the function is not enabled (EN = "0"), the shift does not take place and ENO is also "0" (Figure 13.1).

When the Master Control Relay MCR is activated, output OUT is set to zero if the shift function is processed (EN = "1"). The MCR does not affect the ENO output.

Chapter 15 "Status Bits" explains how the shift functions set the status bits.

Examples

Figure 13.2 gives one example each for different shift functions.

Shift function in a rung

You can connect contacts in series and in parallel prior to input EN and after output ENO.

The shift box itself must only be placed in a branch that leads direct to the left-hand power rail. This branch can also have contacts prior to

Table 13.1 Overview of the Shift Functions

Shift functions	Word Variable	Doubleword variable
Shift left	SHL_W	SHL_DW
Shift right	SHR_W	SHR_DW
Shift with sign	SHR_I	SHR_DI
Rotate left	–	ROL_DW
Rotate right	–	ROR_DW

IF EN == "1"	
THEN	ELSE
OUT := Sfct (IN, N)	
ENO := "1"	ENO := "0"

with Sfct as shift function

Figure 13.1 Function of a Shift Box

input EN and it need not be the uppermost branch. With the direct connection to the left-hand power rail, you can thus connect shift boxes in parallel. When connecting boxes in parallel, you require a coil to terminate the rung. If you have not provided error evaluation, assign a "dummy" operand to the coil, for example, a temporary local data bit.

You can connect shift boxes in series. If the ENO output of the preceding box leads to the EN input of the subsequent box, the subsequent box is always processed. If you want to use the result of the preceding box as the input value for a subsequent box, variables from the temporary local data are a convenient intermediate buffer.

If you arrange several boxes in one rung (parallel to the left-hand power rail and then further in series), the boxes in the uppermost branch are the first to be processed from left to right, followed by the boxes in the second branch from left to right, and so on.

The diskette supplied with the book contains further examples of the representation and arrangement of the shift functions (FB 113 in the "Digital Functions" program).

With incremental programming, you will find the shift functions in the Program Elements Catalog (with VIEW → CATALOG [Ctrl – K] or INSERT → PROGRAM ELEMENTS) under "Shift and Rotate".

13.2 Shift

Shift word variable to the left

The shift function SHL_W shifts the contents of the WORD variable at input IN bitwise to the left by the number of positions specified by the shift number at input N. The bit positions freed up by the shift are filled with zeros. The WORD variable at output OUT contains the result.

The shift number at input N specifies the number of bit positions by which the contents are to be shifted. It can be a constant or a variable. If the shift number = 0, the function is not executed; if it is greater than 15, the output variable contains zero after execution of SHL_W.

Shift doubleword variable to the left

The shift function SHL_DW shifts the contents of the DWORD variable at input IN bitwise to the left by the number of positions specified by

Shifting word variables	The value in memory word MW 130 is shifted to the left by 4 positions and stored in memory word MW 132.	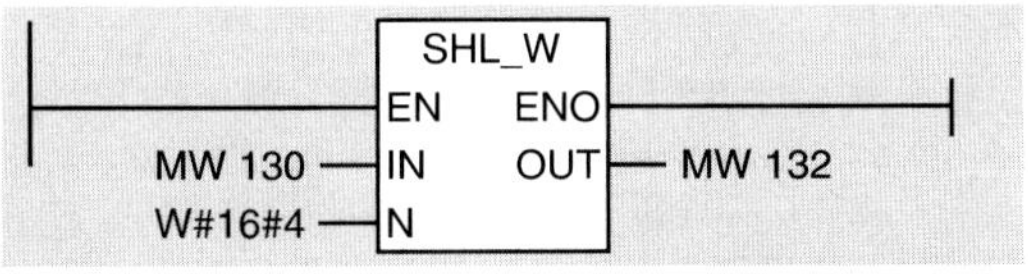 SHL_W EN ENO MW 130 — IN OUT — MW 132 W#16#4 — N
Shifting doubleword variables	The value in the variable "ShiftIn" is shifted to the right by the number of positions specified in "ShiftNum", and then stored in "ShiftOut".	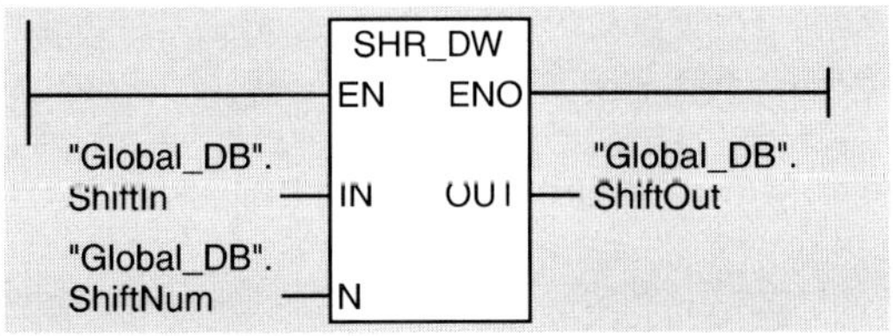 SHR_DW EN ENO "Global_DB". ShiftIn — IN OUT — "Global_DB". ShiftOut "Global_DB". ShiftNum — N
Shifting with sign	The variable #Act_value is shifted 2 positions to the right with sign and transferred to the variable #Indicator.	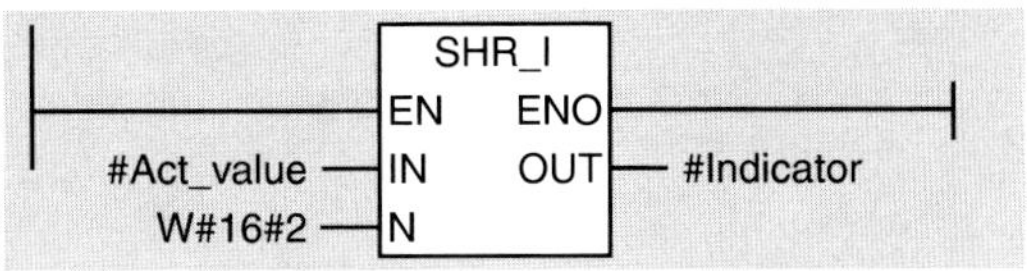 SHR_I EN ENO #Act_value — IN OUT — #Indicator W#16#2 — N

Figure 13.2 Examples of Shift Functions

the shift number at in input N. The bit positions freed up by the shift are filled with zeros. The DWORD variable at output OUT contains the result.

The shift number at input N specifies the number of bit positions by which the contents are to be shifted. It can be a constant or a variable. If the shift number = 0, the function is not executed; if it is greater than 31, the output variable contains zero after execution of SHL_DW.

Shift word variable to the right

The shift function SHR_W shifts the contents of the WORD variable at input IN bitwise to the right by the number of positions specified by the shift number at input N. The bit positions freed up by the shift are filled with zeros. The WORD variable at output OUT contains the result.

The shift number at input N specifies the number of bit positions by which the contents are to be shifted. It can be a constant or a variable. If the shift number = 0, the function is not executed; if it is greater than 15, the output variable contains zero after execution of SHR_W.

Shift doubleword variable to the right

The shift function SHR_DW shifts the contents of the DWORD variable at input IN bitwise to the right by the number of positions specified by the shift number at input N. The bit positions freed up by the shift are filled with zeros. The DWORD variable at output OUT contains the result.

The shift number at input N specifies the number of bit positions by which the contents are to be shifted. It can be a constant or a variable. If the shift number = 0, the function is not executed; if it is greater than 31, the output variable contains zero after execution of SHR_DW.

Shift word variable with sign

The shift function SHR_I shifts the contents of the INT variable at input IN bitwise by to the right by the number of positions specified by the shift number at input IN. The bit positions freed up by the shift are filled with the signal state of bit 31 (this is the sign of a INT number), that is, with "0" in the case of positive numbers and with "1" in the case of negative numbers. The INT variable at output OUT contains the result.

The shift number at input N specifies the number of bit positions by which the contents are to be shifted. It can be a constant or a variable. If the shift number = 0, the function is not executed; if it is greater than 15, all bit positions in the output variable contain the sign after execution of SHR_I.

With a number in data format INT, shifting to the right corresponds to division with an exponential number on base 2. The exponent is the shift number. The result of such a division corresponds to the rounded down whole number.

Shift doubleword variable with sign

The shift function SHR_DI shifts the contents of the DINT variable at input IN bitwise by to the right by the number of positions specified by the shift number at input IN. The bit positions freed up by the shift are filled with the signal state of bit 15 (this is the sign of an DINT number), that is, with "0" in the case of positive numbers and with "1" in the case of negative numbers. The DINT variable at output OUT contains the result.

The shift number at input N specifies the number of bit positions by which the contents are to be shifted. It can be a constant or a variable. If the shift number = 0, the function is not executed; if it is greater than 31, all bit positions in the output variable contain the sign after execution of SHR_DI.

With a number in data format DINT, shifting to the right corresponds to division with an exponential number on base 2. The exponent is the shift number. The result of such a division corresponds to the rounded down whole number.

13.3 Rotate

Rotate doubleword variable to the left

The shift function ROL_DW shifts the contents of the DWORD variable at input IN bitwise to the left by the number of positions specified by the shift number at input N. The bit positions freed up by the shift are filled with the shifted-out bit positions. The DWORD variable at output OUT contains the result.

The shift number at input N specifies the number of bit positions by which the contents are to be shifted. It can be a constant or a variable. If the shift number = 0, the function is not executed; if it is 32, the contents of the input variable are retained and the status bits are set. If the shift number has the value 33, a shift is made by one position, if it has the value 34 the shift is by two positions, and so on (shifting is executed modulus 32).

Rotate doubleword variable to the right

The shift function ROR_DW shifts the contents of the DWORD variable at input IN bitwise to the right by the number of positions specified by the shift number at input N. The bit positions freed up by the shift are filled with the shifted-out bit positions. The DWORD variable at output OUT contains the result.

The shift number at input N specifies the number of bit positions by which the contents are to be shifted. It can be a constant or a variable. If the shift number = 0, the function is not executed; if it is 32, the contents of the input variable are retained and the status bits are set. If the shift number has the value 33, a shift is made by one position, if it has the value 34 the shift is by two positions, and so on (shifting is executed modulus 32).

14 Word Logic Combinations

Word logic combines the values of two variables bitwise according to logic AND, OR or exclusive OR. The combination can be made on word basis or doubleword basis. The functions shown in Table 14.1 are available as word logic combinations.

14.1 Processing a Word Logic Combination

Representation

The box of a word logic combination has two inputs IN1 and IN2 and one output OUT, in addition to the enable input EN and the enable output ENO. The "header" in the box represents the word logic combination executed (for example, WAND_W stands for the wordwise AND combination).

Word logic box
(in the example:
Wordwise AND combination)

WAND_W
EN ENO
IN1 OUT
IN2

The values to be combined are at the inputs IN1 and IN2, and the result of the combination is then at output OUT. The inputs and the output have different data types depending on the word logic combination: WORD in the case of the wordwise (16-bit) combinations and DWORD in the case of the doubleword (32-bit) combinations. The variables applied must be of the same data type as the inputs or the output.

See Section 3.4 "Data Types" for the bit assignments of the data formats.

Function

The word logic combination is always executed if "1" is present at the enable input (power flows into the input EN). If execution of the function is not enabled (EN = "0"), the logic combination does not take place and ENO is also "0" (Figure 14.1).

If the Master Control Relay MCR is active, the output OUT is set to zero when the word logic combination is executed (EN = "1"). The MCR does not influence the ENO output.

The word logic combination generates the result bit for bit. Bit 0 of input IN1 is combined with bit 0 of input IN2; the result is stored in bit 0 of output OUT. The same combination takes place with bit 1, with bit 2, etc. until bit 15 or bit 31. Table 14.2 shows the generation of the result for a single bit.

Chapter 15 "Status Bits" explains how the word logic combinations set the status bits.

Examples

Figure 14.2 shows one example each of the different word logic combinations.

Table 14.1
Overview of the Word Logic Combinations

Word Logic Combination	with a word variable	with a doubleword variable
AND	WAND_W	WAND_DW
OR	WOR_W	WOR_DW
Exclusive OR	WXOR_W	WXOR_DW

IF EN == "1"	
THEN	ELSE
OUT := IN1 Wlog IN2	
ENO := "1"	ENO := "0"

with Wlog as word logic combination

Figure 14.1 Function of a Word Logic Box

Table 14.2
Result Generation in Word Logic Combinations

Content of input IN1	0	0	1	1
Content of input IN2	0	1	0	1
Result with AND	0	0	0	1
Result with OR	0	1	1	1
Result with exclusive OR	0	1	1	0

Word logic combination in a rung

You can arrange contacts in series and in parallel prior to the input EN and after the output ENO.

The word logic box itself must only be placed in a branch that leads direct to the left-hand power rail. This branch can also have contacts before the input EN and it need not be the uppermost branch. With the direct connection to the left-hand power rail, you can thus connect word logic boxes in parallel. When connecting boxes in parallel, you require a coil to terminate the rung. If you have not provided error evaluation, assign a "dummy" operand to the coil, for example, a temporary local data bit.

You can connect word logic boxes in series. If the ENO output of the preceding box leads to the EN input of the subsequent box, the subsequent box is always processed. If you want to use the result of the preceding box as the input value for a subsequent box, variables from the temporary local data are a convenient intermediate buffer.

If you arrange several word logic boxes in one rung (parallel to the left-hand power rail and then further in series), the boxes in the uppermost branch are the first to be processed from left to right, followed by the boxes in the second branch from left to right, and so on.

The diskette supplied with the book contains further examples of the representation and arrangement of the word logic functions (FB 114 in the "Digital Functions" program).

With incremental programming, you will find the conversion functions in the Program Elements Catalog (with VIEW → CATALOG [Ctrl – K] or INSERT → PROGRAM ELEMENTS) under "Word Logic".

14.2 Description of the Word Logic Combinations

AND combination

AND combines the individual bits of the value at input IN1 with the relevant bits of the value at input IN2 according to logic AND. The indi-

Word logic combination according to logic AND	The upper 4 bits in memory word MW 140 are set to "0" and the result is stored in memory word MW 142.	WAND_W EN ENO MW 140 — IN1 OUT — MW 142 W#16#FFF — IN2
Word logic combination according to logic OR	The variables "WLogVal1" and "WLogVal2" are combined bitwise according to logic OR and the result is stored in "WLogRes".	WOR_DW EN ENO "Global_DB".WLogVal1 — IN1 OUT — "Global_DB".WLogRes "Global_DB".WLogVal2 — IN2
Word logic combination according to exclusive OR	The value generated by exclusive OR from the variables #Input and #Mask is stored at variable #Store.	WXOR_W EN ENO #Input — IN1 OUT — #Store #Mask — IN2

Figure 14.2 Examples of Word Logic Combinations

vidual bits in the result word OUT only have signal state "1" if the relevant bits of both values to be combined have signal state" 1".

Since the bits at input IN2 with signal state "0" also set these result bits to "0" regardless of the assignment of these bits at input IN1, we also refer to these bits as being "masked out". This masking out is the primary area of application of the (digital) AND combination.

OR combination

OR combines the individual bits of the value at input IN1 with the relevant bits of the value at input IN2 according to logic OR. The individual bits in the result word OUT only have signal state "0" if the relevant bits of both values to be combined have signal state "0".

Since the bits at input IN2 with signal state "1" also set these result bits to "1" regardless of the assignment of these bits at input IN1, we also refer to these bits being "masked in". This masking in is the primary area of application of the (digital) OR combination.

Exclusive OR combination

Exclusive OR combines the individual bits of the value at input IN1 with the relevant bits of the value at input IN2 according to exclusive OR. The individual bits in the result word OUT only have signal state "1" if only one of the relevant bits of both values to be combined has signal state "1". If a bit at input IN2 has signal state "1", the input from IN1 has the reverse signal state at this position in the result.

In the result, only those bits in both variables that have a different signal state prior to the digital exclusive OR will have signal state "1". Finding the bits assigned with different signal states or "negating" the signal states of individual bits is the main area of application of the (digital) exclusive OR combination.

Program Flow Control

LAD provides you with a variety of possibilities for controlling the flow of the program. You can exit linear program execution within a block or you can structure the program with parameterizable block calls. You can influence program execution depending on values calculated at runtime, or depending on process parameters, or according to your plant status.

- The **status bits** provide information on the result of an arithmetic or math function and on errors (for example, number range violation in a calculation). You can incorporate the signal state of the status bits direct into your program using contacts.
- You can use the **jump functions** to branch within your program either unconditionally or dependent on the RLO.
- A further method of influencing program execution is provided by the **Master Control Relay** (MCR). Originally developed for relay contactor controls, LAD offers a software version of this program control method.
- LAD provides the **block functions** as a means for you to structure your program. You can use functions and function blocks again and again by defining **block parameters**.

Chapter 19 "Block Parameters" contains the examples shown in Chapter 5 "Memory Functions" and 8 "Counter Functions", this time programmed as function blocks with block parameters. These function blocks are then also called in the "Feed" example as local instances.

15 Status Bits

The status bits are binary flags (indicator bits). The CPU uses them for controlling the binary logic operations and sets them in digital processing. You can check these status bits or influence specific bits. The status bits are combined into a word, the status word. However, you cannot access this status word with LAD.

15.1 Description of the Status Bits

Table 15.1 shows the status bits available with LAD. The CPU uses the binary flags for controlling the binary functions; the digital flags indicate primarily results of arithmetic and math functions.

First check

The /FC status bit steers the binary logic operation within a logic control system. A bit logic step always starts with /FC = "0" and a binary check instruction, the first check. The first check sets /FC = "1". The first check corresponds in LAD to the first contact in a network.

A bit logic step ends with a binary value assignment (of, for example, a single coil) or with a conditional jump or a block change. These set /FC = "0".

Result of logic operation (RLO)

The RLO status bit is the intermediate buffer in binary logic operations. In the first check, the CPU transfers the check result to the RLO, combines the check result with the stored RLO at each subsequent check, and stores the result, in turn, in the RLO. You can store the RLO with the SAVE coil in the binary result BR. Memory, timer and counter functions are controlled using the RLO and certain jump functions are executed. The RLO corresponds in LAD to the power flowing in the rung (RLO = "1" is the same as "power flowing").

Status

The STA status bit corresponds to the signal state of the checked binary operand. In the case of memory functions, the value of STA is the same value as the written value or (if no write operation takes place, for example, in the case of RLO = "0" or MCR active), STA corresponds to the value of the addressed (and unmodified) binary operand. With edge evaluations FP or FN, the value of the RLO prior to the edge evaluation is stored in STA. All other binary functions set STA = "1".

The STA status bit has no effect on the processing of the LAD functions.

OR status bit

The OR status bit stores the result of a fulfilled series circuit and indicates to a subsequently processed parallel circuit that the result is already fixed. All other binary functions reset the OR status bit.

Table 15.1 Status Bits

Binary Flags	
/FC	First check
RLO	Result of logic operation (RLO)
STA	Status
OR	OR status bit
BR	Binary result

Digital Flags	
OS	Stored overflow
OV	Overflow
CC0	CC0 (condition code) status bit
CC1	CC1 (condition code) status bit

Overflow

The OV status bit indicates a number range overflow or the use of invalid REAL numbers. The following functions affect the OV status bit: Arithmetic functions, math functions, some conversion functions, REAL comparison functions.

You can check the OV status bit with contacts.

Stored overflow

The OS status bit stores a set OV status bit: When the CPU sets the OV status bit, it also always sets the OS status bit. However, while the next properly executed operation resets OV, OS remains set. This provides you with the opportunity of evaluating, even at a later point in your program, a number range overflow or an operation with an invalid REAL number.

You can check the OS status bit with contacts. A block change resets the OS status bit.

CC0 and CC1 status bits (condition code bits)

The CC0 and CC1 status bits provide information on the result of a comparison function, an arithmetic or math function, a word logic operation or on the shifted out bit in the case of a shift function.

You can check all combinations of the CC0 and CC1 status bits with contacts (see below).

Binary result

LAD uses the BR status bit for implementing the EN/ENO mechanism for boxes. You can also set or reset the BR status bit yourself and check it with contacts.

15.2 Setting the Status Bits

The digital functions affect the CC0, CC1, OV and OS status bits as shown in Table 15.2. You can check these status bits immediately following the function box.

Status bits with INT and DINT calculation

The arithmetic functions with data formats INT and DINT set all digital status bits. A result of zero sets CC0 and CC1 to "0". CC0 = "0" and CC1 = "1" indicates a positive result, CC0 = "1" and CC1 = "0" indicates a negative result. A number range overflow sets OV and OS (please note the other meaning of CC0 and CC1 in the case of overflow). Division by zero is indicated by "1" at all digital status bits.

Status bits with REAL calculation

The arithmetic functions with data format REAL and the math functions set all digital status bits. A result of zero sets CC0 and CC1 to "0". CC0 = "0" and CC1 = "1" indicates a positive result, CC0 = "1" and CC1 = "0" indicates a negative result. A number range overflow sets OV and OS (please note the other meaning of CC0 and CC1 in the case of overflow). An invalid REAL number is indicated with "1" at all digital status bits.

A REAL number is referred to as "denormalized" if it is represented with reduced accuracy. the exponent is then zero; the absolute value of a denormalized REAL number is less than $1.175\,494 \cdot 10^{-38}$. S7-300 CPUs treat denormalized REAL numbers as equal to zero (see also Section 3.4 "Data Types").

Status bits with the conversion functions

Of the conversion functions, the two"s complements affect all digital status bits. In addition, the following conversion functions set status bits OV and OS in the event of an error (number range overflow or invalid REAL number):

- I_BCD and DI_BCD:
  Conversion of INT and DINT to BCD
- CEIL, FLOOR, ROUND, TRUNC:
  Conversion of REAL to DINT

Status bits with shift functions

In the case of the shift functions, the signal state of the last bit to be shifted out is transferred to the CC1 status bit. CC0 and OV are reset.

Status bits with word logic

If the result of the word logic operation is zero (all bits are "0"), CC1 is reset; if at least one bit in the result is "1", CC1 is set. CC0 and OV are reset.

15.3 Evaluating the Status Bit

You can use the normally-open (NO) and the normally-closed (NC) contact to check the digital status bits and the binary result. Figure 15.1

Table 15.2 Setting the Status Bits

INT Calculation				
The result is:	CC0	CC1	OV	OS
< –32 768 (ADD_I, SUB_I)	0	1	1	1
< –32 768 (MUL_I)	1	0	1	1
–32 768 to –1	1	0	0	–
0	0	0	0	–
+1 to +32 767	0	1	0	–
> +32 767 (ADD_I, SUB_I)	1	0	1	1
> +32 767 (MUL_I)	0	1	1	1
32 768 (DIV_I)	0	1	1	1
(–) 65 536	0	0	1	1
Division by zero	1	1	1	1

DINT Calculation				
The result is:	CC0	CC1	OV	OS
< –2 147 483 648 (ADD_DI, SUB_DI)	0	1	1	1
< –2 147 483 648 (MUL_DI)	1	0	1	1
–2 147 483 648 to –1	1	0	0	–
0	0	0	0	–
+1 to +2 147 483 647	0	1	0	–
> +2 147 483 647 (ADD_DI, SUB_DI)	1	0	1	1
> +2 147 483 647 (MUL_DI)	0	1	1	1
2 147 483 648 (DIV_DI)	0	1	1	1
(–) 4 294 967 296	0	0	1	1
Division by zero (DIV_DI, MOD_DI)	1	1	1	1

REAL calculation				
The result is:	CC0	CC1	OV	OS
+ infinite (division by zero)	0	1	1	1
+ normalized	0	1	0	–
± denormalized	0	0	1	1
± zero	0	0	0	–
– normalized	1	0	0	–
– infinite (division by zero)	1	0	1	1
± invalid REAL number	1	1	1	1

Comparison				
The result is:	CC0	CC1	OV	OS
equal to	0	0	0	–
greater than	0	1	0	–
less than	1	0	0	–
invalid REAL number	1	1	1	1

Conversion NEG_I				
The result is:	CC0	CC1	OV	OS
+1 to +32 767	0	1	0	–
0	0	0	0	–
–1 to –32 767	1	0	0	–
(–) 32 768	1	0	1	1

Conversion NEG_D				
The result is:	CC0	CC1	OV	OS
+1 to +2 147 483 647	0	1	0	–
0	0	0	0	–
–1 to –2 147 483 647	1	0	0	–
(–) 2 147 483 648	1	0	1	1

Shift function				
The shifted out bit is:	CC0	CC1	OV	OS
"0"	0	0	0	–
"1"	0	1	0	–
with shift number 0	–	–	–	–

Word logic				
The result is:	CC0	CC1	OV	OS
zero	0	0	0	–
not zero	0	1	0	–

Check	Power flows (check fulfilled) on
>0	Result greater than zero [(CC0=0) & (CC1=1)]
>=0	Result greater than or equal to zero [(CC0=0)]
<0	Result less than zero [(CC0=1) & (CC1=0)]
<=0	Result less than or equal to zero [(CC1=0)]
<>0	Result not equal to zero [(CC0=0) & (CC1=1) ∨ (CC0=1) & (CC1=0)]
==0	Result equal to zero [(CC0=0) & (CC1=0)]
UO	Result invalid (unordered) [(CC0=1) & (CC1=1)]
OV	Number range overflow [OV=1]
OS	Stored overflow [OS=1]
BR	Binary result [BR=1]

Figure 15.1 Evaluating the Status Bits

shows the check with a normally-open contact. The check with a normally-closed contact supplies the reverse check result.

You can handle the NO and NC contacts for evaluating the status bits in exactly the same way as the "normal" contacts. The diskette accompanying the book contains examples of evaluating the status bits (FB 115 in the program "Program Flow Control").

With incremental programming, you will find these checks in the Program Elements Catalog (with VIEW → CATALOG [Ctrl – K] or INSERT → PROGRAM ELEMENTS) under "Status Bits".

15.4 Using the Binary Result

15.4.1 Setting the Binary Result BR

You store the RLO in the binary result using the SAVE coil. If power flows into the SAVE coil, BR is set, otherwise it is reset. You program the SAVE coil in the same way as a "single" coil.

Please note that the SAVE coil does not terminate the logic operation (the /FC status bit is not set to "0"). This means logic operation preceding the SAVE coil is also the preceding logic operation for the next network. If you want to prevent this, program a single coil with a branch parallel to the SAVE coil and assign a "dummy" operand to this coil.

LAD itself also affects the binary result in order to control the enable output ENO (Figure 15.2). If the enable output ENO is switched, its signal state corresponds to that of the BR. In certain cases ("BR corresponds to functions") the executed LAD function sets the binary result as follows:

- ▷ BR := "1" with MOVE, with the shift functions and with the word logic operations
- ▷ BR := OV with the arithmetic and math functions
- ▷ BR := OV or "1" with the conversion functions
- ▷ BR := BR of the called block with block calls

15.4.2 Main Rung

In the LAD programming language, many boxes have an enable input EN and an enable output ENO. If the enable input has "1", the function is processed in the box. When the box is processed correctly, the enable output also has signal state "1". If an error occurs during processing of a box (for example, overflow of an arithmetic function), ENO is set to "0". If EN has signal state "0", ENO is also set to "0".

These characteristics of EN and ENO can be used in order to connect several boxes together in a chain, with the enable output leading to the enable input of the next box (Figure 15.3). This

Is ENO switched ?					
YES			NO		
Is EN switched ?			Is EN switched ?		
YES		NO	YES		NO
Is EN == "1" ?			Is EN == "1" ?		
YES	NO		YES	NO	
BR corresponds to function	BR := "0"	BR corresponds to function	BR := "1"	BR := "0"	BR not affected

Figure 15.2 General Schematic for Setting the Binary Result

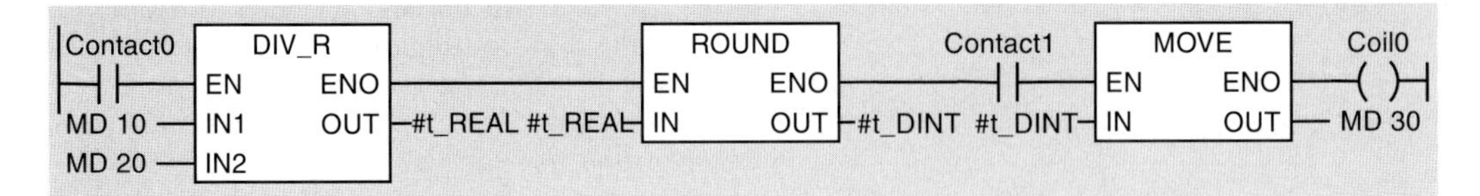

Figure 15.3 Example of a Main Rung

means that, for example, the entire chain can be "switched off" (no boxes are processed if the input I 1.0 in the example has signal state "0") or the rest of the chain is no longer processed if one box signals an error. In the case of boxes arranged in parallel, the enable is switched in parallel.

The input EN and the output ENO are not block parameters but instruction results that the LAD Editor itself generates prior to and following all boxes (also in the case of functions and function blocks). The LAD Editor uses the binary result here to store the signal state at EN while the block is being processed or to check the error signal from the box.

15.4.3 ENO in the case of user-written blocks

LAD provides the call of your own blocks with the enable input EN and the enable output ENO. You can use the enable input EN to call the block conditionally. You can use the ENO output as, for example, a group error signal (signal state "1", if the block has been properly processed; "0", if an error has occurred during processing of the block). All system blocks also signal group errors via BR.

You control the ENO output with the binary result BR. The ENO output has the same signal state that BR has when the block is exited.

For example, BR could be set to "1" at the start of a block. If an error then occurs during processing of the block, for example, a result exceeds the fixed range, so that further processing must be prevented, set the binary result to "0" with the SAVE coil and jump to the block end where the block will be exited (in the event of an error, the condition must supply signal state "0"). Please note that the RET coil sets the BR to "1" if you exit the block via this coil.

16 Jump Functions

You can use jump functions to interrupt the linear flow of the program and continue at another point in the block. This program branching can be executed unconditionally or dependent on the RLO.

16.1 Processing a Jump Function

Representation

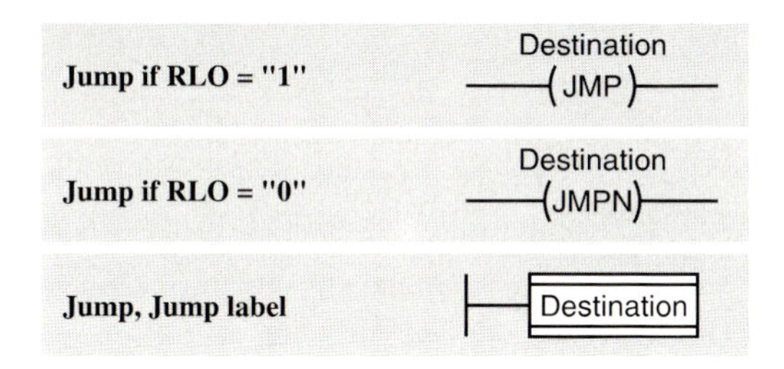

A jump function consists of the jump operation in the form of a coil and a jump label designating the program location at which processing is to continue after the jump. The jump label is above the jump operation.

A jump label consists of up to 4 characters that can include alphanumeric characters and the underscore. It begins with an alpha character; a distinction is made between upper and lower case. A jump label in a box designates the network that is to be processed following completion of the jump operation. The box with the jump label must be at the start of a network ("LABEL" in the Program Elements Catalog).

With incremental programming, you will find the jump functions in the Program Elements Catalog (with VIEW → CATALOG [Ctrl – K] or INSERT → PROGRAM ELEMENTS) under "Logic Control / Jump".

Function

A jump is either always executed (unconditional jump) or it is executed depending on the result of the logic operation (RLO). In the case of a jump dependent on the RLO, you can decide whether the jump is to be executed if RLO = "1" or if RLO = "0".

Jumps can be executed in the forward direction (in the direction of program processing; in the direction of higher network numbers) or in the backward direction. The jump can only take place within a block; that is, the jump destination must be in the same block as the jump function. If you use the Master Control Relay (MCR), the jump label must be in the same MCR zone or in the same MCR area as the jump function.

The jump destination must be designated unambiguously, that is, you must only assign a jump label once in a block. The jump destination can be jumped to from several locations.

LAD stores the designations of the jump labels in the non-execution-relevant section of the block on the data medium of the programming device. Only the jump widths are stored in the work memory of the CPU (in the compiled block). When program modifications are made to blocks on-line at the CPU, these modifications must therefore always be updated on the data medium of the programming device in order to retain the original designations. If this update is not made, or if blocks are transferred from the CPU to the programming device, non-execution-relevant block section will be overwritten or deleted. On the display or the printout, the Editor then generates replacement symbols for the jump labels (M001, M002 etc.).

16.2 Unconditional Jump

The unconditional, always executed jump is the jump function JMP whose coil is connected to the left power rail. This jump function is always

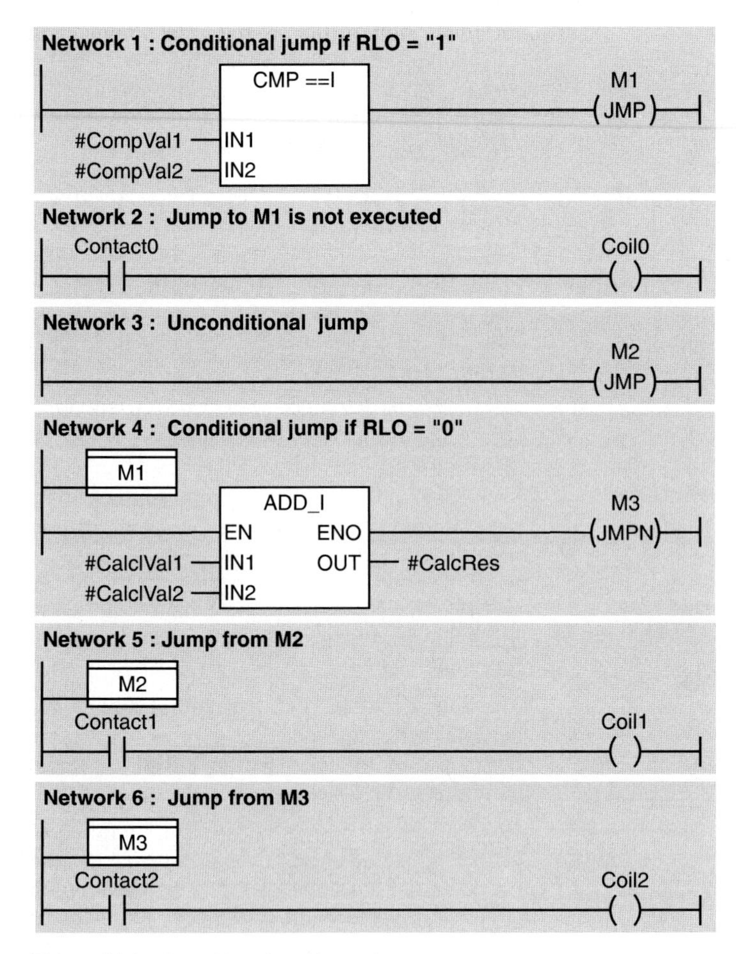

Figure 16.1 Jump Functions Example

executed during processing. The CPU interrupts the linear flow of the program and continues in the network designated by the jump label.

Example (Figure 16.1): In Network 3, there is an unconditional jump to jump label M2. After this network has been processed, the CPU continues program execution in Network 5, at the start of which M2 is located. A jump from another program location is required (for example with jump label M1) in order to process Network 4.

16.3 Jump if RLO = "1"

The conditional jump if RLO = "1" is the jump function JMP, whose coil is not connected direct with the left power rail. The logic operation preceding this coil can be implemented in any way; for example, it can consist of a rung with contacts or of a box. If the RLO = "1" (if the preceding logic operation is fulfilled, power flows into the coil) the CPU interrupts the linear flow of the program and continues in the network designated by the jump label. If the preceding logic operation is not fulfilled (if no power flows into the coil), the CPU continues program execution in the next network.

Example (Figure 16.1): If the comparison in Network 1 is fulfilled, program execution branches to Network 4. If the comparison is not fulfilled, the next network, Network 2 is processed.

16.4 Jump if RLO = "0"

The conditional jump if RLO = "0" is the jump function JMPN, whose coil is not connected direct with the left power rail. The logic operation preceding this coil can be implemented in any way; for example, it can consist of a rung with contacts or of a box. If the RLO = "0" (if the preceding logic operation is not fulfilled and no power flows into the coil), the CPU interrupts the linear flow of the program and continues in the network designated by the jump label. If the preceding logic operation is fulfilled (power flows into the coil), the CPU continues program execution in the next network.

Example (Figure 16.1): If the adder in Network 4 signals an error, the jump to Network 6 (jump label M3) is executed. In the event of error-free processing, Network 5 is the next to be processed.

17 Master Control Relay

With contact controls, a Master Control Relay activates or de-activates a section of the control that can consist of one or more rungs. A de-activated rung

▷ switches all non-retentive coils off and

▷ retains the state of retentive coils or boxes.

You can only change the state of the contactors again when the Master Control Relay (MCR) is active.

Please note that switching off with the "software" Master Control Relay is no substitute for an EMERGENCY OFF or safety facility! Treat Master Control Relay switching in exactly the same way as switching with a memory function!

Activate MCR area	—(MCRA)—
Open MCR zone	—(MCR<)—
Close MCR zone	—(MCR>)—
De-activate MCR area	—(MCRD)—

With the MCRA and MCRD coils, you designate an area in your program in which MCR dependency is to take effect. Within this range, you use the coils (MCR<) and (MCR>) to define one or more zones in which the MCR dependency can be switched on and off. You can also nest the MCR zones. The result of logic operation (RLO) immediately prior to an MCR zone switches MCR dependency on or off within this zone.

17.1 MCR Dependency

MCR dependency affects coils and boxes. If MCR dependency is switched on (corresponds to Master Control Relay switched off)

- a single coil and a midline output set the binary operands to signal state "0" (following the midline output, the RLO is then = "0"; that is, power no longer flows)
- a set and reset coil no longer affect the signal state of the binary operand ("freeze it")
- an SR and RS box no longer affect the signal state of the binary operand ("freeze it")
- a transfer operation writes zero to the digital operand (every function output of a box of digital data type then writes zero to the operand or to the variable)

You switch on MCR dependency in a zone if the RLO is "0" immediately prior to opening the zone (analogous to switching off the Master Control Relay). A logic operation prior to a coil then remains ineffective. If you open an MCR zone with RLO "1" (Master Control Relay switched on), processing within this MCR zone takes place without MCR dependency. MCR dependency is effective only within an MCR zone.

With incremental programming, you will find the MCR functions in the Program Elements Catalog (with VIEW → CATALOG [Ctrl – K] or INSERT → PROGRAM ELEMENTS) under "Program Control".

17.2 MCR Area

To be able to use the characteristics of the Master Control Relay, define an MCR area with the MCRA coil (start) and the MCRD coil (end of the MCR area). MCR dependency is active within an MCR area (but not yet switched on). The MCRA coil and the MCRD coil are connected direct with the left power rail and terminate a rung.

If you call a block within an MCR area, MCR dependency is de-activated in the called block (Figure 17.1). An MCR area only starts again with the MCRA coil. When a block is exited, MCR dependency is set as it was before the block was called, regardless of the MCR dependency with which the block was exited.

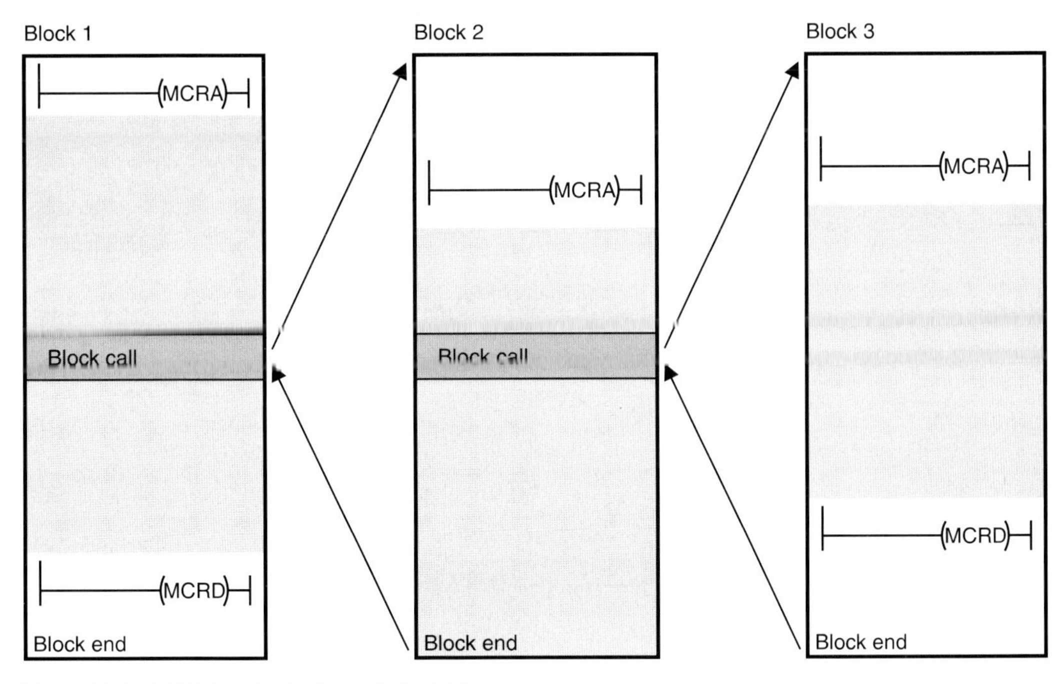

Figure 17.1 MCR Area in the Case of Block Change

17.3 MCR Zone

You define an MCR zone with the MCR<-coil (start) and the MCR>-coil (end of the MCR zone). The MCR<-coil requires a preceding logic operation; the MCR>-coil is connected direct to the left power rail. Both coils terminate a rung. Within this zone, you control MCR dependency with the RLO prior to the MCR<-coil: If power flows into the coil, MCR dependency is switched off ("normal" processing), if power does not flow into the coil, MCR dependency is switched on.

You can open an MCR zone within another MCR zone. The nesting depth for MCR zones is 8; that is, you can open a zone up to eight times before you close a zone.

You control the MCR dependency of a switched on MCR zone with the RLO on opening the zone. However, if MCR dependency is switched on in a "higher-level" zone, you cannot switch off MCR dependency in a "lower-level" MCR zone. The Master Control Relay of the first MCR zone controls the MCR dependency in all switched on zones (Figure 17.2).

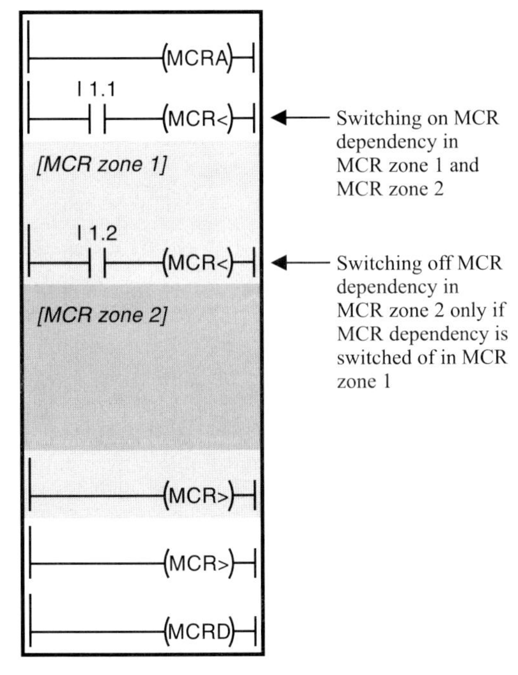

Figure 17.2
MCR Dependency in the Case of Nested MCR Zones

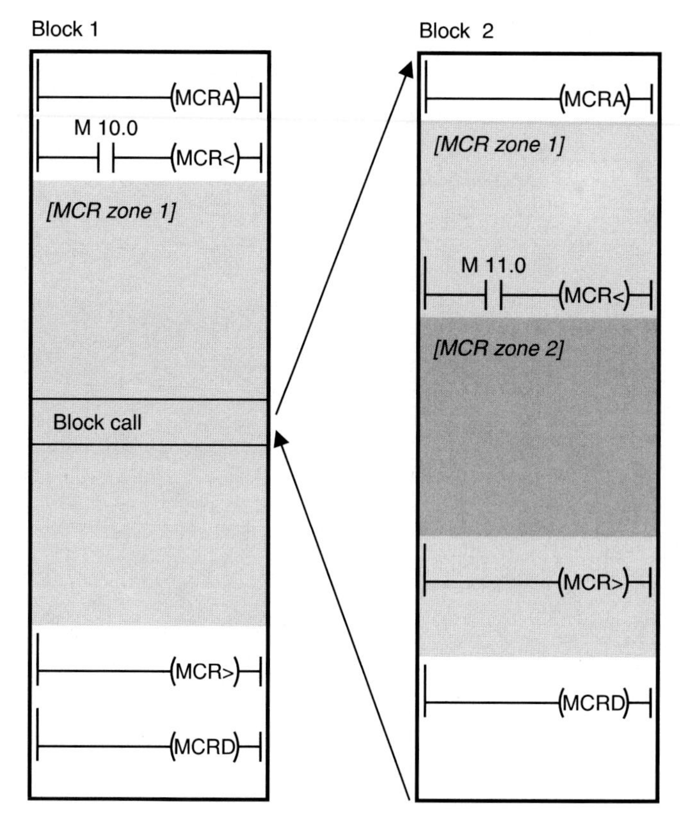

Figure 17.3 MCR Zones in the Case of Block Change

Table 17.1 MCR Dependency in the Case of Nested MCR Zones (Example)

M 10.0	M 11.0	Zone 1	Zone 2
"1"	"1"	No MCR dependency	
"1"	"0"	No MCR dependency	MCR dependency switched on
"0"	"1" or "0"	MCR dependency switched on	

A block call within an MCR zone does not change the nesting depth of an MCR zone. The program in the called block is still in the MCR zone that was open when the block was called (and is controlled form here). However, you must re-activate MCR dependency in a called block by opening the MCR area.

In Figure 17.3, memory bits M 10.0 and M 11.0 control the MCR dependencies. With memory bit M 10.0, you can switch on MCR dependency in both zones (with "0") regardless of the signal state of memory bit M 11.0. If the MCR dependency of zone 1 is switched off with M 10.0 = "1", you can control the MCR dependency of zone 2 with memory bit M 11.0 (Table 17.1).

17.4 Setting and Resetting I/O Bits

Despite MCR dependency being switched on, you can set or reset the bits of an I/O area with the system functions. A requirement for this is that the bits to be controlled are in the process-image output or a process-image output has been defined for the I/O area to be controlled.

Table 17.2 Parameters of the SFCs for Controlling the I/O Bits

SFC	Parameter	Declaration	Data Type	Assignment, Description
79	N	INPUT	INT	Number of bits to be set
	RET_VAL	OUTPUT	INT	Error information
	SA	OUTPUT	POINTER	Pointer to the first bit to be set
80	N	INPUT	INT	Number of bits to be reset
	RET_VAL	OUTPUT	INT	Error information
	SA	OUTPUT	POINTER	Pointer to the first bit to be reset

The system function SFC 79 SET is available for setting the I/O bits, and SFC 80 RSET for resetting (Table 17.2). You call these system functions in an MCR zone. The system functions are only effective if MCR dependency is switched on; if MCR dependency is switched off, the calls of these SFCs remain without effect.

Setting and resetting the I/O bits also simultaneously updates the process-image output. The I/O are affected byte-by-byte. The bits not selected with the SFCs (in the first and in the last byte) retain the signal states as they are currently available in the process-image.

The diskette accompanying the book contains examples of the Master Control Relay and of the system functions SFC 79 and SFC 80 (FB 117 in the program "Program Flow Control").

18 Block Functions

In this chapter, you will learn how to call and terminate code blocks and how to work with operands from data blocks. The next chapter then deals with using block parameters.

18.1 Block Functions for Code Blocks

Block functions for code blocks include instructions for calling and terminating blocks (Figure 18.1). Code blocks are called in the form of a box. If functions or system functions have no block parameters, they can also be called in the form of a coil. In both cases, a preceding logic operation enabling a conditional call (call dependent on conditions) is permissible. The block function RET always requires a preceding logic operation.

In addition to the block change, the call box also contains the transfer of block parameters. When function blocks are called, it also opens the instance data block. The call coil is no more than a change to another block and is only meaningful (and permissible) in the case of functions and system functions.

After a block has been terminated, and following the call function, the CPU continues program execution in the block that made the call (the calling block). If an organization block is terminated, the CPU continues to operate in the operating system.

With incremental programming, you will find the CALL and the RET box in the Program Elements Catalog (with VIEW → CATALOG [Ctrl – K] or INSERT → PROGRAM ELEMENTS) under "Program Control"; you insert block calls with call boxes into your program when you select blocks from "FC/FB/SFC/SFB blocks", "Multiple Instances" or from "Libraries".

18.1.1 Block Calls: General

If a code block is to be processed, it must be "called". Figure 18.2 gives an example of calling function FC 10 in organization block OB 1.

A block call consists of the call box that contains the address of the called box (here: FC 10), the enable input EN, the enable output ENO and any block parameters. Following processing of the call function, the CPU continues program execution in the calling block. The block is processed to the end or until a block end function is encountered. Following this, the CPU returns to the calling block (here: OB 1) and continues processing this block after the call box.

The information the CPU requires to find its way back to the calling block is stored in the block stack (B stack). With every new block call, a new stack element is created that includes the return address, the contents of the data block register and the address of the local data stack of the calling block. If the CPU goes to the Stop state as a

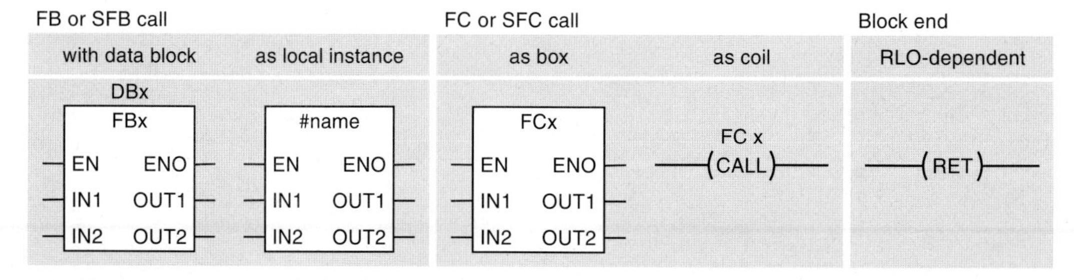

Figure 18.1 Block Functions for Code Blocks

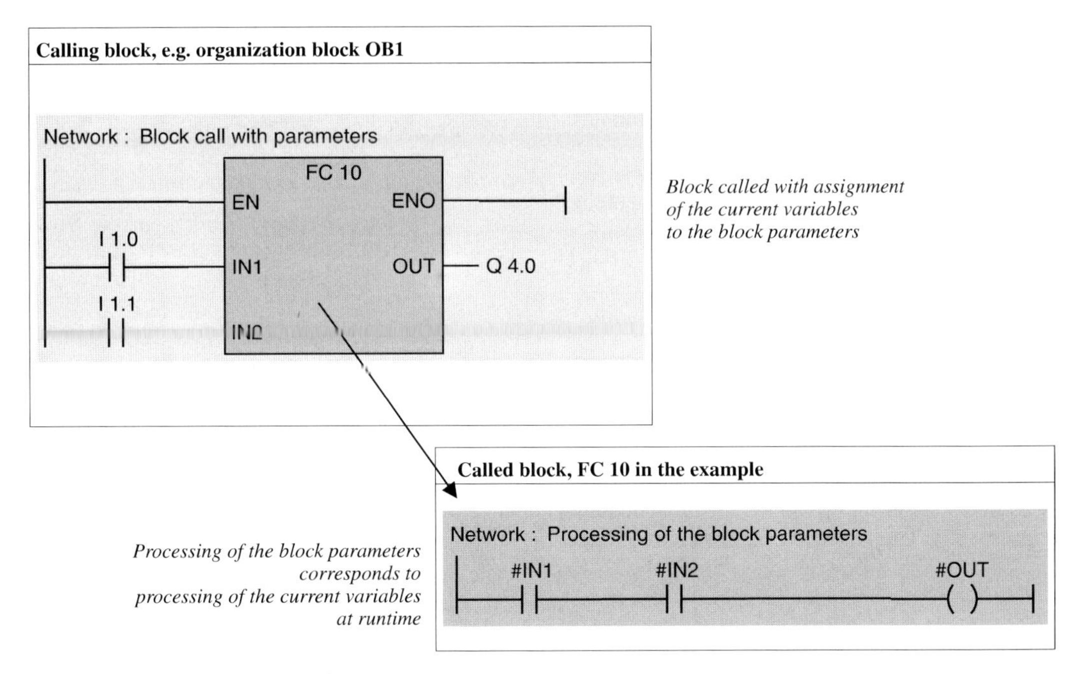

Figure 18.2 Block Call Example

result of an error, you can use the programming device to see from the contents of the B stack which blocks were processed up to the error.

You can transfer data to and from the called block for processing. These data are transferred via block parameters. With the call box, you can also call blocks without block parameters.

18.1.2 Call Box

You use the call box to call FBs, FCs, SFBs and SFCs. (You cannot call organization blocks since they are event-driven and are started by the operating system.)

You can use the EN input to make the block call subject to conditions. If the EN input is connected direct to the left power rail, the call is unconditional; it is always executed. If there is a logic operation (a contact or a rung) preceding EN, the block call is only executed if the preceding logic operation is fulfilled (if power flows into the EN input). The ENO output has the same signal state as the binary result BR on exiting the called block (Figure 18.3).

You label the parameters of the called block, absolutely or symbolically, with the operands current for the call. If a parameter is of data type BOOL, connect a contact or a rung prior to this parameter. A Boolean output parameter cannot be combined further.

You can arrange several call boxes in series, connecting them with each other via EN or ENO. You can only insert a call box in a parallel rung if it is connected direct to the left power rail.

MCR dependency is de-activated when a block is called. The MCR is switched off in the called block regardless of whether the MCR was switched on or off prior to the block call. When exiting a block, MCR dependency assumes the same setting as it had prior to the block call.

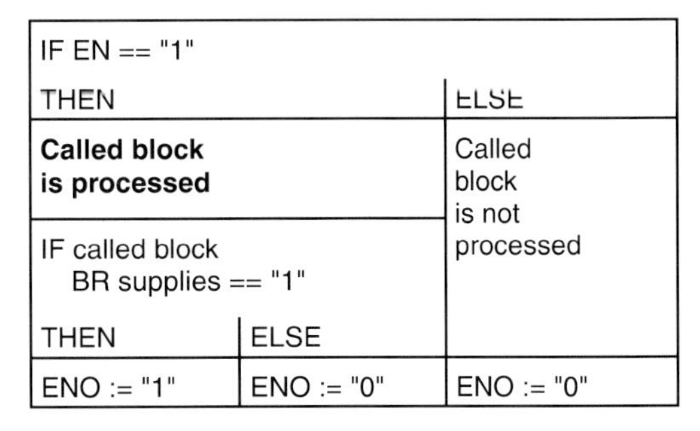

Figure 18.3 Function of a Call Box

Depending on the block parameters, you can modify the contents of the data block register when the block change is made. If the *called* block is a function block, the instance block is always opened in this block via the DI register. If the *calling* block is a function block, the contents of the DI register (the instance data block) are retained after the block call. The contents of the DB register depend on the transferred block parameters, among other things.

Calling function blocks

You call a function block by selecting the relevant function block from the Program Elements Catalog under “FB Blocks”. A requirement is that the function block to be called must already be in the user program. You write the instance data block belonging to the call above the box. Both blocks (function block and instance data block) can have absolute or symbolic addresses.

In the case of function blocks, you do not need to initialize all block parameters at the call. The uninitialized block parameters retain their current value.

You can call function blocks within other function blocks also as local instance. In doing so, the called function block uses the instance data block of the calling function block as the store for its local data. Prior to the call, you declare the local instance in the static local data of the calling function block (the block you are currently programming). The local instance is called by selecting one of the available local instances under “Multiple Instances” in the Program Elements Catalog; it is not necessary to specify an instance data block (see also Section 18.1.5 “Static Local Data”).

Calling functions

You call a function by selecting the relevant function under “FC Blocks” in the Program Elements Catalog. The function can have an absolute or a symbolic address.

When you call functions, you must initialize all available parameters.

Calling functions with a function value takes exactly the same form as calling functions with no function value. Only the first output parameter, corresponding to the function value, has the name RET_VAL.

Calling system blocks

The operating system of the CPU contains system functions SFCs and system function blocks SFBs that you can use. The number and type of system blocks depends on the CPU. You can call all system blocks with the call box.

You call a system function block in the same way as one you have written yourself; set up the associated instance data block in the user memory with the same data type as the SFB.

You call a system function in the same way as a function you have written yourself.

Off-line programming of SFBs and SFCs

System blocks exist only in the operating system of the CPU. If you want to call system blocks during off-line programming, you require a description of the call interface to allow the Editor to initialize the parameters. You will find this interface description under “Built In” in the “StdLib30” library supplied. From here, you can copy the system blocks you require into the off-line user program of your project and then use them at input.

The Program Elements Catalog provides the system blocks currently available in the user program under “SFC Blocks” or “SFB Blocks”. You can find all system blocks known to STEP 7 under “Libraries” → “StdLib30” → “Built In”. You can, for example, select a system block with the mouse and drag it to the currently processed block where it is then called. At the same time, this block (or, more precisely: its interface description) is copied into the user program.

18.1.3 Call Coil

You can call functions and system functions using the call coil. It is a requirement that the called blocks have no block parameters. You can use the call coil if a block is too long or not clear enough for you by simply “breaking down” the block into sections and calling the sections one after the other. One single coil is permitted per network. You can either connect the call coil direct to the left power rail or provide it with a preceding logic operation.

If the call coil is connected direct to the left power rail, it is always executed (unconditional call).

If there is a logic operation preceding the call coil, the call is only executed if the preceding logic operation is fulfilled, that is, if power flows into the call coil. If the preceding logic operation is not fulfilled, the call is not executed and the next network is processed.

When the block change is made, the status bit OS is reset; status bits CC0, CC1 and OV are not affected.

MCR dependency is de-activated when a block is called. The MCR is switched off in the called block regardless of whether the MCR was switched on or off prior to the block call. When exiting a block, MCR dependency assumes the same setting as it had prior to the block call.

Calling a block with the call coil saves the data block register in the B stack; the block end restores its contents when the called block is exited. The global data block current prior to the block call and the instance data block are also open following the block call. If no data block was open prior to the block call (for example, no instance data block in OB 1), no data block is open following the block call either, regardless of which data blocks may be open in the called block.

18.1.4 Block End Function

You can abort processing in a block with the block end function RET.

The block end function is represented as a coil requiring a preceding logic operation. The RET coil must be alone in a network.

If the preceding logic operation is fulfilled (power flows into the RET coil), the block is exited. A return jump is made to the previously processed block in which the block call took place. If an organization block is terminated, the CPU continues to work in the system program.

If the preceding logic operation is not fulfilled (no power flows into the RET coil), the next network in the block is processed (Figure 18.4).

The RET coil simultaneously stores the RLO (whether power flows or not) in the binary result BR, regardless of whether the logic operation was fulfilled or not. The binary result is decisive for controlling the ENO output at the call box (see also Chapter 15 "Status Bits").

IF preceding logic operation == "1"	
THEN	ELSE
The block is exited	The next network is processed
BR := "1"	BR := "0"

Figure 18.4 Function of the RET Coil

18.1.5 Static Local Data

Every function block can have static local data. The static local data are defined in the declaration table under "stat" and stored in the instance data block. The quantity of static local data depends on the data type of the variables and on the length of the data block; this is CPU-specific. Static local data retain their value until the program changes it (exactly like data operands in global data blocks).

Example: The function block "Totalizer" adds an input value to a value stored in the static local data and then stores the total in the static local data. At the next call, the input value is then added to this total, and so on (Figure 18.5).

Total is a variable in the data block "TotalizerData", the instance data block of the "Totalizer" function block (you can define the names of all blocks yourself within the permissible limits in the Symbol Table). The instance data block has the data structure of the function block; in the example, it contains two INT variables with the names *In* and *Total*.

When a function block is called, its instance data block is also specified. In general, the static local data are only processed in the function block itself. However, since they are stored in a data block, you can access static local data at any time in the same way as you can access variables in a global data block, for example, with *"TotalizerData".Total.*

You can generate "multiple instances", that is, you can call a function block as a local instance in another function block. The static local data and also the block parameters of the called function block are then a subset of the static local data of the called block.

Example: In the static local data of the function block "Evaluation", you declare a variable *Memory* that corresponds to the function block "Totalizer" and is of the same data type (Figure 18.5).

FB "Totalizer"

Address	Declaration	Name	Type
+ 0.0	in	In	INT
+ 2.0	stat	Total	INT

ADD_I
EN ENO
#In — IN1 OUT — #Total
#Total — IN2

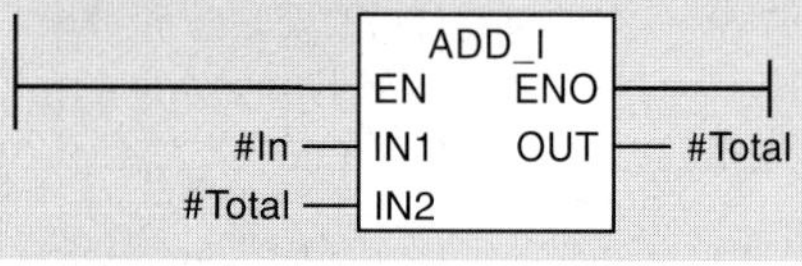

DB "TotalizerData"

Address	Declaration	Name	Type
+ 0.0	in	In	INT
+ 2.0	stat	Total	INT

FB "Evaluation"

Address	Declaration	Name	Type
0.0	in	Add	BOOL
0.1	in	Delete	BOOL
2.0	stat	FM_Add	BOOL
2.1	stat	FM_Del	BOOL
4.0	stat	Memory	Totalizer

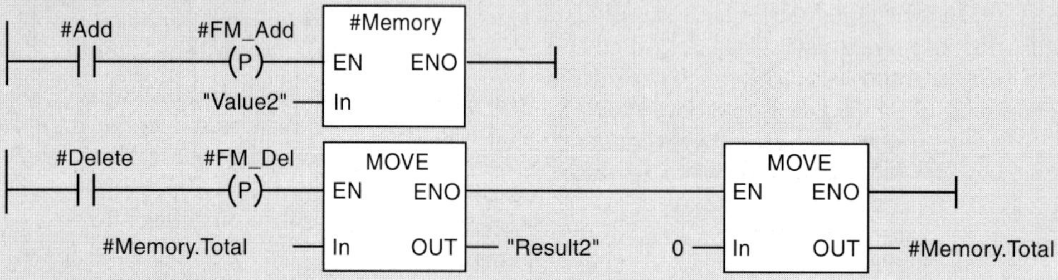

#Add #FM_Add (P) #Memory
EN ENO
"Value2" — In

#Delete #FM_Del (P) MOVE
EN ENO
#Memory.Total — In OUT — "Result2"

MOVE
EN ENO
0 — In OUT — #Memory.Total

DB "EvaluationData"

Address	Declaration	Name	Type
0.0	in	Add	BOOL
0.1	in	Delete	BOOL
2.0	stat	FM_Add	BOOL
2.1	stat	FM_Del	BOOL
4.0	stat:in	Memory.In	INT
6.0	stat	Memory.Total	INT

In the data view, the data block shows all individual variables so that the variables of a local instance appear with their full names.
At the same time, you can see their addresses via which absolute addressing is possible.

Figure 18.5 Example of Static Local Data

Now you can call the function block "Totalizer" via the variable *Memory* but without specifying a data block because the data for *Memory* are stored "block-locally" in the static local data (*Memory* is the local instance of the FB "Totalizer").

You access the static local data of *Memory* in the program of the function block "Evaluation" in the same way as you access structure components, by specifying the structure name (Memory) and the component name (Total).

In addition to the Boolean variable, the instance data block "EvaluationData" also contains the local instance *Memory* whose individual components you can also access as global variables, for example, as *"EvaluationData".Memory.Total.*

You can find the example described here on the diskette accompanying the book in the program "Program Flow Control":

▷ FB 10 "Totalizer"

▷ FB 11 "Evaluation" contains a local instance of FB 10

▷ FB 12 "Local instance" contains the FB 10 and FB 11 calls, each with data block

The remaining block functions are programmed in FB 118. The "Feed" example in Chapter 19 "Block Parameters" gives further applications for local instances.

18.2 Block Functions for Data Blocks

You store your program data in the data blocks. In principle, you can also use the bit memory area for storing data; however, with the data blocks, you have significantly more possibilities with regard to data volume, data structuring and data types. This chapter shows you

- ▷ how to work with data operands,
- ▷ how to call data blocks and
- ▷ how to create, delete and test data blocks at runtime.

You can use data blocks in two versions: as *global data blocks*, that are not assigned to any code block, and as *instance data blocks*, that are assigned to a function block. The data in the global data blocks are, in a manner of speaking, "free" data that every code block can make use of. You yourself determine their volume and structure direct through programming the global data block. An instance data block contains only the data with which the associated function block works; this function block then also determines the structure and storage location of the data in "its" instance data block.

The number and length of data blocks are CPU-specific. The numbering of the data block begins at 1; there is no data block DB 0. You can use each data block either as a global data block or as an instance data block.

You must first create ("set up") the data blocks you use in your program, either by programming, such as code blocks, or at runtime using the system function SFC 22 CREAT_DB.

18.2.1 Two Data Block Registers

Each S7-CPU has two data block registers. These registers contain the numbers of the current data blocks; these are the data blocks with whose operands processing is currently taking place. Before accessing a data block operand, you must open the data block containing the operand. If you use fully-addressed access to data operands (with specification of the data block, see below), you need not be concerned with opening the data blocks and with the assignments of the data block register. The Editor generates the necessary instructions from your specifications.

The Editor uses the first data block register preferably for accessing global data blocks and the second data block register for accessing instance data blocks. For this reason, these registers are given the names "Global data block register" (DB register) and "Instance Data Block Register" (DI register). The handling of the registers by the CPU is absolutely identical. Each data block can be opened via one of the two registers (or also via both simultaneously).

To come to the essential point first: You can affect only the DB register with LAD. If you open a data block with the OPN coil (see below), you always open it via the DB register. With LAD, the instance data block is always opened via the DI register; this happens with the call box when the block is called.

When you load a data word, you must specify which of the two possible open data blocks contains the data word. If the data block has been opened via the DB register, the data word is called DBW; if the data word is in the data block opened via the DI register, it is called DIW. The other data operands are named accordingly (Table 18.1).

18.2.2 Accessing Data Operands

You can use the following methods for accessing data operands:

- ▷ Symbolic addressing with full addressing,
- ▷ Absolute addressing with full addressing and
- ▷ Absolute addressing with part addressing

Table 18.1 Data Operands

Data operand	located in a data block opened via the	
	DB register	DI register
Data bit	DBX y.x	DIX y.x
Data byte	DBB y	DIB y
Data word	DBW y	DIW y
Data double-word	DBD y	DID y

x = Bit address, y = Byte address

Symbolic access to the data operands in global data blocks requires the minimum system knowledge. For absolute access or for using both data block registers, you must observe the notes described below.

Symbolic addressing of data operands

I recommend you use symbolic addressing of data operands as far as possible. Symbolic addressing

- makes it easier to read and understand the program (if meaningful terms are used as symbols),
- reduces write errors in programming (the Editor compares the terms used in the Symbol Table and in the program; "number switching errors" such as DBB 156 and DBB 165 that can occur when using absolute addresses, cannot occur here) and
- does not require programming knowledge at the machine code level (which data block has the CPU opened currently?).

Symbolic addressing uses fully-addressed access (data block together with data operand), so that the data operand always has a unique address.

You determine the symbolic address of a data operand in two steps:

- Assignment of the data block in the Symbol Table
Data blocks are global data that have unique addresses within a program. In the Symbol Table, you assign a symbol (e.g. Motor1) to the absolute address of the data block (e.g. DB 51).
- Assignment of the data operands in the data block
You define the names of the data operands (and the data type) during programming of the data block. The name applies only in the associated block (it is "block-local"). You can also use the same name in another block for another variable.

Fully-addressed access to data operands

In the case of fully-addressed access, you specify the data block together with the data operand. This method of addressing can be symbolic or absolute:

"MOTOR1".ACTVAL
DB 51.DBW 20

MOTOR1 is the symbolic address that you have assigned to a data block in the Symbol Table. ACTVAL is the data operand you defined when programming the data block. The symbolic name "MOTOR1".ACTVAL is just as unique a specification of the data operand as the specification DB 51.DBW 20.

Fully-addressed data access is only possible in conjunction with the global data block register (DB register). In the case of fully-addressed data operands, the Editor first opens the data block via the DB register and then accesses the data operand.

You can use fully-addressed access with all functions permissible for the data type of the addressed data operand. So with block parameters, for example, you can also specify a fully-addressed data operand as the actual parameter.

Absolute addressing of data operands

For absolute addressing of data operands, you must know the addresses at which the Editor places the data operands when setting up. You can find out the addresses by outputting them after programming and compiling the data block. You will then see from the address column the absolute address at which the relevant variable begins. This procedure is suitable for all data blocks, both those you use as global data blocks as well as those you use as instance data blocks (for local instances see Section 18.2.4 below). In this way, you can also see where the Editor stores the block parameters and the static local data in the case of function blocks.

Data operands are addressed bytewise like the bit memory, for example; the functions used on the data operands are also the same as those used with the bit memory.

18.2.3 Opening the Data Block

You use the OPN coil to open a data block via the DB register.

The OPN coil is connected direct to the left power rail and stands alone in a rung. The specified data block is always opened via the DB register. In LAD, it is not possible to open a data block via the DI register with OPN coil. (The DI register uses the call box to open the current instance data block.)

With incremental programming, you will find the OPN coil in the Program Elements Catalog under "Data Block Function".

The open data block must be in the work memory at runtime.

In the networks following the OPN coil, you can use part addressing to access only those data operands that are located in the open data block. If you want to copy from one data block to another, you can, for example go into the temporary local data via an intermediate buffer or (better) use fully-addressed data operands. Please note: A fully-addressed data operand overwrites the DB register with "its" data block.

Example: The value of data word DBW 10 of data block DB 13 is to be transferred to data word DBW 16 of data block DB 14. With the MOVE box, you can only copy the part-addressed data operand within the currently open data block. For copying between data blocks, you require an intermediate buffer, such as a local data word (see Figure 18.6). It is better to use full addressing.

When you open a data block, it remains "valid" until another data block is opened. Under certain circumstances, you may not be able to see this via the Editor, for example, if a block is changed with the call box (see Section 18.2.4).

In the case of a block change with the call box, the contents of the data block register are retained. On returning to the calling block, the block change restores the old contents of the register.

18.2.4 Special Points in Data Addressing

Changing the DB register assignments

With the following functions, the Editor changes the contents of the DB register:

- Full addressing of data operands

Each time data operands are fully addressed, the Editor first opens the data block and then accesses the data operand. The DB register is overwritten each time here. This also applies to supplying block parameters with fully-addressed data operands.

- Accessing block parameters

Access to the following block parameters changes the contents of the DB register: All

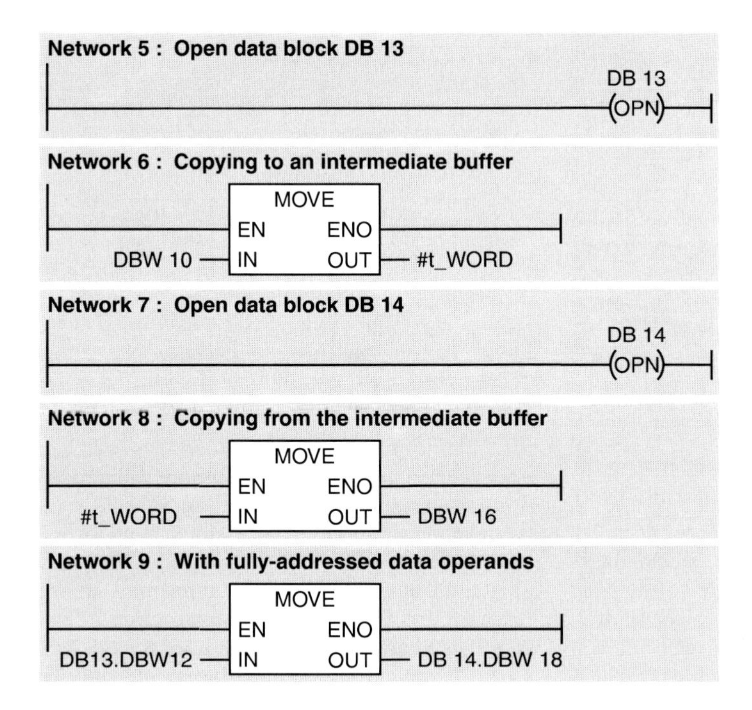

Figure 18.6
Example of Transferring Data Operands

block parameters of complex data type in the case of functions, and in/out parameters of complex data type in the case of function blocks.

• Before the actual function block call, the call box stores the number of the current instance data block in the DB register (by swapping the data block registers) and opens the instance data block for the called function block. In this way, the instance data block associated with a called function block is always open. After the actual block call, the call box swaps data block registers again, so that the current instance data block is available again in the calling function block. In this way, the call box changes the contents of the DB register.

Absolute addressing of instance data operands

In function blocks, the DI register is permanently assigned the number of the current instance data block. All accesses to block parameters or static local data are made via the DI register. The local data address specified in the declaration table of the function only applies if you open the function block with an instance data block; assignment of the data operands then begins at byte zero.

If you call a function block as a local instance, its data are located "in the middle" of the instance data block of the calling block. You can find the absolute address of the local data by displaying the instance data block in data view. Each individual variable, including the variables of the called local instances, is then displayed with name and address.

If you program a function block that is also to be available later as a local instance, you will not yet know the absolute address of the variable at the time of programming. In this case, the contents of the address register AR2 must be added to the variable address – in exactly the same way as the Editor does in symbolic programming. However, this is only possible in the STL programming language.

Changing the assignments of data blocks at a later stage

In programmable controllers, the global operands are assigned permanently to an address. This applies to inputs, outputs, peripheral I/O, memory bits, timers, counters and also blocks. Once it is plugged into the mounting rack, a module has a fixed address that maps to the peripheral I/O operand area as well as inputs and outputs. It makes no sense here to insert or delete an operand, for example, an input byte, with the subsequent shifting of the operands that follow; you would then have to change not only the program but the wiring as well. This behavior has been transferred to the memory bits, timers, counters and blocks. No-one would consider deleting, say, function block FB 35 in the number band with the consequent shifting of all block numbers so that all these blocks would receive a new number.

However, when programming data blocks, it makes sense to insert or delete data operands. Here, it can even be desirable that the subsequent operands (or more precisely, the values of the operands) are shifted. With absolute programming, it is obvious that you must then update the addresses of the (shifted) data operands in the program.

An example: Data block DB 21 "Motor1" is assigned the INT variables *ActValue* (data word DBW 10) and *Setpoint* (DBW 12). If you insert the INT variable *MaxCurrent* prior to the variable *ActValue*, the values shift and with them the variable *ActValue* shifts to data word DBW 12 and *Setpoint* shifts to data word DBW 14.

Assignment before insertion:

	ActValue	*Setpoint*	
DBW 8	DBW 10	DBW 12	DBW 14

Assignment after insertion:

	MaxCurrent	*ActValue*	*Setpoint*
DBW 8	DBW 10	DBW 12	DBW 14

If you have previously addressed data word DBW 10, in order, say, to compare the actual value, you must specify data word DBW 12 after the change.

The same behavior also exists with symbolic addressing! If you create a program in the LAD programming language, the Editor works in incremental input mode. With symbolic addressing of the variables, the Editor stores the absolute address of the associated operand in the program (if, for example, you program "Contact1", the compiled block will have the operand I 1.1). When outputting the program in symbolic representation, the Editor searches the Symbol Table and the declaration tables of the blocks for the stored addresses, and when it has found them, it uses the symbolic name.

To return to the example above, if you have addressed the variable *"Motor1".ActValue* in the program, the Editor writes DB21.DBW10 in the program. If you output the program with symbolic addressing after the change to the data block, *"Motor1".MaxCurrent* will now appear at this point, because the variable *MaxCurrent* is now at address DBW 10.

A change in the data block assignment therefore brings with it the consequence that you must update the addresses in all program sections that access the affected data operands with both *absolute as well as symbolic* addressing.

This also applies to changes in the instance data blocks such as inserting or deleting static local data in function blocks. While all changes to all local variables are correctly updated in the program "block-local" by the Editor when programming code blocks, this does not apply for "external" accesses, when, for example, accessing "foreign" instance data from other blocks. This access then takes place in the same way as an access on global data operands and the addresses must be updated following an operand shift.

18.3 System Functions for Data Blocks

There are three system functions for handling data blocks. Their parameters are described in Table 18.2.

▷ FC 22 CREAT_DB
  Create data block

▷ SFC 23 DEL_DB
  Delete data block

▷ SFC 24 TEST_DB
  Test data block

18.3.1 Creating a Data Block

System function SFC 22 creates a data block in the work memory. As the data block number, the system function takes the lowest free number in the number band given by the input parameters LOW_LIMIT and UP_LIMIT. The numbers specified at these parameters are included in the number band. If both values are the same, the data block is generated with this number. The output parameter DB_NUMBER supplies the number of the actually created data block. With the input parameter COUNT, you specify the length of the data block to be created. The length corresponds to the number of data bytes and must be an even number.

Creating the data block is not the same as calling it. The current data block is still valid. A

Table 18.2 SFCs for Handling Data Blocks

SFC	Name	Decl.	Data Type	Assignment, Description
22	LOW_LIMIT	INPUT	WORD	Lowest number of the data block to be created
	UP_LIMIT	INPUT	WORD	Highest number of the data block to be created
	COUNT	INPUT	WORD	Length of the data block in bytes (even number)
	RET_VAL	OUTPUT	INT	Error information
	DB_NUMBER	OUTPUT	WORD	Number of the created data block
23	DB_NUMBER	INPUT	WORD	Number of the data block to be deleted
	RET_VAL	OUTPUT	INT	Error information
24	DB_NUMBER	INPUT	WORD	Number of the data block to be tested
	RET_VAL	OUTPUT	INT	Error information
	DB_LENGTH	OUTPUT	WORD	Length of the data block (in bytes)
	WRITE_PROT	OUTPUT	BOOL	"1" = write-protected

data block created with the system function contains random data. For meaningful use, data must first be written to a data block created in this way before the data can be read.

In the event of an error, the data block is not created, the output parameter is assigned as undefined and an error number is signaled via the function value.

18.3.2 Deleting a Data Block

System function SFC 23 deletes the data block in RAM (work and load memory) whose number is specified at the input parameter DB_NUMBER. In doing so, the data block can be currently open, even in a lower-level block or on a lower priority level.

In the event of an error, the data block is not deleted and an error number is signaled in the function value.

18.3.3 Testing a Data Block

System function SFC 24 supplies the number of available data bytes for a data block in the work memory (at output parameter DB_LENGTH) and the write-protection ID (at output parameter WRITE_PROT, where signal state "1" signifies write-protected). You specify the number of the selected data block at input parameter DB_NUMBER.

In the event of an error, the output parameter is assigned as undefined and an error number is signaled in the function value.

19 Block Parameters

This chapter describes how to use block parameters. You will learn

▷ how to declare block parameters,

▷ how to work with block parameters,

▷ how to initialize block parameters and

▷ how to "pass on" block parameters.

Block parameters represent the transfer interface between the calling and the called block. All functions and function blocks can be provided with block parameters.

19.1 Block Parameters in General

19.1.1 Defining the Block Parameters

Block parameters make it possible to parameterize the processing instruction in a block, the block function. Example: You want to write a block as an adder that you can use in your program several times with different variables. You transfer the variables as block parameters; in our example, three input parameters and one output parameter (Figure 19.1). Since the adder need not store any values internally, a function is suitable as the block type.

You define a block parameter as an *input parameter* if you only check or load its value in the block program. If you only describe a block parameter (assign, set, reset, transfer), you use an *output parameter*. You must always use an *in/out parameter* if a block parameter is to be both checked and overwritten. The Editor does not check the use of the block parameters.

If you use an output parameter, you must enter it with a value in the block. You should not leave the block before doing this.

19.1.2 Processing the Block Parameters

In the adder program, the names of the block parameters stand as place holders for the later current

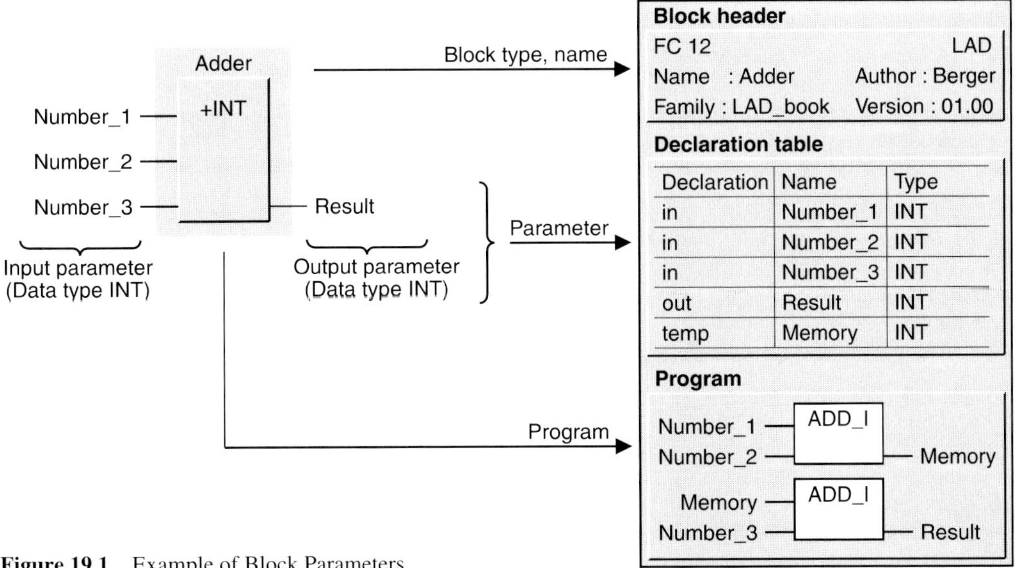

Figure 19.1 Example of Block Parameters

variables. You use the block parameters in the same way as symbolically addressed variables; in the program, they are called *Formal parameters*.

You can call the Adder function several times in your program. With each call, you transfer other values to the adder at the block parameters (Figure 19.2). The values can be constants, operands or variables; they are called *Actual parameters*.

At runtime, the CPU replaces the formal parameters with the actual parameters. The first call in the example adds the contents of memory words MW 30, MW 32 and MW 34 and stores the result in memory word MW 40. The same block with the actual parameters of the second call adds data words DBW 30, DBW 32 and DBW 34 of data block DB 13 and stores the result in data word DBW 40 of data block DB 14.

19.1.3 Declaration of the Block Parameters

You define the block parameters in the declaration section of the block when you program the block. Table 19.1 shows an empty declaration table. In addition to the block parameters (in, out, in_out), you also declare the static local data (stat) and the temporary local data (temp) in this table.

Pre-assignment with an initial value is optional and only makes sense with function blocks if the block parameter is stored as a value. This applies in the case of all block parameters of elementary data type and in the case of input and output parameters of complex data type. A parameter comment can also be given.

The *Block parameter name* can be up to 24 characters in length. It must consist only of alphanumeric characters (without national characters such as the German Umlaut) and the underscore. A distinction is made between upper and lower case. The name must not be a keyword.

For the *Data type of a block parameter* all elementary, combined and user-defined data types are permissible as well as the parameter types (see Chapter 3.4 "Data Types").

STEP 7 stores the names of the block parameters in the non-execution-relevant section of the blocks on the data medium of the programming device. The work memory of the CPU (in the compiled block) contains only the declaration types and the data types. For this reason, program changes made to blocks on-line at the CPU must always be updated on the data medi-

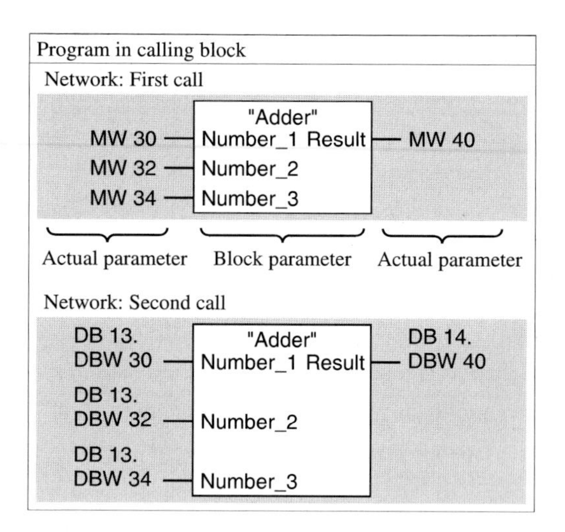

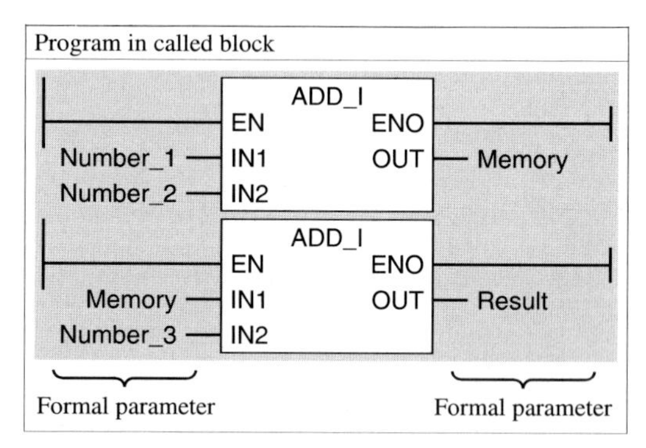

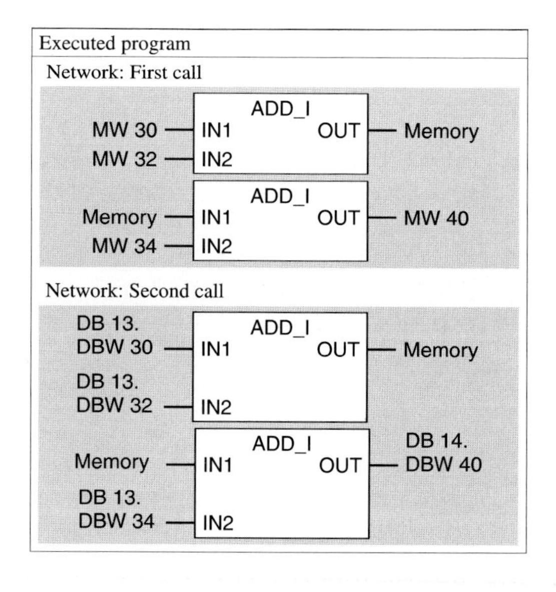

Figure 19.2 Block Call with Block Parameters

Table 19.1 Empty Declaration Table

Address	Declaration	Name	Type	Initial Value	Comment
	in				Input parameter (in the case of FCs and FBs)
	out				Output parameter (in the case of FCs and FBs)
	in_out				In/out parameter (in the case of FCs and FBs)
	stat				Static local data (in the case of FBs)
	temp				Temporary local data (in the case of OBs, FCs and FBs)

um of the programming device, in order to retain the original names. If the update is not made, or if blocks are transferred from the CPU to the programming device, the non-execution-relevant block sections are overwritten or deleted. The Editor then generates replacement symbols for display or printout.

19.1.4 Declaration of the Function Value

The function value in the case of functions is a specially treated output parameter. It has the name RET_VAL (upper case only!) and is defined as the first output parameter. You declare the function value by assigning the name RET_VAL to the first output parameter in the declaration table.

As data type of the function value, all elementary data types are permissible including the data types DATE_AND_TIME, STRING, POINTER, ANY and user-defined data types UDT. ARRAY and STRUCT are not permissible.

As the first output parameter, the function value has no special role to play in the LAD programming language. It only gains significance in the SCL programming language where you can use the block type FUNCTION as a "genuine" function. Here, a function FC can stand in place of an operand in a printout; the function value then represents the value of this function.

19.1.5 Initializing Block Parameters

When calling a block with block parameters, you initialize the block parameters with actual parameters. These can be constants, absolute-addressed operands, fully-addressed data operands or symbolically addressed variables. The actual parameter must be of the same data type as the block parameter.

You must initialize all block parameters of a function at every call. In the case of function blocks, initialization of individual block parameters or all block parameters is optional.

19.2 Formal Parameters

In this chapter, you will learn how to access the block parameters within a block. Table 19.2 shows that it is possible to access block parameters of elementary data types, components of a field or a structure, and timer and counter functions without restriction. Section 19.4 shows you how you can "pass on" block parameters to called blocks.

Access to parameters of complex data types and of parameter types POINTER and ANY is not supported by LAD. However, you can initialize acquired blocks or system blocks that have such parameters with the relevant variable. You can find examples of this in the program "Data Types" on the diskette accompanying the book.

Block parameters with data type BOOL

Block parameters of data type BOOL can be individual binary variables or binary components of fields and structures. You can check input parameters and in/out parameters with contacts or with binary box inputs, and you can influence output parameters and in/out parameters with binary box outputs. After the CPU has used the actual parameter specified as the block parameter, it processes the functions as shown in the relevant chapters.

Block parameters of digital data type

Block parameters of digital data type occupy 8, 16 or 32 bits (all elementary data types except BOOL). They can be individual digital variables

Table 19.2 Access to Block Parameters

Data Types	Permissible with IN	I_O	OUT	Access possible in the block
Elementary data types				
BOOL	x	x	x	Yes
BYTE, WORD, DWORD, CHAR, INT, DINT, REAL, S5TIME, TIME, TOD, DATE	x	x	x	Yes
Complex data types				
DT, STRING	x	x	x	No
ARRAY, STRUCT				
Individual binary components	x	x	x	Yes
Individual digital components	x	x	x	Yes
Complete variables	x	x	x	No
Parameter types				
TIMER	x	–	–	Yes
COUNTER	x	–	–	Yes
BLOCK_FC	x	–	–	Yes
BLOCK_FB	x	–	–	No
BLOCK_DB	x	–	–	Yes
BLOCK_SDB	x	–	–	No
POINTER, ANY	x	x	x[1)]	No

[1)] only with FC functions

or digital components of fields and structures. You apply input and in/out parameters to digital box inputs, and you write output and in/out parameters to digital box outputs.

You exchange values between block parameters with different data types or with different information widths with the MOVE box as described in Chapter 6 “Transfer Functions”.

Block parameters of data type DT and STRING

Direct access to block parameters of data type DT and STRING is not possible. In function blocks, you can “pass on” input and output parameters of data type DT and STRING to parameters of called blocks.

Block parameters of data type ARRAY and STRUCT

Direct access to block parameters of data type ARRAY and STRUCT is possible on a component-wise basis, that is, you can access individual binary or digital components with the relevant operations.

Access to the complete variable (entire field or entire structure) is not possible and neither is access to individual components of combined or user-defined data type. In function blocks, you can “pass on” input and output parameters of data type ARRAY and STRUCT to parameters of called blocks.

Block parameters of user-defined data type

You handle block parameters of user-defined data type in the same way as block parameters of data type STRUCT.

Direct access to block parameters of data type UDT is possible on a component-wise basis, that is, you can access individual binary or digital components with the relevant operations.

Access to the complete variable is not possible and neither is access to individual components of combined or user-defined data type. In function blocks, you can "pass on" input and output parameters of data type UDT to parameters of called blocks.

Block parameters of data type TIMER

You can use block parameters of data type TIMER with all functions, as described in Chapter 7 "Timer Functions". When a timer is started, the time value can also be a block parameter of data type S5TIME.

Block parameters of data type COUNTER

You can use block parameters of data type COUNTER with all functions, as described in Chapter 8 "Counter Functions". When setting a counter, the counter value can also be a block parameter of data type WORD.

Block parameters of data type BLOCK_DB

You can transfer a data block via a block parameter of data type BLOCK_DB. Call this data block with the OPN coil by writing the formal parameters to the OPN coil. When opening a data block via a block parameter, the CPU always uses the global data block register (DB register).

Block parameters of data type BLOCK_FC

You can transfer a function FC via a block parameter of data type BLOCK_FC. Call this function with the CALL coil. You can use the CALL coil with a formal parameter and with or without any preceding logic operation if you are currently programming a function block. If you use the CALL coil with formal parameters in a function, a preceding logic operation is not permissible (absolute call only).

A function FC transferred via a block parameter must not have any block parameters.

Block parameters of data type BLOCK_FB

You can transfer a function block FB via a block parameter of data type BLOCK_FB. Direct access to block parameters of data type BLOCK_FB is not possible with LAD functions.

A function block FB transferred via a block parameter must not have any block parameters.

Block parameters of data type POINTER and ANY

Direct access to block parameters of data type POINTER and ANY is not possible with LAD functions.

19.3 Actual Parameters

When you call a block, you initialize its block parameters with constants, operands or variables with which it is to operate. These are the actual parameters. If you call the block often in your program, you usually use different actual parameters each time it is called.

The actual parameter must agree in data type with the block parameter: You can only apply a binary actual parameter (for example, a memory bit) to a block parameter of data type BOOL; you can only initialize a block parameter of data type ARRAY with an identically dimensioned field variable. Table 19.3 gives an overview of which operands you can use as actual parameters with which data type.

When calling functions, you must initialize all block parameters with actual parameters.

When calling function blocks, it is not a mandatory requirement that you initialize block parameters with actual parameters. If you make no specification at a block parameter, the (old) values stored in the instance data block are used as the actual parameters. These can be, for example, the default values or the values of actual parameters from an earlier call. In/out parameters of complex data type cannot be pre-assigned and neither can any parameter types. You must provide these block parameters with actual parameters at least at the first call.

You can also access the block parameters of the function block direct. Since they are located in a data block, you can handle the block parameters like data operands. Example: A function block with the instance data block "Station_1" controls a binary output parameter with the name *Up*. Following processing in the function block (after its call), you can check the parameter under the symbolic address *"Station_1".Up*, without having initialized the output parameter.

Table 19.3 Initialization with Actual Parameters

Data Type of the Block Parameter	Permissible Actual Parameters
Elementary data type	▷ Simple operands, fully-addressed data operands, constants ▷ Components of fields or structures of elementary data type ▷ Block parameter of the calling block ▷ Components of block parameters of the calling block of elementary data type
Complex data type	▷ Variables or block parameters of the calling block
TIMER, COUNTER and BLOCK_xx	▷ Timers, counters and blocks
POINTER	▷ Simple operands, fully-addressed data operands ▷ Range pointer or DB pointer
ANY	▷ Variables of any data type ▷ ANY pointer

Initializing block parameters with elementary data types

The actual parameters listed in Table 19.4 are permissible as actual parameters of elementary data types.

You can assign either absolute or symbolic addresses to input, output and memory bit operands. Input operands should be placed only at input parameters and output operands at output parameters (however, this is not mandatory). Memory bit operands are suitable for all declaration types. You must apply peripheral inputs only to input parameters and peripheral outputs only to output parameters.

When you use (part-addressed) data operands, you must note that when you access the block parameter (in the *called* block), that the currently open data block is also the "correct" one. Since the Editor may in certain circumstances change the data block when the block is called, part addressing is not recommended for data operands. Use only fully-addressed data operands for this reason.

Temporary local data are usually symbolically addressed. They are located in the L stack of the calling block and are declared in the calling block.

If the calling block is a function block, you can also use its static local data as actual parameters (see Section 19.4 below "Passing On Block Parameters"). Static data are usually symbolically addressed; if you would like to assign absolute addresses via DI operands, please note the information in Section 18.2.4 "Special Points in Data Addressing".

With a block parameter of the BOOL data type, you can apply the constants TRUE (signal state "1") or FALSE (signal state "0"), and with block parameters of digital data type, you can apply all constants corresponding to the data type. Initialization with constants is only permissible with input parameters.

You can also initialize a block parameter of elementary data type with components of fields and structures if the component is of the same data type as the block parameter.

Initializing block parameters of complex data type

Every block parameter can be of the combined type. Variables of the same data type are permissible as actual operands.

For initializing block parameters of data type DT or STRING, individual variables or components of fields or structures of the same data type are permissible.

STRING variables can be of variable length. If you have specified the STRING length when declaring an input or output parameter of a function block, the Editor reserves the specified space in the instance data block; if you have not specified any length, 256 bytes are occupied for

Table 19.4 Actual Parameters of Elementary Data Types

Operands	Permissible with			Binary operand or symbolic name	Digital operand or symbolic name
	IN	I_O	OUT		
Inputs (process image)	x	x	x	I y.x	IB y, IW y, ID y
Outputs (process image)	x	x	x	Q y.x	QB y, QW y, QD y
Memory bits	x	x	x	M y.x	MB y, MW y, MD y
Peripheral inputs	x	–	–	–	PIB y, PIW y, PID y
Peripheral outputs	–	–	x	–	PQB y, PQW y, PQD y
Global data					
Part addressing	x	x	x	DBX y.x	DBB y, DBW y, DBD y
Full addressing	x	x	x	DB z.DBX y.x	DB z.DBB y, etc.
Temporary local data	x	x	x	L y.x	LB y, LW y, LD y
Static local data	x	x	x	DIX y.x	DIB y, DIW y, DID y
Constants	x	–	–	TRUE, FALSE	All digital constants
Components of ARRAY or STRUCT	x	x	x	Complete component name	Complete component name

the STRING variable. The maximum length of the STRING variable that you specify in the declaration must agree between the actual parameter and the formal parameter. Exception: With functions FCs, you specify either no length or the standard length of 254 bytes when declaring a STRING variable; here, you can use STRING variables of all lengths as actual parameters.

For initializing block parameters of data type ARRAY or STRUCT, variables with exactly the same structure as the block parameters are permissible.

Initializing block parameters of user-defined data type

With complex or extensive data structures, the use of user-defined data types (UDTs) is recommended. First, you define the UDT and then you use it, for example, to apply the variable in the data block or to declare the block parameter. Following this, you can use the variable when initializing the block parameter. It is also the case here, that the actual parameter (the variable) must be of the same data type (the same UDT) as the block parameter.

Initializing block parameters of type TIMER, COUNTER and BLOCK_xx

You initialize a block parameter of type TIMER with a timer function, and a block parameter of type COUNTER with a counter function. You can apply only blocks without their own parameters to block parameters of parameter types BLOCK_FC and BLOCK_FB. You initialize BLOCK_DBs with a data block.

Block parameters of types TIMER, COUNTER and BLOCK_xx must only be input parameters.

Initializing block parameters of type POINTER

Pointers (constants) are permissible for block parameters of parameter type POINTER. These pointers are either range pointers (32-bit pointers) or DB pointers (48-bit pointers). The operands are of elementary data type and can also be fully-addressed data operands.

Output parameters of the type POINTER are not permissible with function blocks.

Initializing block parameters of type ANY

Variables of all data types are permissible for block parameters of parameter type ANY. The programming within the called block determines which variables (operands or data types) must be applied to the block parameters, or which variables are feasible. You can also specify a constant in the format of the ANY pointer "P#[data block.]Operand Data type Number" and so define an absolute-addressed area.

Output parameters of type ANY are not permissible with function blocks.

19.4 "Passing On" Block Parameters

"Passing on" block parameters is a special form of access and of initializing block parameters. The block parameters of the calling block are "passed on" to the parameters of the called block. Here, the formal parameter of the calling block then becomes the actual parameter of the called block.

In general, it is also the case here that the actual parameter must be of the same type as the formal parameter (that is, the relevant block parameters must agree in their data types). In addition, you can apply an input parameter of the calling blocks only at an input parameter of the called block, and similarly, an output parameter at an output parameter. You can apply an in/out parameter of the calling block to all declaration types of the called block.

There are restrictions with regard to data types caused by the variations in block parameter storage between functions and function blocks. Block parameters of elementary data type can be passed on without restriction in accordance with the information in the previous paragraph. Complex data types at inputs and output parameters can only be passed on if the calling block is a function block. Block parameters of parameter types TIMER, COUNTER and BLOCK_xx can only be passed on from one input parameter to another if the calling block is a function block. These statements are represented in Table 19.5.

19.5 Examples

19.5.1 Conveyor Belt Example

The example shows the transfer of signal states via block parameters. For this purpose, we use the function of a conveyor belt control explained in Chapter 5 "Memory Functions", programmed here only with coils and contacts. The conveyor belt control is to be located in a function block and all inputs and outputs are to be routed via block parameters, so that the function block can be used repeatedly (for several conveyor belts). Figure 19.3 shows the input and output parameters for the function block as well as the static and temporary local data used.

Distributing the parameters is quite simple in this case: All binary operands that were inputs have become input parameters, all outputs have become output parameters and all memory bits have become static local data.

The function block "Conveyor_Belt" is to control two conveyor belts. For this purpose, it will be called twice; the first time with the inputs and outputs of conveyor belt 1 and the second time with those of conveyor belt 2. For each call, the function block requires an instance data

Table 19.5 Permissible Combinations when Passing On Block Parameters

Calling → called	FC calls FC			FB calls FC			FC calls FB			FB calls FB		
declaration type	E	C	P	E	C	P	E	C	P	E	C	P
Input → Input	x	–	–	x	x	–	x	–	x	x	x	x
Output → Output	x	–	–	x	x	–	x	–	–	x	x	–
In/out → Input	x	–	–	x	–	–	x	–	–	x	–	–
In/out → Output	x	–	–	x	–	–	x	–	–	x	–	–
In/out → In/out	x	–	–	x	–	–	x	–	–	x	–	–

E = Elementary data types
C = Complex data types
P = Parameter types TIMER, COUNTER and BLOCK_xx

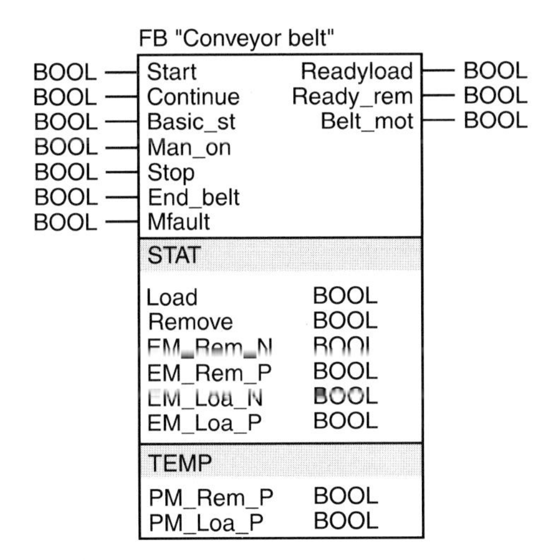

Figure 19.3
Function Block for the Conveyor Belt Example

block where it stores the data for the conveyor belt in each case. The data block for conveyor belt 1 is to be called "BeltData1" and the data block for conveyor belt 2 is to be called "BeltData2".

You can find the executed programming example in the program "Conveyor Example" on the diskette accompanying the book. It shows the programming of function block FB 21 with input parameters, the output parameters and the static local data. You can use any data blocks as instance blocks; in the example, DB 21 is used for "BeltData1" and DB 22 for "BeltData2". In the Symbol Table, these data blocks are of the data type of the function block (FB 21 in the example, if "Conveyor Belt" is the symbol for FB 21).

When you call the function block, you can use the inputs and outputs from the Symbol Table as the actual parameters. In those cases where these global symbols contain special characters, you must place these symbols in quotation marks in the program. The Symbol Table is designed for all three examples of this chapter (Table 19.6).

19.5.2 Parts Counter Example

The example demonstrates the handling of block parameters of elementary data types. The "Parts Counter" example from Chapter 9 "Counter Functions" is the basis of the function. The same function is implemented here as a function block, with all global variables declared either as block parameters or as static local data. The timer and counter functions are controlled here via coils.

Timer and counter functions are transferred via block parameters of parameter types TIMER and COUNTER. These block parameters must be input parameters. The initial values of the counter (Quantity) and the timer function (Dura1 and Dura2) can also be transferred as block parameters, the data type of the block parameters corresponds here to the actual parameters.

The edge memory bits are stored in the static local data and the pulse memory bits are stored in the temporary local data.

You can find the executed programming example in the program "Conveyor Example" on the diskette accompanying the book. It contains function block FB 22 "Parts_counter" and the associated instance data block "CountDat". You can use the inputs, outputs, timer function and counter function from the Symbol Table (table in previous example) as the actual parameters when calling the function block.

19.5.3 Feed Example

The same functions as described in the two previous examples can also be called as local in-

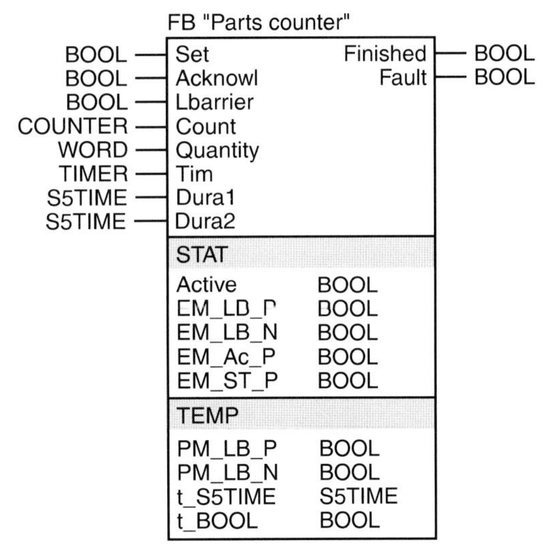

Figure 19.4
Function Block for the Parts Counter Example

Table 19.6 Symbol Table for the Examples Conveyor Belt, Parts Counter and Feed

Symbol	Address	Data type	Comment
Conveyor_belt	FB 21	FB 21	Conveyor belt control
Belt_data1	DB 21	FB 21	Data for conveyor belt 1
Belt_data2	DB 22	FB 21	Data for conveyor belt 2
Parts_counter	FB 22	FB 22	Counter control and monitor
CountDat	DB 29	FB 22	Data for parts_counter
Feed	FB 20	FB 20	Feed with several belts
FeedDat	DB 20	FB 20	Data for Feed
Cycle	OB 1	OB 1	Main program, cyclic execution
Basic_st	I 0.0	BOOL	Set controller to the basic state
Man_on	I 0.1	BOOL	Switch on conveyor belt motors
/Stop	I 0.2	BOOL	Stop conveyor belt motors (zero active)
Start	I 0.3	BOOL	Start conveyor belt
Continue	I 0.4	BOOL	Acknowledgment that parts have been removed
Acknowl	I 0.6	BOOL	Acknowledge fault
Set	I 0.7	BOOL	Set counter, activate monitor
Lbarr1	I 1.0	BOOL	(Light barrier) “End of belt” sensor signal for conveyor belt 1
Lbarr2	I 1.1	BOOL	(Light barrier) “End of belt” sensor signal for conveyor belt 2
Lbarr3	I 1.2	BOOL	(Light barrier) “End of belt” sensor signal for conveyor belt 3
Lbarr4	I 1.3	BOOL	(Light barrier) “End of belt” sensor signal for conveyor belt 4
/Mfault1	I 2.0	BOOL	Motor protection switch conveyor belt 1 (zero-active)
/Mfault2	I 2.1	BOOL	Motor protection switch conveyor belt 2 (zero-active)
/Mfault3	I 2.2	BOOL	Motor protection switch conveyor belt 3 (zero-active)
/Mfault4	I 2.3	BOOL	Motor protection switch conveyor belt 4 (zero-active)
Readyload	Q 4.0	BOOL	Load new parts onto belt
Ready_rem	Q 4.1	BOOL	Remove parts from belt
Finished	Q 4.2	BOOL	Number of parts reached
Fault	Q 4.3	BOOL	Monitor activated
Belt_mot1	Q 5.0	BOOL	Switch on belt motor for conveyor belt 1
Belt_mot2	Q 5.1	BOOL	Switch on belt motor for conveyor belt 2
Belt_mot 3	Q 5.2	BOOL	Switch on belt motor for conveyor belt 3
Belt_mot 4	Q 5.3	BOOL	Switch on belt motor for conveyor belt 4
Quantity	MW 4	WORD	Number of parts
Dura1	MW 6	S5TIME	Monitoring time for light barrier covered
Dura2	MW 8	S5TIME	Monitoring time for light barrier not covered
Count	C 1	COUNTER	Counter for parts
Monitor	T 1	TIMER	Timer function for monitor

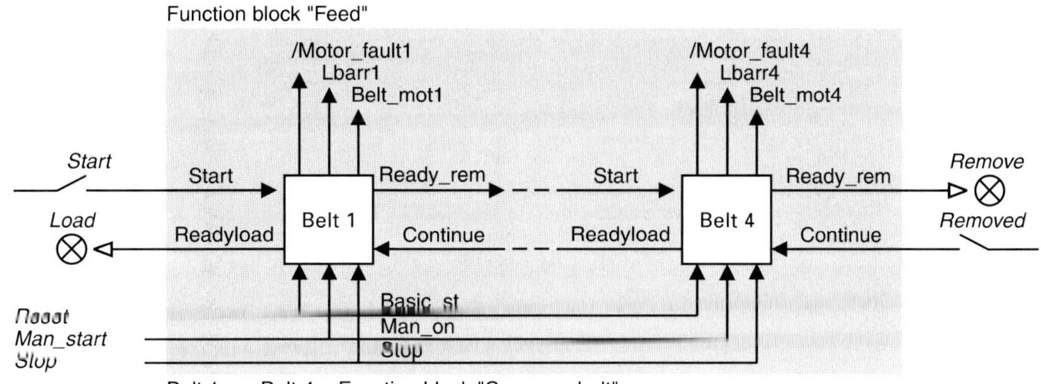

Figure 19.5 Feed Programming Example

stances. In our example, this means that we program a function block "Feed" that is to control four conveyor belts and count the conveyed parts. In this function block, the FB "Conveyor Belt" is called four times and the FB "Parts_counter" is called once. The call does not take place in each case with its own instance data block, but the called FBs are to store their data in the instance data block of the function block "Feed".

Figure 19.5 shows how the individual conveyor belt controls are connected together (the FB "Parts_counter" is not represented here). The start signal is connected to the *Start* input of the controller of belt 1, the *ready_rem* output is connected to the *Start* input of belt 2, etc. Finally, the *ready_rem* output of belt 4 is connected to the *Remove* output of "Feed". The same signal sequence leads in the reverse direction from *Removed* via *Continue* and *Readyload* to *Load*.

Belt_mot, *Lbarr* and */Mfault* are individual signals of the conveyor belts; *Reset*, *Man_start* and *Stop* control all conveyor belts via *Basic_st*, *Man_on* and *Stop*.

The following program for the function block "Feed" is designed in the same way. The input and output parameters of the function block can be seen from the figure. In addition, the digital values for the parts counter *Quantity*, *Dura1* and *Dura2* are designed as input parameters here. We declare the data of the individual conveyor belt controls and the data of the parts counter in the static local data in exactly the same way as for a user-defined data type, i.e. with name and data type. The variable "Belt1" is to receive the data structure of the function block "Conveyor Belt", also the variable "Belt2", etc.; the variable "Check" receives the data structure of the function block "Parts_counter".

The program in the function block starts with the initialization of the signals common to all conveyor belts. Here, we make use of the fact that the block parameters of the function blocks called as local instances are static local data in the current block and can be handled as such. The block parameter *Man_start* of the current function block controls the input parameter *Man_on* of all four conveyor belt controls with a simple assignment. We proceed in the same way with the signals *Stop* and *Reset*. And now the conveyor belt controls are initialized with the common signals. (You can, of course, also initialize these input parameters when the function block is called.)

The subsequent calls of the function blocks for conveyor belt control contain only the block parameters for the individual signals for each conveyor belt and the connection to the block parameters of "Feed". The individual signals are the light barriers, the commands for the belt motor and the motor faults. (We make use here of the fact that when a function block is called, not all block parameters have to be initialized.) We program the connections between the individual belt controllers using assignments.

The FB "Parts_counter" is called as a local instance even if it has no closer connection with the signals of the conveyor belt controls. The instance data block of "Feed" takes the FB data.

Table 19.7 Declaration Section of FB "Feed"

Address	Declaration	Name	Type	Initial Value	Comment
0.0	in	Start	BOOL	FALSE	Start conveyor belts
0.1	in	Removed	BOOL	FALSE	Parts have been removed from belt
0.2	in	Man_start	BOOL	FALSE	Startup conveyor belts manually
0.3	in	Stop	BOOL	FALSE	Stop conveyor belts
0.4	in	Reset	BOOL	FALSE	Set control to the basic setting
2.0	in	Count	COUNTER		Counter for the parts
4.0	in	Quantity	WORD	W#16#200	Number of parts
6.0	in	Tim	TIMER		Timer function for the monitor
8.0	in	Dura1	S5TIME	S5T#5S	Monitoring time for parts
10.0	in	Dura2	S5TIME	S5T#10S	Monitor time for gap
12.0	out	Load	BOOL	FALSE	Load new parts onto belt
12.1	out	Remove	BOOL	FALSE	Remove parts from belt
	in_out				
14.0	stat	Belt1	Conveyor_belt		Control for belt 1
20.0	stat	Belt2	Conveyor_belt		Control for belt 2
26.0	stat	Belt3	Conveyor_belt		Control for belt 3
32.0	stat	Belt4	Conveyor_belt		Control for belt 4
38.0	stat	Check	Parts_counter		Control for counting and monitoring
	temp				

The input parameters *Quantity*, *Dura1* and *Dura2* of "Feed" need to be set only once. This can be done with the default (as in the example) or in the restart program in OB 100 (for example, through direct assignment if these three parameters are treated as global data).

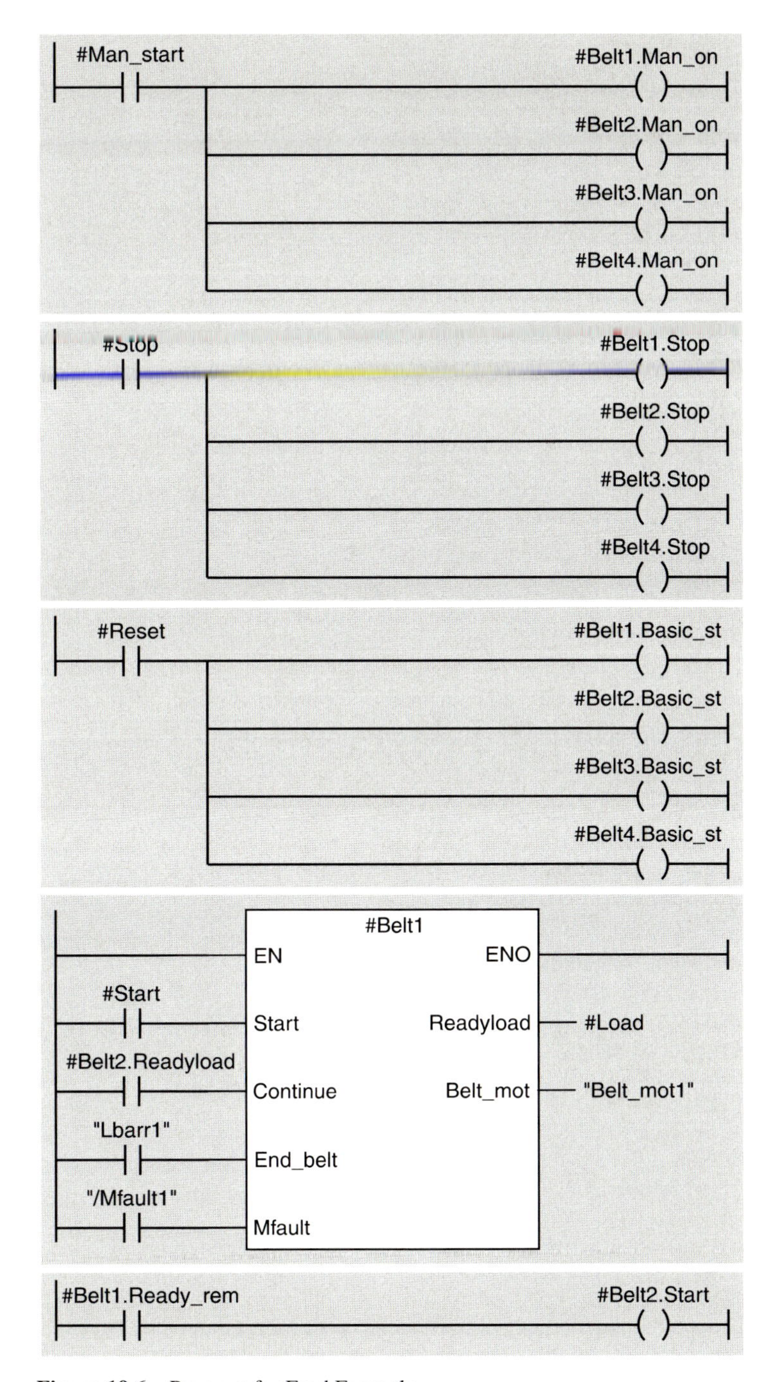

Figure 19.6 Program for Feed Example

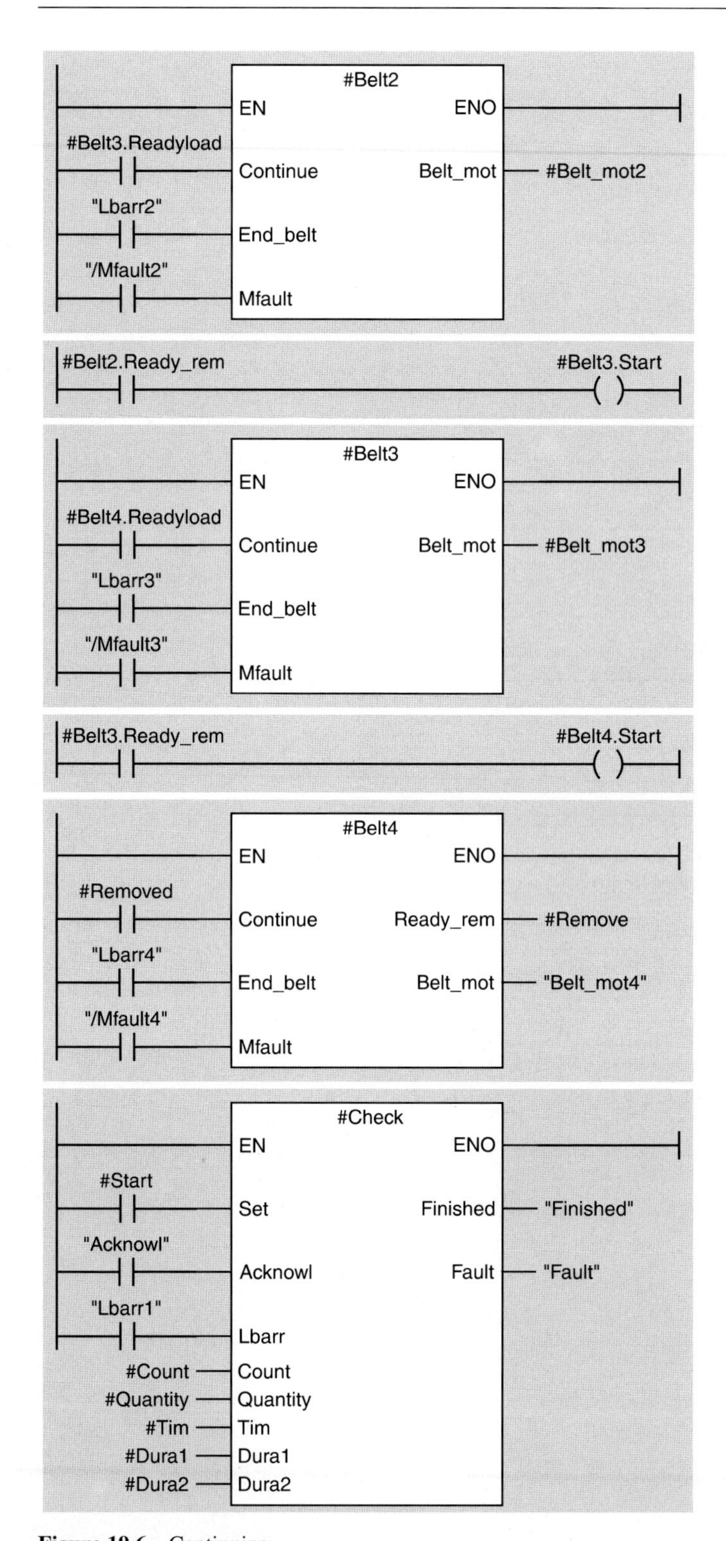

Figure 19.6 Continuing

Program Processing

This section of the book discusses the various methods of program processing.

- The **main program** executes cyclically. After each program pass, the CPU returns to the beginning of the program and executes it again. This is the "standard" method of processing PLC programs.
- Numerous system functions support the utilization of system services, such as controlling the real-time clock or communication via bus systems. In contrast to the static settings made when initializing the CPU, system functions can be used dynamically at program run time.
- The main program can be temporarily suspended to allow **interrupt servicing.** The various types of interrupts (hardware interrupts, watchdog interrupts, time-of-day interrupts, time-delay interrupts, multiprocessor interrupt) are divided into priority classes whose processing priority you may yourself, to a large degree, determine. Interrupt servicing allows you to react quickly to signals from the controlled process or implement periodic control procedures independently of the processing time of the main program.
- Before starting the main program, the CPU initiates a **start-up program** in which you can make specifications regarding program processing, define default values for variables, or initialize modules.
- **Error handling** is also part of program processing. STEP 7 distinguishes between synchronous errors, which occur during processing of a LAD element, and asynchronous errors, which can be detected independently of program processing. In both cases you can adapt the error routine to suit your needs.

20 Main Program

The main program is the cyclically scanned user program; cyclic scanning is the "normal" way in which programs execute in programmable logic controllers. The large majority of control systems use only this form of program execution. If event-driven program scanning is used, it is in most cases only in addition to the main program.

The main program is invoked in organization block OB 1. It executes at the lowest priority level, and can be interrupted by all other types of program processing. The mode selector on the CPU's front panel must be at RUN or RUN-P. When in the RUN-P position, the CPU can be programmed via a programming device. In the RUN position, you can remove the key so that no one can change the operating mode without proper authorization; when the mode selector is at RUN, programs can only be read.

20.1 General Remarks

20.1.1 Program Structure

To analyze a complex automation task means to subdivide that task into smaller tasks or functions in accordance with the structure of the process to be controlled. You then define the individual tasks resulting from this subdividing process by determining the functions and stipulating the interface signals to the process or to other tasks. This breakdown into individual tasks can be done in your program. In this way, the structure of your program corresponds to the subdivision of the automation task.

A subdivided user program can be more easily configured, and can be programmed in sections (even by several people in the case of very large user programs). And not least, subdividing the program simplifies both debugging and service and maintenance.

The structuring of the user program depends on its size and its function. A distinction is made between three different "methods":

- In a *linear program,* the entire main program is in organization block OB 1. Each rung is in a separate network. STEP 7 numbers the networks in sequence. When editing and debugging, you can reference every network direct by its number.

- A *partitioned program* is basically a linear program which is subdivided into blocks. Reasons for subdividing the program might be because it is too long for organization block OB 1 or because you want to make it more readable. The blocks are then called in sequence. You can also subdivide the program in another block the same way you would the program in organization block OB 1. This method allows you to call associated process-related functions for processing from within one and the same block. The advantage of this program structure is that, even though the program is linear, you can still debug and run it in sections (simply by omitting or adding block calls).

- A *structured program* is used when the conceptual formulation is particularly extensive, when you want to reuse program functions, or when complex problems must be solved. Structuring means dividing the program into sections (blocks) which embody self-contained functions or serve a specific functional purpose and which exchange the fewest possible number of signals with other blocks. Assigning each program section a specific (process-related) function will produce easily readable blocks with simple interfaces to other blocks when programmed.

The LAD programming language supports structured programming through functions with which you can create "blocks" (self-contained program sections). Chapter 3, "The LAD Programming Language" discusses under the header "Blocks" the different kinds of blocks and their uses. You will find a detailed description of the functions for calling and ending blocks in Chapter 18, "Block Functions". The blocks receive the signals and data to be processed via the call interface (the block parameters), and forward the results over this same interface. The options for passing parameters are described in detail in Chapter 19, "Block Parameters".

20.1.2 Program Organization

Program organization determines whether and in what order the CPU will process the blocks which you have generated. To organize your program, you program block calls in the desired sequence in the higher-level blocks. You should chose the order in which the blocks are called so that it mirrors the process-related or function-related subdivision of the controlled plant.

Nesting depth

The maximum depth applies for a priority class (for the program in an organization block), and is CPU-dependent. The CPU 314, for example, has a nesting depth of eight, that is, beginning with one organization block (nesting depth 1), you can call seven more blocks in the "horizontal" direction (this is called "nesting"). If more blocks are called, the CPU goes to STOP with a "Block stack overflow" error. Do not forget to include system function block (SFB) calls and system function (SFC) calls when calculating the nesting depth.

A data block call, which is actually only the opening or selecting of a data area, has no effect on the nesting depth of blocks, nor is the nesting depth affected by calling several blocks in succession (linear block calls).

Practice-related program organization

In organization block OB 1, you should call the blocks in the main program in such a way as to roughly organize your program. A program can be organized on either a process-related or function-related basis.

The following points of discussion can give only a rough, very general view with the intention of giving the beginner some ideas on program structuring and on translating his control task into reality. Advanced programmers normally have sufficient experience to organize a program to suit the special control task at hand.

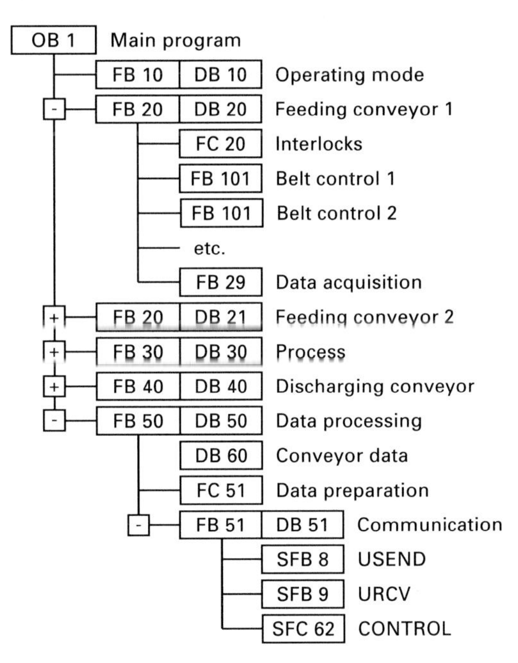

Figure 20.1 Example for Program Structuring

- A *process-related program structure* closely follows the structure of the plant to be controlled. The individual program sections correspond to the individual parts of the plant or of the process to be controlled. Subordinate to this rough structure are the scanning of the limit switches and operator panels and the control of the actuators and display devices (in different parts of the plant). Bit memory or global data are used for signal interchange between different parts of the plant.

- A *function-related program structure* is based on the control function to be executed. Initially, this method of program structuring does not take the controlled plant into account at all. The plant structure first becomes apparent in the subordinate blocks when the control function defined by the rough structure is subdivided further.

- *In practice,* a hybrid of these two concepts is normally used. Figure 20.1 shows an example: A functional structure is mirrored in the operating mode program and in the data processing program which goes above and beyond the plant itself. Program sections Feeding Conveyor 1, Feeding Conveyor 2, Process and Discharging Conveyor are process-related.

The example also shows the use of different types of blocks. The main program is in OB 1; it is in this program that the blocks for the operating modes, the various pieces of plant equipment, and for data processing are called. These blocks are function blocks with an instance data block as data store. Feeding Conveyor 1 and Feeding Conveyor 2 are identically structured; FB 20, with DB 20 as instance data block for Feeding Conveyor 1 and with DB 21 as instance data block for Feeding Conveyor 2, is used for control.

In the conveyor control program, function FC 20 processes the interlocks; it scans inputs or bit memories and controls FB 20's local data. Function block FB 101 contains the control program for a conveyor belt, and is called once for each belt. The call is a local instance, so that its local data are in instance data block DB 20. The same applies for the data acquisition program in FB 29.

The data processing program in FB 50, which uses DB 50, processes the data acquired with FB 29 (and other blocks), located in global data block DB 60. Function FC 51 prepares these data for transfer. The transfer is controlled by FB 51 (with DB 51), in which system blocks SFB 8, SFB 9 and SFB 62 are called. Here, too, the SFBs save their instance data in "higher-level" data block DB 51.

20.2 Scan Cycle Control

20.2.1 Process Image Updating

Before the operating system calls organization block OB 1, it first loads the process-image input table (the operand area for inputs is loaded with the current signal states from the process). As a rule, the user program then works with these signal states and changes signal states in the process-image output table. When organization block OB 1 terminates, the operating system transfers the signal states from the process-image output table to the output modules (now, for the first time, the process is supplied with the processed data).

The process image is part of the CPU's internal system memory (Chapter 1.4, "CPU Memory Areas"). It begins at I/O address 0 and ends at the address stipulated by the relevant CPU. Normally, all digital modules lie in the process image address area, while all analog modules have addresses outside this area. If the CPU has free address allocation, you can use the configuration table to direct any module over the process image or address it outside the process image area.

Subprocess images

On the S7-400, the process image is divided into eight subprocess images whose sizes you can stipulate when you initialize the CPU. System functions SFC 26 UPDAT_PI for the process-image input table and SFC 27 UPDAT_PO for the process-image output table are used to update these subprocess images (Table 20.1). You can update individual subprocess images by calling these SFCs at any time and at any location in the program. For instance, you can define a subprocess image for a priority class (a program scanning level) and then have it updated at the beginning and end of the relevant organization block when that priority class is processed.

The automatic updating of the entire process image by the operating system is programmable. You can completely disable automatic updating and allow only subprocess image updating or, by specifying the number 0 in the SFC call, you can specify the time at which you want the entire process image updated.

The updating of a process image can be interrupted by calling a higher priority class.

Table 20.1 Parameters for the SFCs for Process Image Updating

Parameter Name	SFC		Declaration	Data Type	Contents, Description
PART	26	27	INPUT	BYTE	Number of the subprocess image (0 to 8)
RET_VAL	26	27	OUTPUT	INT	Error information
FLADDR	26	27	OUTPUT	WORD	On an access error: the address of the first byte to cause the error

If an error occurs during updating of a process image, it is reported via the SFC's function value. If an error occurs while the operating system is executing an automatic process image update, organization block OB 85, "Program run errors", is invoked.

20.2.2 Scan Cycle Monitoring Time

Program scanning in organization block OB 1 is monitored by the so-called "scan cycle monitor" or "scan cycle watchdog". The default value for the scan cycle monitoring time is 150 ms. You can change this value in the range from 1 ms to 6 s by initializing the CPU accordingly.

If the main program takes longer to scan than the specified scan cycle monitoring time, the CPU calls OB 80 ("Time-out"). If OB 80 has not been programmed, the CPU goes to STOP.

The scan cycle monitoring time includes the full scan time for OB 1. It also includes the scan times for higher priority classes which interrupt the main program (in the current cycle). Communication processes carried out by the operating system, such as GD communication or PG access to the CPU (block status!), also increase the runtime of the main program. The increase can be reduced in part by the way you parameterize the CPU ("cycle performance" parameter block). The CPU's cyclic memory test (S7-300) also increases the scan cycle time. You can limit or disable this test by initializing the CPU accordingly.

An SFC 43 RE_TRIGR system function call restarts the scan cycle monitoring time; the timer restarts with the new value. SFC 43 has no parameters.

Operating system run times

The scan cycle time also includes the operating system run times. These are composed of the following:

- System control of cyclic scanning ("no-load cycle"), fixed value
- Updating of the process image; dependent on the number of bytes to be updated
- Updating of the timers; dependent on the number of timers to be updated
- Communication load; dependent on the communication just completed (PG-CPU or CPU-CPU link); can be restricted to a value between 5% and 50% (default is 20%) of the scan cycle time by parameterizing the CPU accordingly.

All values at operating system runtime are properties of the relevant CPU.

20.2.3 Minimum Scan Cycle Time, Background Scanning

On the S7-400, you may specify a minimum scan cycle time. If the main program (including interrupts) takes less time, the CPU waits until the specified minimum scan cycle time has expired before beginning the next cycle by recalling OB 1.

The default value for the minimum scan cycle time is 0 ms, that is to say, the function is disabled. You can set a minimum scan cycle time of between 1 ms and 6 s in tab "Cycle performance" when you initialize the CPU.

Background Processing OB 90

In the interval between the actual end of the cycle and expiry of the minimum cycle time, the CPU executes organization block OB 90 "Background processing" (Figure 20.2). OB 90 is executed "in slices". When the operating system

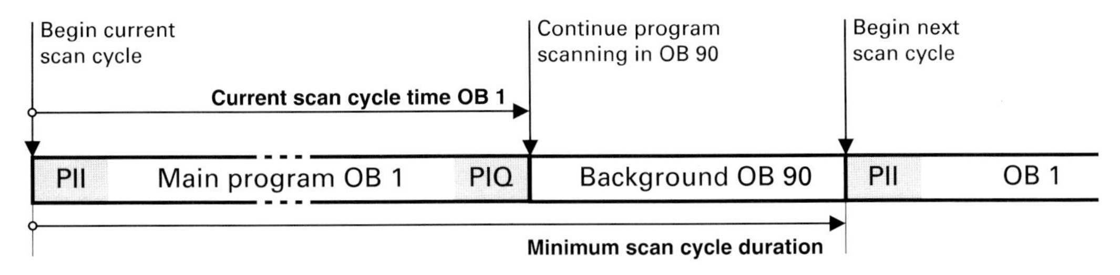

Figure 20.2 Background Processing with OB 90

Table 20.2 Start Information for Background Processing with OB 90

Name	Data Type	Description	Contents
OB90_EV_CLASS	BYTE	Event class	B#16#11 = Call default OB
OB90_STRT_INF	BYTE	Start information	See text
OB90_PRIORITY	BYTE	Priority	B#16#1D = 29
OB90_OB_NUMBR	BYTE	OB number	B#16#5A = 90
OB90_RESERVED_1	BYTE	Reserved	–
OB90_RESERVED_2	BYTE	Reserved	–
OB90_RESERVED_3	INT	Reserved	–
OB90_RESERVED_4	INT	Reserved	–
OB90_RESERVED_5	INT	Reserved	–
OB90_DATE_TIME	DT	Event occurrence	Instant of OB request

calls OB 1, execution of OB 90 is interrupted; it is then resumed at the point of interruption when OB 1 has terminated. OB 90 can be interrupted after each statement; any system block called in OB 90, however, is first scanned in its entirety. The length of a "slice" depends on the current scan cycle time of OB 1. The closer OB 1's scan time is to the minimum scan cycle time, the less time remains for executing OB 90. The program scan time is not monitored in OB 90.

OB 90 is scanned only in RUN mode. It can be interrupted by interrupt and error events, just like OB 1. You can activate and deactivate background processing when parameterizing the CPU by entering a value of at least 20, or 0, in the "Local data" tab for priority class 29. The start information in the temporary local data (Table 20.2) also tells which events cause OB 90 to execute from the beginning:

▷ B#16#91 After a restart (complete or warm restart);

▷ B#16#92 After a block processed in OB 90 was deleted or replaced;

▷ B#16#93 After (re)loading of OB 90 in RUN mode;

▷ B#16#95 After the program in OB 90 was scanned and a new background cycle begins.

20.2.4 Response Time

Before the main program is processed, the CPU's operating system loads the signal states of the digital input modules into the process-image input table. During program execution, the signal states of the inputs are then scanned and logically combined as per the functions in the user program; the outputs are set. After the program has terminated, the operating system transfers the signal states in the process-image output table to the output modules. A program cycle is then complete. Once the operating system has finished what it has to do, a new (user) program cycle begins with the updating of the process-image input table.

The introduction of a process image has given the programmable controller a response time which is dependent on the program execution time (scan cycle time). The response time lies between one and two scan cycles, as the following example explains.

When a limit switch is activated, for instance, it changes its signal state from "0" to "1". The programmable controller detects this change during the subsequent updating of the process image, and sets the inputs allocated to the limit switch to "1". The program evaluates this change by resetting an output, for example, in order to switch off the corresponding motor. The new signal state of the output that was reset

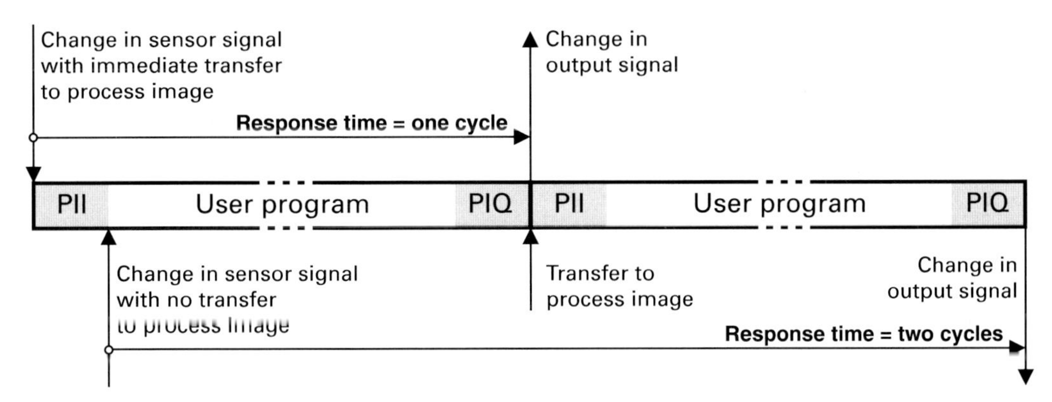

Figure 20.3 Response Times of Programmable Logic Controllers

is transferred at the end of the program scan; only then is the corresponding bit reset on the digital output.

In a best-case situation, the process image is updated immediately following the change in the limit switch's signal. It would then take only one cycle for the relevant output to respond (Figure 20.3). In the worst-case situation, updating of the process image has just been completed when the limit switch signal changes. It would then be necessary to wait approximately one cycle for the programmable controller to detect the signal change and set the input. After yet another cycle, the program can respond.

When so considered, the user program's execution time contains all procedures in one program cycle (including, for instance, the servicing of interrupts, the functions carried out by the operating system, such as updating timers, controlling the MPI interface and updating the process images).

The response time to a change in an input signal can thus be between one and two cycles. Added to the response time are the delays for the input modules, the switching times of contactors, and so on.

In some instances, you can reduce the response times by addressing the I/Os direct or calling program sections on an event-driven basis.

20.2.5 Scan Cycle Statistics

If your programmer is connected on-line with an active CPU, you can invoke the menu command PLC → MODULE INFORMATION to open a dialog box containing several tabs. The "Scan Cycle Time" tab shows the current scan cycle time as well as the shortest and longest scan cycle times. Also included in the display are the specified minimum scan cycle time and the scan cycle monitoring time.

The OB 1 start information (see below) also shows you the shortest and longest cycle times and the scan time of the preceding cycle. You can process this information direct in your program.

20.3 Program Functions

In addition to parameterizing the CPU via STEP 7, you can also select a number of program functions dynamically at runtime via the integrated system functions.

20.3.1 Real-Time Clock

The following system functions can be used to control the CPU's real-time clock:

- ▷ SFC 0 SET_CLK
  Set date and time
- ▷ SFC 1 READ_CLK
  Read date and time
- ▷ SFC 48 SNC_RTCB
  Synchronize CPU clocks

You will find a list of system function parameters in Table 20.3.

When several CPUs are connected to one another in a subnetwork, parameterize one of the CPUs' clocks as "master clock". When parameterizing the CPU, also enter the synchronization

Table 20.3 SFC Parameters for the Real-Time Clock

SFC	Parameter Name	Declaration	Data Type	Contents, Description
0	PDT	INPUT	DT	Date and time (new)
	RET_VAL	OUTPUT	INT	Error information
1	RET_VAL	OUTPUT	INT	Error information
	CDT	OUTPUT	DT	Date and time (current)
48	RET_VAL	OUTPUT	INT	Error information

interval after which all clocks in the subnetwork are to be automatically synchronized to the master clock.

Call SFC 48 SNC_RTCB in the CPU with the master clock. This call synchronizes all clocks in the subnetwork independently of automatic synchronization. When you set a master clock with SFC 0 SET_CLK, all other clocks in the subnetwork are automatically synchronized to this value.

20.3.2 Read System Clock

A CPU's system Clock Starts running on power-up or on a complete restart. As long as the CPU is executing the restart routine or is in RUN mode, the system clock keeps running. When the CPU goes to STOP, the current system time is "frozen". If you initiate a warm restart on an S7-400 CPU, the system Clock Starts running again, using the saved value as its starting time.

The system time has data format TIME, whereby it can assume only positive values (TIME#0ms to TIME#24d20h31m23s647ms). In the event of an overflow, the Clock starts again at 0. An S7-300 CPU updates the system clock every 10 milliseconds, an S7-400 every millisecond.

You can read the current system time with system function SFC 64 TIME_TCK. The RET_VAL parameter contains the system time in the TIME data format.

You can use the system clock, for example, to read out the current CPU runtime or, by computing the difference, to calculate the time between two SFC 64 calls. The difference between two values in TIME format is computed using DINT subtraction.

20.3.3 Run-Time Meter

A run-time meter in a CPU counts the hours. You can use the run-time meter for such tasks as determining the CPU runtime or ascertaining the runtime of devices connected to that CPU.

The number of run-time meters per CPU depends on the CPU. When the CPU is at STOP, the run-time meter also stops running; when the CPU is restarted (complete or warm restart), the run-time meter begins again with the previous value. When a run-time meter reaches 32767 hours, it stops and reports an overflow. A run-time meter can be set to a new value or reset to zero only via an SFC call.

The following system functions are available to control a run-time meter:

▷ SFC 2 SET_RTM
Set run-time meter

▷ SFC 3 CTRL_RTM
Start or stop run-time meter

▷ SFC 4 READ_RTM
Read run-time meter

Table 20.4 shows the parameters for this system functions.

The NR parameter stands for the number of the run-time meter, and has the data type BYTE. It can be initialized using a constant or a variable (as can all input parameters of elementary data type). The PV parameter (data type INT) is used to set the run-time meter to an initial value. SFC 3's-S-parameter starts (with signal state "1") or stops (with signal state "0") the selected run-time meter. CQ indicates whether the run-time meter was running (signal state "1") or stopped (signal state "0") when scanned. The CV parameter records the hours in INT format.

Table 20.4 Parameters of the SFCs for the Run-Time Meter

SFC	Parameter	Declaration	Data Type	Contents, Description
2	NR	INPUT	BYTE	Number of the run-time meter (B#16#01 to B#16#08)
	PV	INPUT	INT	New value for the run-time meter
	RET_VAL	OUTPUT	INT	Error information
3	NR	INPUT	BYTE	Number of the run-time meter (B#16#01 to B#16#08)
	S	INPUT	BOOL	Start (with "1") or stop (with "0") run-time meter
	RET_VAL	OUTPUT	INT	Error information
4	NR	INPUT	BYTE	Number of the run-time meter (B#16#01 to B#16#08)
	RET_VAL	OUTPUT	INT	Error information
	CQ	OUTPUT	BOOL	Run-time meter running ("1") or stopped ("0")
	CV	OUTPUT	INT	Current value of the run-time meter

20.3.4 Compressing CPU Memory

Multiple deletion and reloading of blocks, which often occurs during on-line block modification, can result in gaps in the CPU's work memory and in the RAM load memory which decrease the amount of usable space in memory. When you call the "Compress" function, you start a CPU program which fills these gaps by pushing the blocks together. You can initiate the "Compress" function via a programming device connected to the CPU or by calling system function SFC 25 COMPRESS. The parameters for system function SFC 25 are listed in Table 20.5.

The compression procedure is distributed over several program cycles. The SFC returns BUSY = "1" to indicate that it is still in progress, and DONE = "1" to indicate that it has completed the compression operation. The SFC cannot compress when an externally initiated compression is in progress, when the "Delete Block" function is active, or when programming device functions are accessing the block to be shifted (for instance the Block Status function).

Note that blocks of a particular CPU-specific maximum length cannot be compressed, so that gaps would still remain in CPU memory. Only the Compress function initiated via the programming device while the CPU is at STOP closes all gaps.

20.3.5 Waiting and Stopping

The system function SFC 47 WAIT halts the program scan for a specified period of time; the system function SFC 46 STP terminates the program scan, and the CPU goes to STOP.

SFC 47 WAIT has input parameter WT of data type INT in which you can specify the waiting time in microseconds (µs). The maximum waiting time is 32767 microseconds; the minimum waiting time corresponds to the execution time of the system function, which is CPU dependent.

SFC 47 can be interrupted by higher-priority events. On the S7-300, this increases the waiting time by the scan time of the higher-priority interrupt routine.

SFC 46 STP has no parameters.

Table 20.5 Parameters for SFC 25

SFC	Parameter	Declaration	Data Type	Contents, Description
25	RET_VAL	OUTPUT	INT	Error information
	BUSY	OUTPUT	BOOL	Compression still in progress (with "1")
	DONE	OUTPUT	BOOL	Compression completed (with "1")

20.3.6 Multiprocessing Mode

The S7-400 enables multiprocessing. As many as four appropriately designed CPUs can be operated in one rack on the same P bus and K bus. Each CPU has a CPU number by which it is referenced in multiprocessing mode. You can assign this number when you configure the hardware ("Multicomputing" tab). The configuration data for all the CPUs must be loaded into the PLC, even when you make changes to only one CPU.

After assigning parameters to the CPUs, you must assign each module in the station to a CPU. This is done by parameterizing the module in the "Addresses" tab under "CPU Allocation". At the same time that you assign the module's address area, you also allocate the module's interrupts to this CPU. With VIEW → FILTER → CPU NO.X-MODULES, you can emphasize the modules allocated to a CPU in the configuration tables.

The CPUs in a multiprocessing network all have the same operating mode. This means

- ▷ They must all be parameterized with the same restart mode;
- ▷ They all go to RUN simultaneously;
- ▷ They all go to HALT when you debug in single-step mode in one of the CPUs;
- ▷ They all go to STOP as soon as one of the CPUs goes to STOP.

When one rack in the station fails, organization block OB 86 is called in each CPU.

The user program in these CPUs execute independently of one another; they are not synchronized.

An SFC 35 MP_ALM call starts organization block OB 60 "Multiprocessor interrupt" in all CPUs simultaneously (see section 21.6, "Multiprocessor Interrupt").

20.4 Data Interchange Using System Functions

This section deals with system blocks for data interchange

- ▷ With the distributed I/O;
- ▷ Between CPUs via global data communication;
- ▷ Between two partners in the same station;
- ▷ Between two partners in different stations;
- ▷ Via system function blocks (SFB communication).

The interface description for the system blocks described in this section can be found in library "StdLib30" under "Built In".

Section 22.5 "Initializing Modules" describes how to initialize modules or transfer data records in RUN mode using system functions.

20.4.1 System Functions for Distributed I/O

With PROFIBUS-DP, information is transferred in packets. It is sometimes necessary to transfer a particular area in one packet as associated (consistent) data. Variables, which can comprise as much as a double word, always contain consistent data. The data consistency of longer variables (arrays, structures) or data areas is 32 bytes on the S7-300 and 122 bytes on the S7-400. This data consistency is ensured when you use system functions SFC 14 DPRD_DAT and SFC 15 DPWR_DAT. Each station of distributed I/O can send so-called diagnostic data, which provides information on the status of the station.

The following SFCs can be used to address a standard distributed I/O slave:

- ▷ SFC 13 DPNRM_DG
  Read diagnostic data from a standard DP slave
- ▷ SFC 14 DPRD_DAT
  Read user data from a standard DP slave
- ▷ SFC 15 DPWR_DAT
  Write user data to a standard DP slave

The parameters for these SFCs are listed in Table 20.6.

The RECORD parameter describes the area in which the data that are read are to be stored or from which data are to be read. Actual parameters may be variables of data type ARRAY or STRUCT or an ANY pointer of data type BYTE (for example:
P#DBzDBX*y.x* BYTE *nnn*).

• Read diagnostic data
SFC 13 DPNRM_DG reads diagnostic data from a DP slave in the form stipulated by EN 50170 Volume 2, PROFIBUS. The read procedure is initiated with REQ = "1", and is terminated when BUSY = "0" is returned. Function

Table 20.6 Parameters for SFCs Used to Reference the Distributed I/O

SFC	Parameter	Declaration	Data Type	Contents, Description
13	REQ	INPUT	BOOL	Read request with REQ = "1"
	LADDR	INPUT	WORD	Configured diagnostic address (from the I area)
	RET_VAL	OUTPUT	INT	Error information
	RECORD	OUTPUT	ANY	Destination area for the diagnostic data read
	BUSY	OUTPUT	BOOL	Read still in progress when BUSY = "1"
14	LADDR	INPUT	WORD	Configured start address (from the I area)
	RET_VAL	OUTPUT	INT	Error information
	RECORD	OUTPUT	ANY	Destination area for useful data read
15	LADDR	INPUT	WORD	Configured start address (from the Q area)
	RECORD	INPUT	ANY	Source area for useful data to be written
	RET_VAL	OUTPUT	INT	Error information

value RET_VAL then contains the number of bytes read. Depending on the slave, diagnostic data may comprise from 6 bytes to 240 bytes. If there are more than 240 bytes, the first 240 bytes are transferred and the relevant overflow bit is then set in the data.

• Read user data
SFC 14 DPRD_DAT reads consistent user data (data which are associated through their contents) from a standard DP slave. You specify the length of the consistent data when you initialize the DP slave.

• Write user data
SFC 15 DPWR_DAT writes consistent user data (data which are associated through their contents) to a standard DP slave. You specify the length of the consistent data when you initialize the DP slave.

20.4.2 System Functions for GD Communication

In S7-400 systems, you can also control GD communication in your program. Additionally or alternatively to the cyclic transfer of global data, you can send or receive a GD packet with the following SFCs:

▷ SFC 60 GD_SND
Send GD packet

▷ SFC 61 GD_RCV
Receive GD packet

The parameters for these SFCs are listed in Table 20.7. The prerequisite for the use of these SFCs is a configured global data table. After compiling this table, STEP 7 shows you, in the "GD Identifier" column, the numbers of the GD

Table 20.7 SFC Parameters for GD Communication

SFC	Parameter	Declaration	Data Type	Contents, Description
60	CIRCLE_ID	INPUT	BYTE	Number of the GD circle
	BLOCK_ID	INPUT	BYTE	Number of the GD packet to be sent
	RET_VAL	OUTPUT	INT	Error information
61	CIRCLE_ID	INPUT	BYTE	Number of the GD circle
	BLOCK_ID	INPUT	BYTE	Number of the GD packet to be received
	RET_VAL	OUTPUT	INT	Error information

circles and GD packets which you need for parameter assignments (for instance, GD 1.2.3 corresponds to circle 1, packet 2, element 3).

The size of a GD packet may not exceed 22 bytes of net data on an S7-300 or 54 bytes on an S7-400, in both instances in the case of a single element of that length. An element is either an operand or an operand area, such as MB 0:15, which corresponds to memory bytes MB 0 to MB 15. The data consistency applies to an element, whereby the size of a consistent operand area is CPU-specific (8 bytes on the S7-300, 16 bytes on the S7-412/413, 32 bytes on the S7-414/416). If an operand area exceeds the length applicable for data consistency, blocks with consistent data in the applicable length are formed, beginning with the first byte.

If you want to ensure data consistency for the entire GD data packet when transferring with SFCs 60 and 61, you must disable or delay higher-priority interrupts and asynchronous errors on both the Send and Receive side during processing of the SFC 60 or SFC 61 (see section 21.7, "Handling Interrupt Events").

Do not forget about the system resources when using the SFCs for sending GD packets; under some circumstances, you will have to simulate the scan rates per program.

20.4.3 System Functions for Data Interchange within a Station

This section deals with system functions for internal station connections not configured in the connection table ("Communication via Non-Configured Connections"). These functions address nodes directly allocated to the CPU, for instance via PROFIBUS-DP. The node identification is derived from the I/O address. Use the LADDR parameter to specify the module start address and the IOID parameter to indicate whether this address is located in the input or output area.

These system functions establish the required communication links dynamically and – if specified – break the connections when the job has been executed. If a connection cannot be set up due to lack of resources in either the transmitter or the receiver, "temporary lack of resources" is signaled. The transfer must then be retried. Between two communication partners there can be no more than one connection in each direction.

By modifying the block parameters at run time, you can utilize a system function for different communication connections. An SFC may not interrupt itself. You may modify a program section in which one of these SFCs is used only in STOP mode: afterwards, you must execute a complete restart.

These SFCs transfer a maximum of 76 bytes of user data. A CPU's operating system combines the user data into blocks consistent within themselves, without regard to the direction of transfer. In S7-300 systems, these blocks are 8 bytes long, in systems with a CPU 412/413 they are 16 bytes long, and in systems with a CPU 414/415 they are 32 bytes long. When two CPUs exchange data, the block size on the "passive" CPU is decisive for data consistency.

The following system functions handle data transfers between two CPUs in the same station:

▷ SFC 72 I_GET
Read data

▷ SFC 73 I_PUT
Write data

▷ SFC 74 I_ABORT
Disconnect

The parameters for these SFCs are listed in Table 20.8.

The SD, RD and VAR_ADDR parameters describe the area from which the data to be transferred are to be read or to which the receive data are to be written. Actual parameters may be operands, variables or data areas addressed with an ANY pointer. The Send and Receive data are not checked for identical data types.

A job is initiated with REQ = "1" and BUSY = "0" ("first call"). While the job is in progress, BUSY is set to "1". Changes to the REQ parameter no longer have any effect. When the job is completed, BUSY is reset to "0". If REQ is still "1", the job is immediately restarted.

- Read data

When the read procedure has been initiated, the operating system in the partner CPU assembles and sends the requested data. An SFC call transfers the Receive data to the destination area. RET_VAL then shows the number of bytes transferred. If CONT is = "0", the communica-

Table 20.8 SFC Parameters for Internal Station Communication

Parameter	For SFC			Declaration	Data Type	Contents, Description
REQ	72	73	74	INPUT	BOOL	Initiate job with REQ = "1"
CONT	72	73	–	INPUT	BOOL	CONT = "1": Connection remains intact after job terminates
IOID	72	73	74	INPUT	BYTE	B#16#54 = Input area, B#16#55 = Output area
LADDR	72	73	74	INPUT	WORD	Module start address
VAR_ADDR	72	73	–	INPUT	ANY	Data area in partner CPU
SD	–	73	–	INPUT	ANY	Data area in own CPU, which contains the Send data
RET_VAL	72	73	74	OUTPUT	INT	Error information
BUSY	72	73	74	OUTPUT	BOOL	Job in progress when BUSY = "1"
RD	72	–	–	OUTPUT	ANY	Data area in own CPU, which will take the Receive data

tion link is broken. If CONT is = "1", the link is maintained. The data are also read when the communication partner is in STOP mode.

• Write data
When the write procedure has been initiated, the operating system transfers all data from the source area to an internal buffer on the first call, and sends them to the partner in the link. There, the receiver writes the data into data area VAR_ADDR. BUSY is then set to "0". The data are also written when the receiving partner is at STOP.

• Disconnect
REQ = "1" breaks a connection to the specified communication partner. With I_ABORT, you can break only those connections established in the same station with I_GET or I_PUT.

20.4.4 System Functions for Data Interchange Between Two Stations

This section covers system functions for external station connections not configured in the connection table ("Communication via Non-configured Connections"). These functions address nodes on the same MPI subnet. The node identification is derived from the MPI address (DEST_ID parameter).

These system functions set up the required communication links dynamically and – if specified – break them when the job has been executed. If a connection cannot be established because of a lack of resources in either the sender or the receiver, "temporary lack of resources" is reported. The transfer must then be retried. Between two communication partners, there can be only one connection in each direction.

By modifying the block parameters at run time, you can utilize a system function for different communication links. An SFC may not interrupt itself. You may modify a program section in which one of these SFCs is used only in STOP mode; a complete restart must then be executed.

These SFCs transfer a maximum of 76 bytes of user data. A CPU's operating system combines the user data into blocks consistent within themselves, without regard to the direction of transfer. In S7-300 systems, these blocks have a length of 8 bytes, in systems with a CPU 412/413 a length of 16 bytes, and in systems with a CPU 414/416 a length of 32 bytes. If two CPUs exchange data via X_GET or X_PUT, the block size of the "passive" CPU is decisive to data consistency. In the case of a SEND/RECEIVE connection, all data are consistent.

On a transition from RUN to STOP, all active connections (all SFCs except X_RECV) are cleared.

The following system functions handle data transfers between partners in different stations:

▷ SFC 65 X_SEND
Send data

▷ SFC 66 X_RCV
Receive data

▷ SFC 67 X_GET
Read data

▷ SFC 68 X_PUT
Write data

▷ SFC 69 X_ABORT
Disconnect

The parameters for these SFCs are listed in Table 20.9

The SD, RD and VAR_ADDR parameters describe the area from which the data to be sent are to be read or to which the Receive data are to be written. Actual parameters may be operands, variables, or data areas addressed with an ANY pointer. Send and Receive data are not checked for matching data types. When the Receive data are irrelevant, a "blank" ANY pointer (NIL pointer) as RD parameter in X_RCV is permissible.

A job is initiated with REQ = "1" and BUSY = "0" ("first call"). While the job is in progress, BUSY is set to "1"; changes to the REQ parameter now no longer have any effect. When the job terminates, BUSY is set back to "0". If REQ is still "1", the job is immediately restarted.

• Send data
When the Send request has been submitted, the operating system transfers all data from the source area to an internal buffer on the first call, and transfers the data to the partner CPU. BUSY is "1" for the duration of the send procedure. When the partner has signaled that it has fetched the data, BUSY is set to "0" and the send job terminated. If CONT is = "0", the connection is broken and the respective CPU resources are available to other communication links. If CONT is = "1", the connection is maintained. The REQ_ID parameter makes it possible for you to assign an ID to the Send data which you can evaluate with SFC X_RCV.

• Receive data
The Receive data are placed in an internal buffer. Multiple packets can be put in a queue in the chronological order of their arrival. Use EN_DT = "0" to check whether or not data were received; if so, NDA is "1" , RET_VAL shows the number of bytes of Receive data, and

Table 20.9 SFC Parameters for External Station Communication

Parameter	For SFC					Declaration	Data Type	Contents, Description
REQ	65	–	67	68	69	INPUT	BOOL	Job initiation with REQ = "1"
CONT	65	–	67	68	–	INPUT	BOOL	CONT = "1": Connection is maintained when job is completed
DEST_ID	65	–	67	68	69	INPUT	WORD	Partner's node identification (MPI address)
REQ_ID	65	–	–	–	–	INPUT	DWORD	Job identification
VAR_ADDR	–	–	67	68	–	INPUT	ANY	Data area in partner CPU
SD	65	–	–	68	–	INPUT	ANY	Data area in own CPU, which contains the Send data
EN_DT	-	66	–	–	–	INPUT	BOOL	When "1": Accept Receive data
RET_VAL	65	66	67	68	69	OUTPUT	INT	Error information
BUSY	65	–	67	68	69	OUTPUT	BOOL	Job in progress when BUSY = "1"
REQ_ID	–	66	–	–	–	OUTPUT	DWORD	Job identification
NDA	–	66	–	–	–	OUTPUT	BOOL	When "1": Data received
RD	–	66	67	–	–	OUTPUT	ANY	Data area in own CPU, which will accept the Receive data

REQ_ID is the same as the corresponding parameter in SFC X_SEND. When EN_DT = "1", the SFC transfers the first (oldest) packet to the destination area; NDA is then "1" and RET_VAL shows the number of bytes transferred. If EN_DT = "1" but there are no data in the internal queue, NDA is "0". On a complete restart, all data packets in the queue are rejected. In the event of a broken connection or a restart, the oldest entry in the queue, if already "queried" with EN_DT = "0", is retained; otherwise, it is rejected like the other queue entries

- Read data

When the read procedure has been initiated, the operating system in the partner CPU assembles and sends the data requested under VAR_-ADDR. On an SFC call, the Receive data are entered in the destination area. RET_VAL then shows the number of bytes transferred. If CONT is "0", the communication link is broken. If CONT is "1", the connection is maintained. The data are then read even when the communication partner is in STOP mode.

- Write data

When the write procedure has been initiated, the operating system transfers all data from the source area to an internal buffer on the first call, and sends the data to the partner CPU. There, the partner CPU's operating system writes the Receive data to the VAR_ADDR data area. BUSY is then set to "0". The data are written even when the communication partner is in STOP mode.

- Disconnect

REQ = "1" breaks an existing connection to the specified communication partner. X_ABORT can be used to break only those connections established in the CPU's own station with X_SEND, X_GET or X_PUT.

20.4.5 SFB Communication

The prerequisite for communication via system function blocks is a configured connection table ("Communication via Configured Connections"). The connection table is used to define and describe communication links.

A communication link is specified by a connection ID for each communication partner. STEP 7 assigns the connection IDs when it compiles the connection table. Use the "local ID" to initialize the SFB in the local or "own" module and the "remote ID" to initialize the SFB in the partner module.

The same logical connection can be used for different Send/Receive requests. A distinction must be made by adding a job ID to the connection ID in order to stipulate the association between Send and Receive block.

Two-way data interchange

The following SFBs are available for two-way data interchange:

▷ SFB 8 USEND
Uncoordinated sending of a data packet of CPU-specific length

▷ SFB 9 URCV
Uncoordinated receiving of a data packet of CPU-specific length

▷ SFB 12 BSEND
Sending of a data block of up to 64 Kbytes in length

▷ SFB 13 BRCV
Receiving of a data block of up to 64 Kbytes in length

SFB 8 and SFB 9 or SFB 12 and SFB 13 must always be used as a pair. The parameters for these SFBs are listed in Table 20.10.

A positive edge at the REQ (request) parameter starts the data exchange, a positive edge at the R (reset) parameter aborts it. A "1" in the EN_R (enable receive) parameter signals that the partner is ready to receive. Initialize the ID parameter with the connection ID, which STEP 7 enters in the connection table for both the local and the remote partner (the two IDs may differ). R_ID allows you to choose a specifiable but unique job ID which must be identical for the Send and the Receive block. This allows several pairs of Send and Receive blocks to share a single logical connection (as each has a unique ID). The block transfers the actual values of the ID and R_ID parameters to its instance data block on the *first call.* The first call establishes the communication relationship (for this instance) until the next complete restart.

A "1" in the DONE or NDR parameter signals the block that the job terminated without error. An error, if any, is flagged in the ERROR parameter. A value other than zero in the STATUS parameter indicates either a warning (ERROR =

Table 20.10 SFB Parameters for Sending and Receiving Data

Parameter	For SFB				Declaration	Data Type	Contents, Description
REQ	8	–	12	–	INPUT	BOOL	Start data exchange
EN_R	–	9	–	13	INPUT	BOOL	Ready to receive
R	–	–	12	–	INPUT	BOOL	Abort data exchange
ID	8	9	12	13	INPUT	WORD	Connection ID
R_ID	8	9	12	13	INPUT	DWORD	Job ID
DONE	8	–	12	–	OUTPUT	BOOL	Job terminated
NDR	–	9	–	13	OUTPUT	BOOL	New data accepted
ERROR	8	9	12	13	OUTPUT	BOOL	Error occurred
STATUS	8	9	12	13	OUTPUT	WORD	Job status
SD_1	8	–	12	–	IN_OUT	ANY	First Send area
SD_2	8	–	–	–	IN_OUT	ANY	Second Send area
SD_3	8	–	–	–	IN_OUT	ANY	Third Send area
SD_4	8	–	–	–	IN_OUT	ANY	Fourth Send area
RD_1	–	9	–	13	IN_OUT	ANY	First Receive area
RD_2	–	9	–	–	IN_OUT	ANY	Second Receive area
RD_3	–	9	–	–	IN_OUT	ANY	Third Receive area
RD_4	–	9	–	–	IN_OUT	ANY	Fourth Receive area
LEN	–	–	12	13	IN_OUT	WORD	Data block length in bytes

"0") or an error (ERROR = "1"). You must evaluate the DONE, NDR, ERROR and STATUS parameters after *every* block call.

• SFB 8 USEND and SFB 9 URCV: The SD_n and RD_n parameters are used to specify the variable or the area you want to transfer. Send area SD_n must correspond to the respective Receive area RD_n. Use the parameters without gaps, beginning with 1. No values need be specified for unneeded parameters (like an FB, not all SFB parameters need be assigned values). The first time SFB 9 is called, a Receive mailbox is generated; on all subsequent calls, the Receive data must fit into this mailbox.

• SFB 12 BSEND and SFB 13 BRCV: Enter a pointer to the first byte of the data area in parameter SD_1 or RD_1 (the length is not evaluated); the number of bytes of Send or Receive data is in the LEN parameter. Up to 64 Kbytes may be transferred; the data are transferred in blocks (sometimes called frames), and the transfer itself is asynchronous to the user program scan.

Unilateral data interchange

The following SFBs are available for unilateral data interchange:

▷ SFB 14 GET
Read data up to a CPU-specific maximum length

▷ SFB 15 PUT
Write data up to a CPU-specific maximum length

The operating system in the partner CPU collects the data read with SFB 14; the operating system in the partner CPU distributes the data written with SFB 15. A Send or Receive (user) program in the partner CPU is not required. Table 20.11 lists the parameters for these SFBs.

Table 20.11 SFB Parameters for Reading and Writing Data

Parameter	For SFB		Declaration	Data Type	Contents, Description
REQ	14	15	INPUT	BOOL	Start data exchange
ID	14	15	INPUT	WORD	Connection ID
NDR	14	–	OUTPUT	BOOL	Accept new data
DONE	–	15	OUTPUT	BOOL	Job terminated
ERROR	14	15	OUTPUT	BOOL	Error occurred
STATUS	14	15	OUTPUT	WORD	Job status
ADDR_1	14	15	IN_OUT	ANY	First data area in partner CPU
ADDR_2	14	15	IN_OUT	ANY	Second data area in partner CPU
ADDR_3	14	15	IN_OUT	ANY	Third data area in partner CPU
ADDR_4	14	15	IN_OUT	ANY	Fourth data area in partner CPU
RD_1	14	–	IN_OUT	ANY	First Receive area
RD_2	14	–	IN_OUT	ANY	Second Receive area
RD_3	14	–	IN_OUT	ANY	Third Receive area
RD_4	14	–	IN_OUT	ANY	Fourth Receive area
SD_1	–	15	IN_OUT	ANY	First Send area
SD_2	–	15	IN_OUT	ANY	Second Send area
SD_3	–	15	IN_OUT	ANY	Third Send area
SD_4	–	15	IN_OUT	ANY	Fourth Send area

A positive edge at parameter REQ (request) starts the data interchange. Set the ID parameter to the connection ID entered by STEP 7 in the connection table.

A "1" in the DONE or NDR parameter signals the block that the job terminated without error. An error, if any, is flagged with a "1" in the ERROR parameter. A value other than zero in the STATUS parameter is indicative of either a warning (ERROR = "0") or an error (ERROR = "1"). You must evaluate the DONE, NDR, ERROR and STATUS parameters after *every* block call.

Use the ADDR_n parameter to specify the variable or the area in the partner CPU from which you want to fetch or to which you want to send the data. The areas in ADDR_n must coincide with the areas specified in SD_n or RD_n. Use the parameters without gaps, beginning with 1. Unneeded parameters need not be specified (as in an FB, an SFB does not have to have values for all parameters).

Transferring print data

SFB 16 PRINT allows you to transfer a format description and data to a printer via a CP 441 communications processor. Table 20.12 lists the parameters for this SFB.

A positive edge at the REQ parameter starts the data exchange with the printer specified by the ID and PRN_NR parameters. The block signals an error-free transfer by setting DONE to "1". An error, if any, is flagged by a "1" in the ERROR parameter. A value other than zero in the STATUS parameter is indicative of either a warning (ERROR = "0") or an error (ERROR = "1"). You must evaluate the DONE, ERROR and STATUS parameters after *every* block call.

Enter the characters to be printed in STRING format in the FORMAT parameter. You can integrate as many as four format descriptions for variables in this string, defined in parameters SD_1 to SD_4. Use the parameters without gaps, beginning with 1; do not specify values

Table 20.12 Parameters for SFB 16 PRINT

Parameter	Declaration	Data Type	Contents, Description
REQ	INPUT	BOOL	Start data exchange
ID	INPUT	WORD	Connection ID
DONE	OUTPUT	BOOL	Job terminated
ERROR	OUTPUT	BOOL	Error detected
STATUS	OUTPUT	WORD	Job status
PRN_NR	IN_OUT	BYTE	Printer number
FORMAT	IN_OUT	STRING	Format description
SD_1	IN_OUT	ANY	First variable
SD_2	IN_OUT	ANY	Second variable
SD_3	IN_OUT	ANY	Third variable
SD_4	IN_OUT	ANY	Fourth variable

Table 20.13 SFB Parameters for Partner Control

Parameter	For SFB			Declaration	Data Type	Contents, Description
REQ	19	20	21	INPUT	BOOL	Start data exchange
ID	19	20	21	INPUT	WORD	Connection ID
DONE	19	20	21	OUTPUT	BOOL	Job terminated
ERROR	19	20	21	OUTPUT	BOOL	Error detected
STATUS	19	20	21	OUTPUT	WORD	Job status
PI_NAME	19	20	21	IN_OUT	ANY	Program name (P_PROGRAM)
ARG	19	–	21	IN_OUT	ANY	Irrelevant
IO_STATE	19	20	21	IN_OUT	BYTE	Irrelevant

for unneeded parameters. You can transfer up to 420 bytes (the sum of FORMAT and all variables) per print request.

Control functions

The following SFBs are available for controlling the communication partner:

▷ SFB 19 START
Execute a complete restart in the partner controller

▷ SFB 20 STOP
Switch the partner controller to STOP

▷ SFB 21 RESUME
Execute a warm restart in the partner controller

These SFBs are for unilateral data exchange. The parameters for these SFBs are listed in Table 20.13.

A positive edge at the REQ parameter starts the data exchange. Enter as ID parameter the connection ID which STEP 7 entered in the connection table.

With a "1" in the DONE parameter the block signals that the job terminated without error. An error, if any, is flagged by a "1" in the ERROR parameter. A value other than zero in the STA-

TUS parameter is indicative of either a warning (ERROR = "0") or an error (ERROR = "1"). You must evaluate the DONE, ERROR and STATUS parameters after *every* block call.

Specify as PI_NAME an array variable with the contents "P_PROGRAM" (ARRAY [1..9] OF CHAR). The ARG and IO_STATE parameters are currently irrelevant, and need not be assigned a value.

• SFB 19 START executes a complete restart of the partner CPU. Prerequisite is that the partner CPU is at STOP and that the mode selector is positioned to either RUN or RUN-P.

• SFB 20 STOP sets the partner CPU to STOP. A requirement for error-free execution of this job request is that the partner CPU not be at STOP when the request is issued.

• SFB 21 RESUME executes a warm restart of the partner CPU. Requirements are that the partner CPU is at STOP, that the mode selector is set to either RUN or RUN-P, and that a warm restart is permissible at this time.

Monitoring functions

The following system blocks are available for monitoring functions:

▷ SFB 22 STATUS
Query partner status

▷ SFB 23 USTATUS
Receive partner status

▷ SFC 62 CONTROL
Query status of an SFB instance

Table 20.14 lists the parameters for the SFBs, Table 20.15 those for SFC 62.

A positive edge at the REQ (request) parameter starts the data exchange. A "1" in the EN_R (enable receive) parameter signals readiness to receive. Enter in the ID parameter the connection ID which STEP 7 entered in the connection table.

A "1" in the NDR parameter signals the block that the job terminated without error. An error, if any, is flagged with a "1" in the ERROR parameter. A value other than zero in the STATUS parameter is indicative of either a warning (ERROR = "0") or an error (ERROR = "1"). You must evaluate the NDR, ERROR and STATUS parameters after *every* block call.

• SFB 22 STATUS fetches the status of the partner CPU and displays it in the PHYS (physical status), LOG (logical status) and LOCAL (operating status if the partner is an S7-CPU) parameters.

• SFB 23 USTATUS receives the status of the partner device, which it sends, unprompted, in the event of a change (can be configured in the connection table: select connection, EDIT → OBJECT PROPERTIES, "Send status messages"). The device status is displayed in the PHYS (physical status), LOG (logical status) and LOCAL (operating status if the partner is an S7-CPU) parameters.

• SFC 62 CONTROL determines the status of an SFB instance and the associated connection in the local controller. In the I_DB parameter, enter the SFB's instance data block. If the SFB is called as local instance, specify the number of

Table 20.14 SFB Parameters for Querying Status

Parameter	For SFB		Declaration	Data Type	Contents, Description
REQ	22	–	INPUT	BOOL	Start data exchange
EN_R	–	23	INPUT	BOOL	Ready to receive
ID	22	23	INPUT	WORD	Connection ID
NDR	22	23	OUTPUT	BOOL	Accept new data
ERROR	22	23	OUTPUT	BOOL	Error detected
STATUS	22	23	OUTPUT	WORD	Job status
PHYS	22	23	IN_OUT	ANY	Physical status
LOG	22	23	IN_OUT	ANY	Logical status
LOCAL	22	23	IN_OUT	ANY	Status of an S7 CPU as partner

Table 20.15 Parameters for SFC 62 CONTROL

Parameter	Declaration	Data Type	Contents, Description
EN_R	INPUT	BOOL	Ready to receive
I_DB	INPUT	BLOCK_DB	Instance data block
OFFSET	INPUT	WORD	Number of the local instance
RET_VAL	OUTPUT	INT	Error information
ERROR	OUTPUT	BOOL	Error detected
STATUS	OUTPUT	WORD	Status word
I_TYP	OUTPUT	BYTE	Block type identifier
I_STATE	OUTPUT	BYTE	Current status identifier
I_CONN	OUTPUT	BOOL	Connection status ("1" = connection exists)
I_STATUS	OUTPUT	WORD	STATUS parameter for SFB instance

the local instance in the OFFSET parameter (0 when no local instance, one for the first local instance, 2 for the second, and so on).

20.5 Start Information

The CPU's operating system forwards start information to organization block OB 1, as it does to every organization block, in the first 20 bytes of temporary local data. You can generate the declaration for the start information yourself or you can use information from standard library "StdLib30" under "Std OBs".

20.5.1 Start Information for OB 1

Table 20.16 shows this start information, the default symbolic designation, and the data types. You can change the designation at any time and choose more suitable names. Even if you don't use the start information, you must reserve the first 20 bytes of temporary local data for this purpose (for instance in the form of a 20-byte array).

In SIMATIC S7, all event messages have a fixed structure which is specified by the event class. The start information for OB 1, for example, reports event B#16#11 as calling of a standard OB. From the contents of the next byte you can tell whether the main program is in the first cycle after power-up and is therefore calling, for instance, initialization routines in the cyclic program.

The priority and OB number of the main program are fixed. With three INT values, the start information provides information on the cycle time of the last scan cycle and on the minimum and maximum cycle times since the last power-up. The last value, in DATE_AND_TIME format, indicates when the priority control program received the event for calling OB 1.

Note that direct reading of the start information for an organization block is possible only in that organization block because that information consists of temporary local data. If you require the start information in blocks which lie on deeper levels, call system function SFC 6 RD_SINFO at the relevant location in the program.

20.5.2 Reading Out Start Information

System function SFC 6 RD_SINFO makes the start information on the current organization block (that is, the OB at the top of the call tree) and on the start-up OB last executed available to you even in a deeper call level (Table 20.17). SFC 6 RD_SINFO can not only be called at any location in the main program but in every priority class, even in an error organization block or in the start-up routine.

Output parameter TOP_SI contains the first 12 bytes of start information on the current OB, and output parameter START_UP_SI contains the first 12 bytes of start information on the last start-up OB executed. There is no time stamp in either case. If SFC 6 is called in the start-up routine, both parameters contain the same start information.

Table 20.16 Start Information for OB 1

Name	Data Type	Description	Contents
OB1_EV_CLASS	BYTE	Event class	B#16#11 = Call standard OB
OB1_STRT_INFO	BYTE	Start information	B#16#01 = 1st cycle after complete restart B#16#02 = 1st cycle after warm restart B#16#03 = Every other cycle
OB1_PRIORITY	BYTE	Priority	B#16#01
OB1_OB_NUMBR	BYTE	OB number	B#16#01
OB1_RESERVED_1	BYTE	Reserved	–
OB1_RESERVED_2	BYTE	Reserved	–
OB1_PREV_CYCLE	INT	Previous scan cycle time	in ms
OB1_MIN_CYCLE	INT	Minimum scan cycle time	in ms
OB1_MAX_CYCLE	INT	Maximum scan cycle time	in ms
OB1_DATE_TIME	DT	Event occurrence	cyclic

Table 20.17 Parameters for SFC 6 RD_SINFO

Parameter	Declaration	Data Type	Contents, Description
RET_VAL	OUTPUT	INT	Error information
TOP_SI	OUTPUT	STRUCT	Start information for the current OB
START_UP_SI	OUTPUT	STRUCT	Start information for the last OB started

21 Interrupt Handling

Interrupt handling is always *event-driven*. When such an event occurs, the operating system interrupts scanning of the main program and calls the routine allocated to this particular event. When this routine has executed, the operating system resumes scanning of the main program at the point of interruption. Such an interruption can take place after every operation (statement).

Applicable events may be interrupts and errors. The order in which virtually simultaneous interrupt events are handled is regulated by a priority scheduler. Each event has a particular servicing priority. In S7-300 systems, the priorities are fixed; in S7-400 systems, you yourself can determine the priority class for an interrupt. Several interrupt events can be combined into priority classes.

Every routine associated with an interrupt event is written in an organization block in which additional blocks can be called. A higher-priority event interrupts execution of the routine in an organization block with a lower priority. You can affect the interruption of a program by higher-priority events using system functions.

21.1 General Remarks

SIMATIC S7 provides the following interrupt events (interrupts):

▷ Hardware interrupt
Interrupt from a module, either via an input derived from a process signal or generated on the module itself

▷ Watchdog interrupt
An interrupt generated by the operating system at periodic intervals

▷ Time-of-day interrupt
An interrupt generated by the operating system at a specific time of day, either once only or periodically

▷ Time-delay interrupt
An interrupt generated after a specific amount of time has passed; a system function call determines the instant at which this time period begins

▷ Multiprocessor interrupt
An interrupt generated by another CPU in a multiprocessor network

Other interrupt events are the synchronous errors which may occur in conjunction with program scanning and the asynchronous errors, such as the diagnostic interrupt. The handling of these events is discussed in Chapter 23, “Error Handling”.

21.1.1 Priorities

An event with a higher priority interrupts a program being processed with lower priority because of another event. The main program has the lowest priority (priority class 1), asynchronous errors the highest (priority class 26), apart from the start-up routine. All other events are in the intervening priority classes. In S7-300 systems, the priorities are fixed; in S7-400 systems, you can change the priorities by parameterizing the CPU accordingly.

An overview of all priority classes, together with the default organization blocks for each, is presented in section 3.1.2, “Priority Classes”.

21.1.2 Disabling Interrupts

The organization blocks for event-driven program scanning can be disabled and enabled with system functions SFC 39 DIS_IRT and SFC 40 EN_IRT and delayed and enabled with SFC 41 DIS_AIRT and SFC 42 EN_AIRT (see section 21.7, “Interrupt Handling”).

21.1.3 Current Signal States

When an event-driven program is called, the process images are not automatically updated.

If you want to work with the input signal states current at the time of the call (and not with the signal states from the start of the cyclic program), you must scan the input modules directly (with the MOVE box via peripheral input area PI) or, in S7-400 systems, you must define a subprocess image and have it updated at the beginning of the event-driven program.

If you want to set or reset outputs immediately in the event-driven program (instead of at the end of the main program), you must address the output modules directly (with the MOVE box via peripheral output area PQ) or, in S7-400 systems, you must define a subprocess image and have it updated at the end of the event-driven program. Direct addressing of the peripheral output area automatically updates the process-image output table.

Note that updating of the inputs and outputs in the event-driven program is asynchronous to scanning of the main program in both cases. For example, this may result in an input or output having a different signal state in the main program prior to interrupt servicing than after interrupt servicing, which in turn would mean that you might have different input or output signal states in a single main program scan cycle.

21.1.4 Start Information, Temporary Local Data

Table 21.1 provides an overview of the start information for the event-driven organization blocks. In S7-300 systems, the available temporary local data have a fixed length of 256 bytes. In S7-400 systems, you can specify the length per priority class by parameterizing the CPU accordingly (parameter block "priority classes"), whereby the total may not exceed a CPU-specific maximum. Note that the minimum number of bytes for temporary local data for the priority class used must be 20 bytes so as to be able to accommodate the start information. Specify zero for unused priority classes.

21.2 Hardware Interrupts

Hardware interrupts are used to enable the immediate detection in the user program of events in the controlled process, making it possible to respond with an appropriate interrupt handling routine. STEP 7 provides organization blocks OB 40 to OB 47 for servicing hardware interrupts; which of these eight organization blocks are actually available, however, depends on the CPU.

Table 21.1 Start Information for Interrupt Organization Blocks

Byte	Multiprocessor Interrupt OB 60	Hardware Interrupts OB 40 to OB 47	Watchdog Interrupts OB 30 to OB 38	Time-Delay Interrupt OB 20 to OB 23	Time-of-Day Interrupts OB 10 to OB 17
0	Event class	Event class	Event class	Event class	Event class
1	Start event	Start event	Start event	Start event	Start event
2	Priority class	Priority class	Priority class	Priority class	Priority class
3	OB number	OB number	OB number	OB number	OB number
4	–	–	–	–	–
5	–	Address Identifier	–	–	–
6..7	Job identifier (INT)	Module start address (WORD)	Phase offset in ms (WORD)	Job identifier (WORD)	Interval (WORD)
8..9	–	Hardware interrupt	–	Expired delay	–
10.. 11	–	information (DWORD)	Time cycle in ms (INT)	(TIME)	–
12.. 19	Event instant (DT)	Event instant (DT)	Event instant (DT)	Event instant (DT)	Event instant (DT)

Hardware interrupt handling is programmed in the hardware configuration data. With system functions SFC 55 WR_PARM, SFC 56 WR_DPARM and SFC 57 PARM_MOD, you can (re)parameterize the modules with hardware interrupt capability even in RUN mode.

21.2.1 Generating a Hardware Interrupt

A hardware interrupt is generated on the modules with this capability. This could, for example, be a digital input module that detects a signal from the process or a function module that generates a hardware interrupt because of an activity taking place on the module.

By default, hardware interrupts are disabled. A parameter is used to enable servicing of a hardware interrupt (static parameter), and you can specify whether the hardware interrupt should be generated for a coming event, a leaving event, or both (dynamic parameter). Dynamic parameters are parameters which you can modify at runtime using SFCs.

The hardware interrupt is acknowledged on the module when the organization block containing the service routine for that interrupt has finished executing.

Resolution on the S7-300

If an event occurs during execution of a hardware interrupt OB which itself would trigger generation of the same hardware interrupt, that hardware interrupt will be lost when the event that triggered it is no longer present following acknowledgment. It makes no difference whether the event comes from the module whose hardware interrupt is currently being serviced or from another module.

A diagnostic interrupt can be generated while a hardware interrupt is being serviced. If another hardware interrupt occurs on the same channel between the time the first hardware interrupt was generated and the time that interrupt was acknowledged, the loss of the latter interrupt is reported via a diagnostic interrupt to system diagnostics.

Resolution on the S7-400

If during execution of a hardware interrupt OB an event occurs on the same channel on the same module which would trigger the same hardware interrupt, that interrupt is lost. If the event occurs on another channel on the same module or on another module, the operating system restarts the OB as soon as it has finished executing.

21.2.2 Servicing Hardware Interrupts

Checking interrupt information

The start address of the module that triggered the hardware interrupt is in bytes 6 and 7 of the hardware interrupt OB's start information. If this address is an input address, byte 5 of the start information contains B#16#54; otherwise it contains B#16#55. If the module in question is a digital input module, bytes 8 to 11 contain the status of the inputs; for any other type of module, these bytes contain the interrupt status of the module.

Interrupt handling in the start-up routine

In the start-up routine, the modules do not generate hardware interrupts. Interrupt handling begins with the transition to RUN mode. Any hardware interrupts pending at the time of the transition are lost.

Error handling

If a hardware interrupt is generated for which there is no hardware interrupt OB in the user program, the operating system calls OB 85 (program execution error). The hardware interrupt is acknowledged. If OB 85 has not been programmed, the CPU goes to STOP.

Hardware interrupts deselected when the CPU was parameterized cannot be serviced, even when the OBs for these interrupts have been programmed. The CPU goes to STOP.

Disabling, delaying and enabling

Calling of the hardware interrupt OBs can be disabled and enabled with system functions SFC 39 DIS_IRT and SFC 40 EN_IRT, and delayed and enabled with SFC 41 DIS_AIRT and SFC 42 EN_AIRT.

21.2.3 Configuring Hardware Interrupts with STEP 7

Hardware interrupts are programmed in the hardware configuration data. Open the selected CPU with EDIT → OBJECT PROPERTIES and choose the "Interrupts" tab in the dialog box.

In S7-300 systems, the default priority for OB 40 is 16, and cannot be changed. In S7-400 systems, you can choose a priority between 2 and 24 for every possible OB (on a CPU-specific basis); priority 0 deselects execution of an OB. You should never assign the same priority twice because interrupts can be lost when more than 12 interrupt events with the same priority occur simultaneously.

You must also enable the triggering of hardware interrupts on the respective modules. To this purpose, these modules are parameterized much the same as the CPU.

When it saves the hardware configuration, STEP 7 writes the compiled data to the *System Data* object in off-line user program *Blocks;* from here, you can load the parameterization data into the CPU while the CPU is in STOP mode. The parameterization data for the CPU go into force immediately following loading; the parameterization data for the modules take effect after the next start-up.

21.3 Watchdog Interrupts

A watchdog interrupt is an interrupt which is generated at periodic intervals and which initiates execution of a watchdog interrupt OB. A watchdog interrupt allows you to execute a particular program periodically, independently of the processing time of the cyclic program.

In STEP 7, organization blocks OB 30 to OB 38 have been set aside for watchdog interrupts; which of these nine organization blocks are actually available depends on the CPU used.

Watchdog interrupt handling is set in the hardware configuration data when the CPU is parameterized.

21.3.1 Handling Watchdog Interrupts

Triggering watchdog interrupts in an S7-300 system

In an S7-300 system, the organization block for servicing watchdog interrupts is OB 35, which has the priority 12. You can set the interval in the range from 1 millisecond to 1 minute, in 1-millisecond increments, by parameterizing the CPU accordingly.

Triggering watchdog interrupts in an S7-400 system

You define a watchdog interrupt when you parameterize the CPU. A watchdog interrupt has three parameters: the interval, the phase offset, and the priority. You can set all three. Specifiable values for interval and phase offset are from 1 millisecond to 1 minute, in 1-millisecond increments; the priority may be set to a value between 2 and 24 or zero, depending on the CPU (zero means the watchdog interrupt is not active).

STEP 7 provides the organization blocks listed in Table 21.2, in their maximum configurations.

Table 21.2 Defaults for Watchdog Interrupts

OB	Time Interval	Phase	Priority
30	5 s	0 ms	7
31	2 s	0 ms	8
32	1 s	0 ms	9
33	500 ms	0 ms	10
34	200 ms	0 ms	11
35	100 ms	0 ms	12
36	50 ms	0 ms	13
37	20 ms	0 ms	14
38	10 ms	0 ms	15

Phase offset

The phase offset can be used to stagger the execution of watchdog interrupt handling routines despite the fact that these routines are timed to a multiple of the same interval. Use of the phase offset would put less of a time load on the main program.

The start time of the time intervals and the phase offset is the instant of transition from STARTUP to RUN. The call instant for a watchdog interrupt OB is thus the time interval plus the phase offset. Figure 21.1 shows an example of this. No phase offset is set for time interval 1, time interval 2 is twice as long as time interval 1. Because of time interval 2's phase offset, the OBs for time interval 2 and those for time interval 1 are not called simultaneously. The lower-priority OB thus need not wait, and can precisely maintain its time interval.

Performance characteristics during startup

Watchdog interrupts cannot be serviced in the start-up OB. The time intervals do not begin until a transition is made to RUN mode.

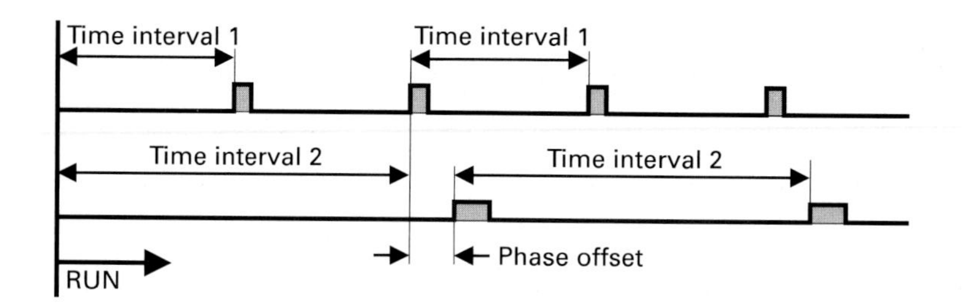

Figure 21.1
Example of Phase Offset for Watchdog Interrupts

Error characteristics

When the same watchdog interrupt is generated again while the associated watchdog interrupt handling OB is still executing, the operating system calls OB 80 (timing error). If OB 80 has not been programmed, the CPU goes to STOP.

The operating system saves the watchdog interrupt that was not serviced, servicing it at the next opportunity. Only one unserviced watchdog interrupt is saved per priority class, regardless of how many unserviced watchdog interrupts accumulate.

Watchdog interrupts that were deselected when the CPU was parameterized cannot be serviced, even when the corresponding OB is available. The CPU goes to STOP in this case.

Disabling, delaying and enabling

Calling of the watchdog interrupts can be disabled and enabled with system functions SFC 39 DIS_IRT and SFC 40 EN_IRT and delayed and enabled with SFC 41 DIS_AIRT and SFC 42 EN_AIRT.

21.3.2 Configuring Watchdog Interrupts with STEP 7

Watchdog interrupts are configured via the hardware configuration data. Simply open the selected CPU with EDIT → OBJECT PROPERTIES and choose the "Cyclic Interrupt" tab from the dialog box.

In S7-300 systems, the processing priority is permanently set to 12. In S7-400 systems, you may set a priority between 2 and 24 for each possible OB (CPU-specific); priority 0 deselects the OB to which it is assigned. You should not assign a priority more than once, as interrupts might be lost if more than 12 interrupt events with the same priority occur simultaneously.

The interval for each OB is selected under "Execution", the delayed call instant under "Phase Offset".

When it saves the hardware configuration, STEP 7 writes the compiled data to the *System Data* object in the off-line user program *Blocks*. From here, you can load the parameter assignment data into the CPU while the CPU is at STOP; the data take effect immediately.

21.4 Time-of-Day Interrupts

Time-of-day interrupts are used when you want to run a program at a particular time, either once only or periodically, for instance daily. In STEP 7, organization blocks OB 10 to OB 17 are provided for servicing time-of-day interrupts; which of these eight organization blocks are actually available depends on the CPU used.

You can configure the time-of-day interrupts in the hardware configuration or control them at runtime via the program using system functions. The prerequisite for proper handling of the time-of-day interrupts is a correctly set real-time clock on the CPU.

21.4.1 Handling Time-of-Day Interrupts

General remarks

To start a time-of-day interrupt, you must first set the start time, then activate the interrupt. You can perform the two activities separately via the hardware configuration data or using SFCs. Note that when activated via the hardware configuration data, the time-of-day interrupt is started automatically following parameterization of the CPU.

You can start a time-of-day interrupt in two ways:

- Single-shot: the relevant OB is called once only at the specified time, or

- Periodically: depending on the parameter assignments, the relevant OB is started every minute, hourly, daily, weekly, monthly or yearly.

Following a single-shot time-of-day interrupt OB call, the time-of-day interrupt is canceled. You can also cancel a time-of-day interrupt with SFC 29 CAN_TINT.

If you want to once again use a canceled time-of-day interrupt, you must set the start time again, then reactivate the interrupt.

You can check the status of a time of day interrupt with SFC 31 QRY_TINT.

Performance characteristics during startup

During a complete restart, the operating system clears all settings made with SFCs. Settings made via the hardware configuration data are retained. On a warm restart, the CPU resumes servicing of the time-of-day interrupts in the first complete scan cycle of the main program.

You can check the status of the time-of-day interrupts in the start-up OB by calling SFC 31, and subsequently cancel or reset and reactivate the interrupts. The time-of-day interrupts are serviced only in RUN mode.

Error characteristics

If a time-of-day interrupt OB is called but was not programmed, the operating system calls OB 85 (program execution error). If OB 85 was not programmed, the CPU goes to STOP.

Time-of-day interrupts that were deselected when the CPU was parameterized cannot be serviced, even when the relevant OB is available. The CPU goes to STOP.

If you activate a time-of-day interrupt on a single-shot basis, and if the start time has already passed (from the real-time clock's point of view), the operating system calls OB 80 (timing error). If OB 80 is not available, the CPU goes to STOP.

If you interrupt a time-of-day interrupt on a periodic basis, and if the start time has already passed (from the real-time clock's point of view), the time-of-day interrupt OB is executed the next time that time period comes due.

If you set the real-time clock ahead, whether for the purpose of correction or synchronization, thus skipping over the start time for the time-of-day interrupt, the operating system calls OB 80 (timing error). The time-of-day interrupt OB is then executed precisely once.

If you set the real-time clock back, whether for the purpose of correction or synchronization, an activated time-of-day interrupt OB will no longer be executed at the instants which are already past.

If a time-of-day interrupt OB is still executing when the next (periodic) call occurs, the operating system invokes OB 80 (timing error). When OB 80 and the time-of-day interrupt OB have executed, the time-of-day interrupt OB is restarted.

Periodic time-of-day interrupts must correspond to an actual date. For example, monthly repetitions on the 31st day of each month will be carried out only in months having 31 days.

Disabling, delaying and enabling

Time-of-day interrupt OB calls can be disabled and enabled with SFC 39 DIS_IRT and SFC 40 EN_IRT, and delayed and enabled with SFC 41 DIS_AIRT and SFC 42 EN_AIRT.

21.4.2 Configuring Time-of-Day Interrupts with STEP 7

The time-of-day interrupts are configured via the hardware configuration data. Open the selected CPU with EDIT → OBJECT PROPERTIES and choose the "Time-of-Day Interrupts" tab from the dialog box.

In S7-300 systems, the processing priority is permanently set to 2. In S7-400 systems, you can set a priority between 2 and 24, depending on the CPU, for each possible OB; priority 0 deselects a particular OB. You should not assign a priority more than once, as interrupts might be lost when more than 12 interrupt events with the same priority occur simultaneously.

The "Active" option activates automatic starting of the time-of-day interrupt. The "Execution" option screens a list which allows you to choose whether you want the OB to execute on a single-shot basis or at specific intervals. The final parameter is the start time (date and time).

When it saves the hardware configuration, STEP 7 writes the compiled data to the *System Data* object in the off-line user program *Blocks*. From here, you can load the parameter assignment data into the CPU while the CPU is at STOP; these data then go into force immediately.

Table 21.3 SFC Parameters for Time-of-Day Interrupts

SFC	Parameter	Declaration	Data Type	Contents, Description
28	OB_NR	INPUT	INT	Number of the OB to be called at the specified time on a single-shot basis or periodically
	SDT	INPUT	DT	Start date and start time in the format DATE_AND_TIME
	PERIOD	INPUT	WORD	Period on which start time is based: W#16#0000 = Single-shot W#16#0201 = Every minute W#16#0401 = Hourly W#16#1001 = Daily W#16#1201 = Weekly W#16#1401 = Monthly W#16#1801 = Yearly
	RET_VAL	OUTPUT	INT	Error information
29	OB_NR	INPUT	INT	Number of the OB whose start time is to be deleted
	RET_VAL	OUTPUT	INT	Error information
30	OB_NR	INPUT	INT	Number of the OB to be activated
	RET_VAL	OUTPUT	INT	Error information
31	OB_NR	INPUT	INT	Number of the OB whose status is to be checked
	RET_VAL	OUTPUT	INT	Error information
	STATUS	OUTPUT	WORD	Status of the time-of-day interrupt

21.4.3 System Functions for Time-of-Day Interrupts

The following system functions can be used for time-of-day interrupt control:

▷ SFC 28 SET_TINT
Set time-of-day interrupt

▷ SFC 29 CAN_TINT
Cancel time-of-day interrupt

▷ SFC 30 ACT_TINT
Activate time-of-day interrupt

▷ SFC 31 QRY_TINT
Check time-of-day interrupt

The parameters for these system functions are listed in Table 21.3.

• Set time-of-day interrupt
You determine the start time for a time-of-day interrupt by calling system function SFC 28 SET_TINT. SFC 28 sets only the start time; to start the time-of-day interrupt OB, you must activate the time-of-day interrupt with SFC 30 ACT_TINT. Specify the start time in the SDT parameter in the format DATE_AND_TIME, for instance DT#1997-06-30-08:30. The operating system ignores seconds and milliseconds and sets these values to zero. Setting the start time will overwrite the old start time value, if any. An active time-of-day interrupt is canceled, that is, it must be reactivated.

• Activate tine-of-day interrupt
A time-of-day interrupt is activated by calling system function SFC 30 ACT_TINT. When a TOD interrupt is activated, it is assumed that a time has been set for the interrupt. If, in the case of a single-shot interrupt, the start time is already past, SFC 30 reports an error. In the case of a periodic start, the operating system calls the relevant OB at the next applicable time. Once a single-shot time-of-day interrupt has been serviced, it is, for a all practical purposes, canceled. You can reset and reactivate it (for a different start time) if desired.

• Cancel time-of-day interrupt
You can delete a start time, thus deactivating the time-of-day interrupt, with system function SFC 29 CAN_TINT. The respective OB is no longer called. If you want to use this same time-of-day interrupt again, you must first set the start time, then activate the interrupt.

• Check time-of-day interrupt
You can check the status of a time-of-day interrupt by calling system function SFC 31

Table 21.4
STATUS Parameter for SFC 31 QRY_TINT

Bit	Meaning when bit = "1"
0	TOD interrupt disabled by operating system
1	New TOD interrupt rejected
2	TOD interrupt not activated and not expired
3	–
4	TOD interrupt OB loaded
5	No disable
6 ...	–

QRY_TINT. The required information is returned in the STATUS parameter (see Table 21.4).

21.5 Time-Delay Interrupts

A time-delay interrupt allows you to implement a time delay independently of the standard timers. In STEP 7, organization blocks OB 20 to OB 23 are set aside for time-delay interrupts; which of these four organization blocks are actually available depends on the CPU used.

The priorities for time-delay interrupt OBs are programmed in the hardware configuration data; system functions are used for control purposes.

21.5.1 Handling Time-Delay Interrupts

General remarks

A time-delay interrupt is started by calling SFC 32 SRT_DINT; this system function also passes the delay interval and the number of the selected organization block to the operating system. When the delay interval has expired, the OB is called.

You can cancel servicing of a time-delay interrupt, in which case the associated OB will no longer be called.

You can check the status of a time-delay interrupt with SFC 34 QRY_DINT.

Performance characteristics during startup

On a complete restart, the operating system deletes all programmed settings for time-delay interrupts. On a warm restart, the settings are retained until processed in RUN mode, whereby the "residual cycle" is counted as part of the start-up routine.

You can start a time-delay interrupt in the start-up routine by calling SFC 32. When the delay interval has expired, the CPU must be in RUN mode in order to be able to execute the relevant organization block. If this is not the case, the CPU waits to call the organization block until the start-up routine has terminated, then calls the time-delay interrupt OB before the first network in the main program.

Error characteristics

If no time-delay interrupt OB was programmed, the operating system calls OB 85 (program execution error). If there is no OB 85 in the user program, the CPU goes to STOP.

If the delay interval has expired and the associated OB is still executing, the operating system calls OB 80 (timing error) or goes to STOP if there is no OB 80 in the user program.

Time-delay interrupts which were deselected during CPU parameterization cannot be serviced, even when the respective OB has been programmed. The CPU goes to STOP.

Disabling, delaying and enabling

The time-delay interrupt OBs can be disabled and enabled with system functions SFC 39 DIS_IRT and SFC 40 EN_IRT, and delayed and enabled with SFC 41 DIS_AIRT and SFC 42 EN_AIRT.

21.5.2 Configuring Time-Delay Interrupts with STEP 7

Time-delay interrupts are configured in the hardware configuration data. Simply open the selected CPU with EDIT → OBJECT PROPERTIES and choose the "Interrupts" tab from the dialog box.

In S7-300 systems, the priority is permanently preset to 3. In S7-400 systems, you can choose a priority between 2 and 24, depending on the CPU, for each possible OB; choose priority 0 to deselect an OB. You should not assign a priority more than once, as interrupts could be lost if more than 12 interrupt events with the same priority occur simultaneously.

When it saves the hardware configuration, STEP 7 writes the compiled data to the *System Data* object in the off-line user program *Blocks*. From here, you can transfer the parameter assignment data while the CPU is at stop; the data take effect immediately.

21.5.3 System Functions for Time-Delay Interrupts

A time-delay interrupt can be controlled with the following system functions:

▷ SFC 32 SRT_DINT
Start time-delay interrupt

▷ SFC 33 CAN_DINT
Cancel time-delay interrupt

▷ SFC 34 QRY_DINT
Check time-delay interrupt

The parameters for these system functions are listed in Table 21.5.

• Start time-delay interrupt
A time-delay interrupt is started by calling system function SFC 32 SRT_DINT. The SFC call is also the start time for the programmed delay interval. When the delay interval has expired, the CPU calls the programmed OB and passes the time delay value and a job identifier in the start information for this OB. The job identifier is specified in the SIGN parameter for SFC 32; you can read the same value in bytes 6 and 7 of the start information for the associated time-delay interrupt OB. The time delay is set in increments of 1 ms. The accuracy of the time delay is also 1 ms. Note that execution of the time-delay interrupt OB may itself be delayed if organization blocks with higher priorities are being processed when the time-delay interrupt OB is called. You can overwrite a time delay with a new value by re-calling SFC 32. The new time delay goes into force with the SFC call.

• Cancel time-delay interrupt
You can call system function SFC 33 CAN_-DINT to cancel a time-delay interrupt, in which case the programmed organization block is not called.

• Check time-delay interrupt
System function SFC 34 QRY_DINT informs you about the status of a time-delay interrupt. You select the time-delay interrupt via the OB number, and the status information is returned in the STATUS parameter (Table 21.6).

Table 21.6
STATUS Parameter in SFC 23 QRY_DINT

Bit	Meaning when bit = "1"
0	Time-delay interrupt disabled by operating system
1	New time-delay interrupt rejected
2	Time-delay interrupt activated and not expired
3	–
4	Time-delay OB loaded
5	No disable
6 ...	–

Table 21.5 SFC Parameters for Time-Delay Interrupts

SFC	Parameter	Declaration	Data Type	Contents, Description
32	OB_NR	INPUT	INT	Number of the OB to be called when the delay interval has expired
	DTIME	INPUT	TIME	Delay interval; permissible: T#1ms to T#1m
	SIGN	INPUT	WORD	Job identification in the respective OB's start information when the OB is called (arbitrary characters)
	RET_VAL	OUTPUT	INT	Error information
33	OB_NR	INPUT	INT	Number of the OB to be canceled
	RET_VAL	OUTPUT	INT	Error information
34	OB_NR	INPUT	INT	Number of the OB whose status is to be checked
	RET_VAL	OUTPUT	INT	Error information
	STATUS	OUTPUT	WORD	Status of the time-delay interrupt

21.6 Multiprocessor Interrupt

The multiprocessor interrupt allows a synchronous response to an event in all CPUs in multiprocessor mode. A multiprocessor interrupt is triggered using SFC 35 MP_ALM. Organization block OB 60, which has a fixed priority of 25, is the OB used to service a multiprocessor interrupt.

General remarks

An SFC 35 MP_ALM call initiates execution of the multiprocessor interrupt OB. If the CPU is in single-processor mode, OB 60 is started immediately. In multiprocessor mode, OB 60 is started simultaneously on all participating CPUs, that is to say, even the CPU in which SFC 35 was called waits before calling OB 60 until all the other CPUs have indicated that they are ready.

The multiprocessor interrupt is not programmed in the hardware configuration data; it is already present in every CPU with multicomputing capability. Despite this fact, however, a sufficient number of local data bytes (at least 20) must still be reserved in the CPU's "Local Data" tab under priority class 25.

Performance characteristics during startup

The multiprocessor interrupt is triggered only in RUN mode. An SFC 35 call in the start-up routine terminates after returning error 32 929 (W#16#80A1) as function value.

Error characteristics

If OB 60 is still in progress when SFC 35 is recalled, the system function returns error code 32 928 (W#16#80A0) as function value. OB 60 is not started in any of the CPUs.

The unavailability of OB 60 in one of the CPUs at the time it is called or the disabling or delaying of its execution by system functions has no effect, nor does SFC 35 report an error.

Disabling, delaying and enabling

The multiprocessor OB can be disabled and enabled with system functions SFC 39 DIS_IRT and SFC 40 EN_IRT, and delayed and enabled with system functions SFC 41 DIS_AIRT and SFC 42 EN_AIRT.

System functions for the multiprocessor interrupt

A multiprocessor interrupt is triggered with system function SFC 35 MP_ALM. Its parameters are listed in Table 21.7.

The JOB parameter allows you to forward a job identifier. The same value can be read in bytes 6 and 7 of OB 60's start information in all CPUs.

21.7 Handling Interrupts

The following system functions are available for handling interrupts and asynchronous errors:

▷ SFC 39 DIS_IRT
Disable interrupts

▷ SFC 40 EN_IRT
Enable disabled interrupts

▷ SFC 41 DIS_AIRT
Delay interrupts

▷ SFC 42 EN_AIRT
Enable delayed interrupts

These system functions affect all interrupts and all asynchronous errors. Table 21.8 lists the parameters for these system functions. System functions SFC 36 to SFC 38 are provided for handling synchronous errors.

21.7.1 Disabling Interrupts

System function SFC 39 DIS_IRT disables servicing of new interrupts and asynchronous errors. All new interrupts and asynchronous errors are rejected. If an interrupt or asynchronous error occurs following a Disable, the associated organization block is not executed; if the OB does not exist, the CPU does *not* go to STOP.

The Disable remains in force for all priority classes until it is revoked with SFC 40 EN_IRT. After a complete restart, all interrupts and asynchronous errors are enabled.

Table 21.7 Parameters for SFC 35 MP_ALM

Parameter	Declaration	Data Type	Contents, Description
JOB	INPUT	BYTE	Job identification in the range B#16#00 to B#16#0F
RET_VAL	OUTPUT	INT	Error information

Table 21.8 SFC Parameters for Interrupt Handling

SFC	Parameter	Declaration	Data Type	Contents, Description
39	MODE	INPUT	BYTE	Disable mode (see text)
	OB_NR	INPUT	INT	OB number (see text)
	RET_VAL	OUTPUT	INT	Error information
40	MODE	INPUT	BYTE	Enable mode (see text)
	OB_NR	INPUT	INT	OB number (see text)
	RET_VAL	OUTPUT	INT	Error information
41	RET_VAL	OUTPUT	INT	(New) number of delays
42	RET_VAL	OUTPUT	INT	Number of delays remaining

The MODE and OB_NR parameters are used to specify which interrupts and asynchronous errors are to be disabled. MODE = B#16#00 disables all interrupts and asynchronous errors. MODE = B#16#01 disables an interrupt class whose first OB number is specified in the OB_NR parameter. For example, MODE = B#16#01 and OB_NR = 40 disables all hardware interrupts; OB = 80 would disable all asynchronous errors. MODE = B#16#02 disables the interrupt or asynchronous error whose OB number you entered in the OB_NR parameter.

Regardless of a Disable, the operating system enters each new interrupt or asynchronous error in the diagnostic buffer.

21.7.2 Enabling Disabled Interrupts

System function SFC 40 EN_IRT enables the interrupts and asynchronous errors disabled with SFC 39 DIS_IRT. An interrupt or asynchronous error occurring after the Enable will be serviced by the associated organization block; if that organization block is not in the user program, the CPU goes to STOP (except in the case of OB 81, the organization block for power supply errors).

The MODE and OB_NR parameters specify which interrupts and asynchronous errors are to be enabled. MODE = B#16#00 enables all interrupts and asynchronous errors. MODE = B#16#01 enables an interrupt class whose first OB number is specified in the OB_NR parameter. MODE = B#16#02 enables the interrupt or asynchronous error whose OB number you entered in the OB_NR parameter.

21.7.3 Delaying Interrupt Events

System function SFC 41 DIS_AIRT delays the servicing of higher-priority new interrupts and asynchronous errors. Delay means that the operating system saves the interrupts and asynchronous errors which occurred during the delay and services them when the delay interval has expired. Once SFC 41 has been called, the program in the current organization block (in the current priority class) will not be interrupted by a higher-priority interrupt; no interrupts or asynchronous errors are lost.

A delay remains in force until the current OB has terminated its execution or until SFC 42 EN_AIRT is called.

You can call SFC 41 several times in succession. The RET_VAL parameter shows the number of calls. You must call SFC 42 exactly the same number of times as SFC 41 in order to reenable the interrupts and asynchronous errors.

21.7.4 Enabling Delayed Interrupts

System function SFC 42 EN_AIRT reenables the interrupts and asynchronous errors delayed with SFC 41. You must call SFC 42 exactly the same number of times as you called SFC 41 (in the current OB). The RET_VAL parameter shows the number of delays still in force; if RET_VAL is = 0, the interrupts and asynchronous errors have been reenabled.

If you call SFC 42 without first having called SFC 41, RET_VAL contains the value 32896 (W#16#8080).

22 Start-Up Characteristics

22.1 General Remarks

22.1.1 Operating Modes

Before the CPU begins processing the main program following power-up, it executes a start-up routine. STARTUP is one of the CPU's operating modes, as is STOP or RUN. This chapter describes the CPU's activities on a transition from and to STARTUP and in the start-up routine itself.

Following power-up ①, the CPU is in the STOP mode (Figure 22.1). If the keyswitch on the CPU's front panel is at RUN or RUN-P, the CPU switches to STARTUP mode ②, then to RUN mode ③. If an "unrecoverable" error occurs while the CPU is in STARTUP or RUN mode or if you position the keyswitch to STOP, the CPU returns to the STOP mode ④ ⑤.

The user program is tested in "single-step" operation in the HOLD mode (S7-400 only and STL only). You can switch to this mode from both RUN and STARTUP, and return to the original mode when you abort the test ⑥ ⑦. You can also set the CPU to the STOP mode from the HOLD mode ⑧.

When you parameterize the CPU, you can define start-up characteristics with the "Startup" tab such as the maximum permissible amount of time for the Ready signals from the modules following power-up, or whether the CPU is to start up when the configuration data do not coincide with the actual configuration, or what mode the CPU startup is to be in.

SIMATIC S7 has two start-up modes: complete restart and warm restart. On a complete restart, the main program is always processed from the beginning. A warm restart resumes the main program at the point of interruption, and "finishes" the cycle. S7-300 systems support complete restart; S7-400 systems support both modes.

You can scan a program on a single-shot basis in STARTUP mode. STEP 7 provides organization blocks OB 100 (complete restart) and OB 101 (warm restart) expressly for this purpose. Sample applications are the parameterization of modules unless this was already taken care of by the CPU, and the programming of defaults for your main program.

22.1.2 HOLD Mode

In HOLD mode, the output modules are disabled. Writing to the modules affects the module memory, but does not switch the signal states "out" to the module outputs. The modules are not reenabled until you exit the HOLD mode.

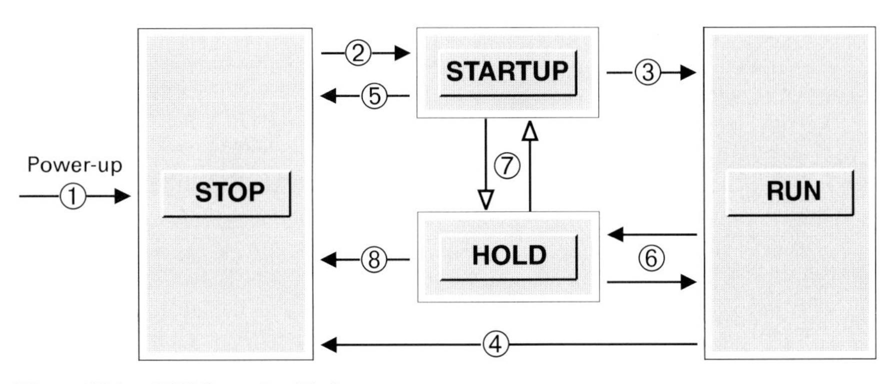

Figure 22.1 CPU Operating Modes

In HOLD mode, everything having to do with timing is discontinued. This includes, for example, the processing of timers, clock memory and run-time meters, cycle time monitoring and minimum scan cycle time, and the servicing of time-of-day and time-delay interrupts. Exception: the real-time clock continues to function normally.

Every time the progression is made to the next statement in test mode, the timers for the duration of the single step run a little further, thus simulating a dynamic behavior similar to "normal" program scanning.

In HOLD mode, the CPU is capable of passive communication, that is, it can receive global data or take part in the one-way exchange of data.

If the power fails while the CPU is in HOLD mode, battery-backed CPUs go to STOP on power recovery. CPUs without backup batteries execute an automatic complete restart.

22.1.3 Disabling the Output Modules

In the STOP and HOLD modes, all output modules are disabled (OD (output disable) signal). Disabled output modules output a zero signal or, if they have the capability, the substitute value. Via a variable table, you can control outputs on the modules with the "Isolate PQ" function, even in STOP mode.

During startup, the output modules remain disabled. Only when the cyclic scan begins are the output modules enabled. The signal states in a module's memory (not the process image!) are then applied to the outputs.

While the output modules are disabled, module memory can be set either with direct access (MOVE box with operand area PQ) or via the transfer of the process-image output table. If the CPU removes the Disable signal, the signal states in module memory are applied to the external outputs.

On a complete restart (OB 100), the process images and the module memory are cleared. If you want to scan inputs in OB 100, you must load the signal states from the module using direct access. You can then set the inputs (transfer them, for instance, with the MOVE box from operand area PI to operand area I), then work with the inputs. If you want certain outputs to be set on a transition from a complete restart to the cyclic program (prior to calling OB 1), you must use direct access to address the output modules. It is not enough to just set the outputs (in the process image), as the process-image output table is not transferred at the end of the complete restart routine.

On a warm restart, the "old" process-image input and process-image output tables, which were valid prior to power-down or STOP, are used in OB 101 and in the remainder of the cycle. At the end of that cycle, the process-image output table is transferred to module memory (but not yet switched through to the external outputs, since the output modules are still disabled). You now have the option of parameterizing the CPU to clear the process-image output table and the module memory at the end of the warm restart. Before switching to OB 1, the CPU revokes the Disable signal so that the signal states in the module memory are applied to the external outputs.

22.1.4 Start-Up Organization Blocks

On a *complete restart,* the CPU calls organization block OB 100 on a single-shot basis prior to processing the main program. If there is no OB 100, the CPU begins cyclic program scanning immediately.

On a *warm restart,* the CPU calls organization block OB 101 on a single-shot basis before processing the main program. If there is no OB 101, the CPU begins scanning at the point of interruption.

The start information in the temporary local data has the same format for both organization blocks; Table 22.1 shows the start information for OB 100. The reason for the restart is shown in the restart request.

- B#16#81 Manual complete restart (OB 100)
- B#16#82 Automatic complete restart (OB 100)
- B#16#83 Manual warm restart (OB 101)
- B#16#84 Automatic warm restart (OB 101)

The number of the stop event and the additional information define the restart more precisely (tells you, for example, whether a manual complete restart was initiated via the mode switch). With this information, you can develop an appropriate event-related start-up routine.

Table 22.1 Start Information for the Start-Up OBs

Byte	Name	Data Type	Description
0	OB100_EV_CLASS	BYTE	Event class
1	OB100_STRTUP	BYTE	Restart request
2	OB100_PRIORITY	BYTE	Priority class
3	OB100_OB_NUMBR	BYTE	OB number
4	OB100_RESERVED_1	BYTE	Reserved
5	OB100_RESERVED_2	BYTE	Reserved
6..7	OB100_STOP	WORD	Number of the stop event
8..11	OB100_STRT_INFO	DWORD	Additional information on the current restart
12..19	OB100_DATE_TIME	DT	Date and time event occurred

22.2 Power-Up

22.2.1 STOP Mode

STOP mode is reached in the following cases:

▷ After the CPU has been switched on

▷ When the mode selector is set from RUN to STOP

▷ When an "unrecoverable" error occurs during program processing

▷ When system function SFC 46 STP is executed

▷ When requested by a communication function (stop request from the programming device or via communication function blocks from another CPU).

The CPU enters the reason for the STOP in the diagnostic buffer. In this mode, you can also read the CPU information with a programming device in order to localize the problem.

In STOP mode, the user program is not scanned. The CPU retrieves the settings – either the values which you entered in the hardware configuration data when you parameterized the CPU or the defaults – and sets the modules to the specified initial state.

In STOP mode, the CPU can receive global data via GD communication and carry out passive one-way communication functions. The real-time clock keeps running.

You can parameterize the CPU in STOP mode, for instance you can also set the MPI address, transfer or modify the user program, and execute a CPU memory reset.

22.2.2 Memory Reset

A memory reset sets the CPU to the "initial state". You can initiate a memory reset with the mode selector or with a programming device only in STOP mode.

The CPU erases the entire user program both in work memory and in RAM load memory. System memory (for instance bit memory, timers and counters) is also erased, regardless of retentivity settings.

The CPU sets the parameters for all modules, including its own, to their default values. The MPI parameters are an exception. They are not changed so that a CPU whose memory has been reset can still be addressed on the MPI bus. A memory reset also does not affect the diagnostic buffer, the real-time clock, or the run-time meters.

If a memory card with Flash EPROM is inserted, the CPU copies the user program from the memory card to work memory. The CPU also copies any configuration data it finds on the memory card.

22.2.3 Retentivity

A memory area is retentive if its contents are retained even when the mains power is switched off as well as on a transition from STOP to RUN following power-up. Retentive memory areas may be those for bit memory, timers, counters and, on the S7-300, data areas. The volume depends on the CPU. You can specify the retentive areas via the "Retentivity" tab when you parameterize the CPU.

The settings for retentivity are in the system data blocks (SDBs) in load memory, that is, on the memory card. If the memory card is a RAM card, you must operate the programmable controller with a backup battery to save the retentivity settings permanently.

When you use a backup battery, the signal states of the memory bits, timers and counters specified as being retentive are retained. The user program and the user data remain unchanged. It makes no difference whether the memory card is a RAM or a Flash EPROM card.

If the memory card is a Flash EPROM card and there is no backup battery, an S7-300 and an S7-400 respond differently. In S7-300 controllers, the signal states of the retentive memory bits, timers and counters are retained, in S7-400 controllers they are not. The contents of retentive data blocks also remain unchanged in S7-300 controllers. The remaining data blocks in an S7-300 controller and all data blocks in an S7-400 controller are copied from the memory card to work memory, as are the code blocks. The only data blocks whose contents are retained are those on the memory card. Data blocks generated with system function SFC 22 CREAT_DB are not retentive. After startup, the data blocks have the contents that are on the memory card, that is, the contents with which they were programmed.

22.3 Types of Start-Up

22.3.1 STARTUP Mode

The CPU executes a start-up in the following cases:

▷ When the mains power is switched on (via STOP)

▷ When the mode selector is set from STOP to RUN or RUN-P

▷ On a request from a communication function (request from a programming device or via communication function blocks from another CPU)

A *manual* startup is initiated via the keyswitch or a communication function, an *automatic* startup by switching on the mains power.

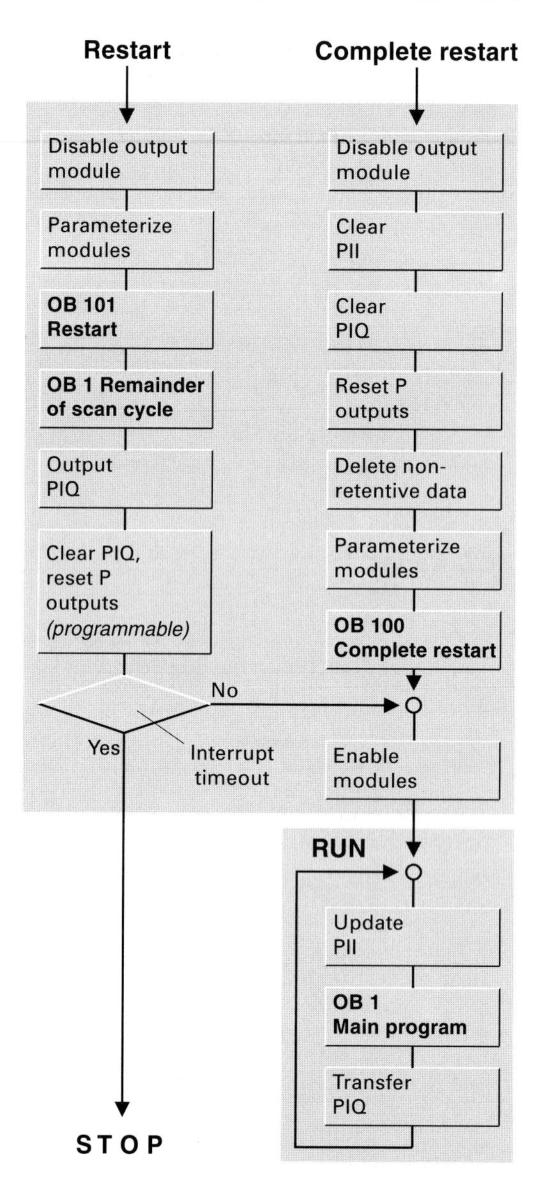

Figure 22.2 CPU Activities During Start-Up

The start-up routine may be as long as required, and there is no time limit on its execution; the scan cycle monitor is not active. During the execution of the start-up routine, no interrupts will be serviced (exceptions being errors, which are handled as they would be in RUN mode). In the start-up routine, the CPU updates the timers, the run-time meters and the real-time clock. Figure 22.2 shows the activities carried out by the CPU during a startup.

A start-up routine can be aborted, for instance when the mode selector is actuated or when there is a power failure. The aborted start-up routine is then executed from the beginning when the power is switched on. If a complete restart is aborted, it must be executed again. If a warm restart is aborted, it is possible to execute either a complete restart or a warm restart.

22.3.2 Complete Restart

On a complete restart, the CPU sets both itself and the modules to the programmed initial state, erases the non-retentive data, calls OB 100, and then executes the main program in OB 1 from the beginning.

Complete manual restart

A complete manual restart is initiated in the following instances:

• Via the mode switch on the CPU on a transition from STOP to RUN or RUN-P; on the S7-400, the start-up switch is in the CRST position.

• Via a communication function from a PG or with an SFB from another CPU; the mode selector must be in the RUN or RUN-P position.

A complete manual restart can always be initiated unless the CPU requests a memory reset.

Complete automatic restart

An automatic complete restart is initiated by switching on the mains power. The restart is executed if

▷ the CPU was not at STOP when the power was switched off

▷ the mode switch is at RUN or RUN-P

▷ the CPU was interrupted by a power outage while executing a complete restart

▷ no automatic warm restart was specified (S7-400 only)

In all other cases, the CPU remains at STOP when the power is switched on. The startup switch (S7-400) has no effect in the case of an automatic complete restart.

When operated without a backup battery, the CPU executes an automatic non-retentive complete restart. The CPU starts the memory reset automatically, then copies the user program from the memory card to work memory. The memory card must be a Flash EPROM.

22.3.3 Warm Restart

A warm restart is possible only on an S7-400.

On a STOP or power outage, the CPU saves all interrupts as well as the internal CPU registers that are important to the processing of the user program. On a warm restart, it can therefore resume at the location in the program at which the interruption occurred. This may be the main program, or it may be an interrupt or error handling routine. All ("old") interrupts are saved and will be serviced.

The so-called "residual cycle", which extends from the point at which the CPU resumes the program following a warm restart to the end of the main program, counts as part of the restart. No (new) interrupts are serviced. The output modules are disabled, and are in their initial state.

A warm restart is permitted only if there were no changes in the user program while the CPU was at STOP, such as modification of a block.

By parameterizing the CPU accordingly, you can specify how long the interruption may be for the CPU to still be able to execute a warm restart (from 100 milliseconds to 1 hour). If the interruption is longer, only a complete restart is allowed. The length of the interruption is the amount of time between exiting the RUN mode (STOP or power-down) and reentering the RUN mode (following execution of OB 101 and the residual cycle).

Manual warm restart

A manual warm restart is initiated

• by moving the mode selector from STOP to RUN or RUN-P when the startup switch is at WRST

• via a communication function from a programming device or with an SFB from another CPU; the mode selector must be at RUN or RUN-P.

A manual warm restart is possible only when the warm restart disable was revoked in the "Startup" tab when the CPU was parameterized. The cause of the STOP must have been a manual activity, either via the mode selector or through a communication function; only then can a manual warm restart be executed while the CPU is at STOP.

Table 22.2 Parameters for the SFCs Used to Determine the Module Address

SFC	Parameter	Declaration	Data Type	Contents, Description
5	SUBNETID	INPUT	BYTE	Area identifier
	RACK	INPUT	WORD	Number of the rack
	SLOT	INPUT	WORD	Number of the slot
	SUBSLOT	INPUT	BYTE	Number of the submodule
	SUBADDR	INPUT	WORD	Offset in the module's user data address area
	RET_VAL	OUTPUT	INT	Error information
	IOID	OUTPUT	BYTE	Area identifier
	LADDR	OUTPUT	WORD	Logical address of the channel
50	IOID	INPUT	BYTE	Area identifier
	LADDR	INPUT	WORD	A logical module address
	RET_VAL	OUTPUT	INT	Error information
	PEADDR	OUTPUT	ANY	WORD field for the PI addresses
	PECOUNT	OUTPUT	INT	Number of PI addresses returned
	PAADDR	OUTPUT	ANY	WORD field for the PQ addresses
	PACOUNT	OUTPUT	INT	Number of PQ addresses returned
49	IOID	INPUT	BYTE	Area identifier
	LADDR	INPUT	WORD	A logical module address
	RET_VAL	OUTPUT	INT	Error information
	AREA	OUTPUT	BYTE	Area identifier
	RACK	OUTPUT	WORD	Number of the rack
	SLOT	OUTPUT	WORD	Number of the slot
	SUBADDR	OUTPUT	WORD	Offset in the module's user data address area

Automatic warm restart

An automatic warm restart is initiated by switching on the mains power. The CPU executes an automatic warm restart only in the following instances:

▷ If it was not at STOP when switched off

▷ If the mode selector was at RUN or RUN-P when the CPU was switched on

▷ If an automatic warm restart was specified per parameter

▷ If the backup battery is inserted and in working order.

The position of the startup switch is irrelevant to an automatic warm restart.

22.4 Ascertaining a Module Address

You can determine module addresses with the following SFCs:

▷ SFC 5 GADR_LGC
Determine logical dual-port RAM address

▷ SFC 50 RD_LGADR
Determine all logical module addresses

▷ SFC 49 LGC_GADR
Determine slot

Table 22.2 shows the parameters for these SFCs.

The SFCs have IOID and LADDR as common parameters for the logical address (= address in the I/O area). IOID is either B#16#54, which stands for the peripheral inputs (PIs) or B#16#55,

Table 22.3 Description of SFC 49 LGC_GADR's Output Parameters

AREA	System	Meaning of RACK, SLOT and SUBADDR
0	S7-400	RACK = Number of the rack
1	S7-300	SLOT = Number of the slot
		SUBADDR = Address offset to the start address
2	Distributed I/O	RACK, SLOT and SUBADDR irrelevant
3	S5-P area	RACK = Number of the rack
4	S5-Q area	SLOT = Slot number of the adapter casing
5	S5-IM3 area	SUBADDR = Address in the S5 area
6	S5-IM4 area	

which stands for the peripheral outputs (PQs). LADDR contains an I/O address in the PI or PQ area which corresponds to the specified channel. If the channel is 0, it is the module start address.

The hardware configuration data must specify an allocation between logical address (module start address) and slot address (location of the module in a rack or a station for distributed I/O) for the addresses ascertained with these SFCs.

• Ascertain the logical address of a channel
System function SFC 5 GADR_LGC returns the logical address of a channel when you specify the slot address ("geographic" address). Enter the number of the subnet in the SUBNETID parameter if the module belong to the distributed I/O or B#16#00 if the module is plugged into a rack. The RACK parameter specifies the number of the rack or, in the case of distributed I/O, the number of the station. If the module has no submodule slot, enter B#16#00 in the SUBSLOT parameter. SUBADDR contains the address offset in the module's user data (W#16#0000, for example, stands for the module start address).

• Ascertain all logical addresses for a module
On the S7-400, you can assign addresses for a module's user data bytes which are not contiguous (available soon). SFC 50 RD_LGADR returns all logical addresses for a module when you specify an arbitrary address from the user data area. Use the PEADDR and PAADDR parameters to define an area of WORD components (a word-based ANY pointer, for example P#*DBzDBXy.x* WORD *nnn)*. SFC 50 then shows you the number of entries returned in these areas in the RECOUNT and PACOUNT parameters.

• Ascertain the slot address of a module
SFC 49 LGC_GADR returns the slot address of a module when you specify an arbitrary logical module address. The value in the AREA parameter specifies the system in which the module is operated (Table 22.3).

22.5 Parameterizing Modules

The following system functions are available for parameterizing modules:

▷ SFC 55 WR_PARM
Write dynamic parameters

▷ SFC 56 WR_DPARM
Write predefined parameters

▷ SFC 57 PARM_MOD
Parameterize module

▷ SFC 58 WR_REC
Write data record

▷ SFC 59 RD_REC
Read data record

Use system functions SFC 58 WR_REC and SFC 59 RD_REC to transfer arbitrary data records from a module with the appropriate capability. The parameters for these and a number of other system functions are listed in Table 22.4.

Table 22.5 shows the available data records which can be transferred with the aforementioned system functions. Depending on the data record, the information transferred may comprise up to 240 bytes.

Table 22.4 Parameters for System Functions Used for Data Transfer

Parameter for SFC					Parameter Name	Decl.	Data Type	Contents, Description
55	56	57	58	59	REQ	INPUT	BOOL	"1" = Write request
55	56	57	58	59	IOID	INPUT	BYTE	B#16#54 = Peripheral inputs (PIs) B#16#55 = Peripheral outputs (PQs)
55	56	57	58	59	LADDR	INPUT	WORD	Module start address
55	56	–	58	59	RECNUM	INPUT	BYTE	Data record number
55	–	–	58	–	RECORD	INPUT	ANY	Data record
55	56	57	58	59	RET_VAL	OUTPUT	INT	Error information
55	56	57	58	59	BUSY	OUTPUT	BOOL	Transfer still in progress when "1"
–	–	–	–	59	RECORD	OUTPUT	ANY	Data record

General remarks on parameterizing modules

Some S7 modules can be parameterized, that is to say, values may be set on the module which deviate from the defaults. To specify parameters, open the module in the hardware configuration and fill in the tabs in the dialog box. When you transfer the *System Data* object in the *Blocks* container to the PLC, you are also transferring the module parameters.

The CPU transfers the module parameters to the module automatically

- ▷ On startup (complete restart or warm restart)
- ▷ When a module has been plugged into a configured slot (S7-400)
- ▷ Following the "return" of a rack or a distributed I/O station.

The module parameters are subdivided into static parameters and dynamic parameters. You can set both parameter types off-line in the hardware configuration. You can also modify the dynamic parameters at runtime using SFC calls. In the startup routine, the parameters set on the modules using SFCs are overwritten by the parameters set (and stored on the CPU) via the hardware configuration.

The parameters for the signal modules are in two data records: the static parameters in data record 0 and the dynamic parameters in data record 1 in the system data blocks (SDBs). You can transfer both data records to the module with SFC 57 PARM_MOD, data record 0 or 1 with SFC 56 WR_DPARM, and only data record 1 with SFC 55 WR_PARM. The data records must be in the SDBs.

After parameterization of an S7-400 module, the specified values do not go into force until bit 2 ("Operating mode") in byte 2 of diagnostic data record 0 has assumed the value "RUN" (can be read with SFC 59 RD_REC).

As far as addressing for data transfer is concerned, use the *lowest* module start address (LADDR parameter) together with the identifier indicating whether you have defined this address as input or output (IOID parameter). If

Table 22.5 Available Data Records

Record No.	Contents for Read	Contents for Write
0	Diagnostic data	Parameters
1	Diagnostic data	Parameters
2 to 127	User data	User data
128 to 255	Diagnostic data	Parameters

you assigned the same start address to both the input and output area, use the identifier for input. Use the I/O identifier regardless of whether you want to execute a Read or a Write operation.

Use the RECORD parameter with the data type ANY to define an area of BYTE components. This may be a variable of type ARRAY, STRUCT or UDT, or an ANY pointer of type BYTE (for example P#DBzDBX*y.x* BYTE *nnn)*. If you use a variable, it must be a "complete" variable; individual array or structure components are not permissible.

Writing dynamic parameters

System function SFC 55 WR_PARM transfers the data record addressed by RECORD to the module specified by the IOID and LADDR parameters. Specify the number of the data record in the RECNUM parameter. It is a requirement that the data record is in the correct SDB system data block, and that the data record is not static.

When the job is initiated, the SFC reads the entire data record; the transfer may be distributed over several program scan cycles. The BUSY parameter is "1" during the transfer.

Writing predefined parameters

System function SFC 56 WR_DPARM transfers the data record with the number specified in the RECNUM parameter from the relevant SDB system data block to the module identified by the IOID and LADDR parameters.

The transfer may be distributed over several program scan cycles; the BUSY parameter is "1" during the transfer.

Parameterizing a module

System function SFC 57 PARM_MOD transfers all the data records programmed when the module was parameterized via the hardware configuration.

The transfer may be distributed over several program scan cycles; the BUSY parameter is "1" during the transfer.

Writing a data record

SFC 58 WR_REC transfers the data record addressed by the RECORD parameter and the number RECNUM to the module defined by the IOID and LADDR parameters. A "1" in the REQ parameter starts the transfer. When the job is initiated, the SFC reads the complete data record.

The transfer may be distributed over several program cycles; the BUSY parameter is "1" during the transfer.

Reading a data record

When the REQ parameter is "1", SFC 59 RD_REC reads the data record addressed by the RECNUM parameter from the module and places it in destination area RECORD. The destination area must be longer than or at least as long as the data record. If the transfer is completed without error, the RET_VAL parameter contains the number of bytes transferred.

The transfer may be distributed over several program scan cycles; the BUSY parameter is "1" during the transfer.

S7-300s delivered prior to February 1997: the SFC reads as much data from the specified data record as the destination area can accommodate. The size of the destination area may not exceed that of the data record.

23 Error Handling

The CPU reports errors or faults detected by the modules or by the CPU itself in different ways:

▷ Errors in arithmetic operations (overflow, invalid REAL number) by setting status bits (status bit OV, for example, for a numerical overflow)

▷ Errors detected while processing LAD functions (synchronous errors) by calling organization blocks OB 121 and OB 122

▷ Errors in the programmable controller which do not relate to program scanning (asynchronous errors) by calling organization blocks OB 80 to OB 87

The CPU signals the occurrence of an error or fault, and in some cases the cause, by setting error LEDs on the front panel. In the case of unrecoverable errors (such as invalid OP code), the CPU goes direct to STOP.

With the CPU in STOP mode, you can use a programming device and the CPU information functions to read out the contents of the block stack (B stack), the interrupt stack (I stack) and the local data stack (L stack) and then draw conclusions as to the cause of error.

The system diagnostics can detect errors/faults on the modules, and enters these errors in a diagnostic buffer. Information on CPU mode transitions (such as the reasons for a STOP) are also placed in the diagnostic buffer. The contents of this buffer are retained on STOP, on a memory reset, and on power failure, and can be read out following power recovery and execution of a start-up routine using a programming device.

23.1 Synchronous Errors

The CPU's operating system generates a synchronous error when an error occurs in immediate conjunction with program scanning. In the case of a programming error, OB 121 is called,

Table 23.1 Start Information for the Synchronous Error OBs

Variable Name	Data Type	Description, Contents
OB12x_EV_CLASS	BYTE	B#16#25 = Call programming error OB 121 B#16#29 = Call access error OB 122
OB12x_SW_FLT	BYTE	Error code (see text)
OB12x_PRIORITY	BYTE	Priority class in which the error occurred
OB12x_OB_NUMBR	BYTE	OB number (B#16#79 or. B#16#80)
OB12x_BLK_TYPE	BYTE	Type of block interrupted (S7-400 only) OB: B#16#88, FB: B#16#8E, FC: B#16#8C
OB121_RESERVED_1 OB122_MEM_AREA	BYTE	Addition to error code (see text)
OB121_FLT_REG OB122_MEM_ADDR	WORD	OB 121: Error source (see text) OB 122: Address at which the error occurred
OB12x_BLK_NUM	WORD	Number of the block in which the error occurred (S7-400 only)
OB12x_PRG_ADDR	WORD	Error address in the block that caused the error (S7-400 only)
OB12x_DATE_TIME	DT	Time at which programming error was detected

while OB 122 is called in the event of an access error. If a synchronous error OB has not been programmed, the CPU goes to STOP. Table 22.3 shows the start information for both synchronous error organization blocks.

A synchronous error OB has the same priority as the block in which the error occurred. It is for this reason that it is possible to access the registers of the interrupted block in the synchronous error OB, and it is also for this reason that the program in the synchronous error OB (under certain circumstances with modified content) can return the registers to the interrupted block. Note that when a synchronous error OB is called, its 20 bytes of start information are also pushed onto the L stack for the priority class that caused the error, as are the other temporary local data for the synchronous error OB and for all blocks called in this OB.

In the case of a CPU 416, another synchronous error OB can be called in an error OB. The block nesting depth for a synchronous error OB is 3 for S7-400 CPUs and 4 for S7-300 CPUs.

You can disable and enable a synchronous error OB call with system functions SFC 36 MSK_FLT and SFC 37 DMSK_FLT.

23.1.1 Programming Errors

If the operating system detects a programming error at runtime, it calls OB 121.

The error code in variable OB121_SW_FLT is described in detail in section 23.2.1, "Error Filter". Bits 4 to 7 of variables OB121_RESERVED_1 and OB122_MEM_AREA (in OB 122) contain the access mode in the event of a read or write error

1 Bit access
2 Byte access
3 Word access
4 Doubleword access

and bits 0 to 3 contain the address area

0 I/O area PI or PQ
1 Process-image input area I
2 Process-image output area Q
3 Bit memory area M
4 Global data block DB
5 Instance data block DI
6 Temporary local data L
7 Temporary local data for the preceding block V

Example: Error code B#16#24 means that a synchronous error occurred while accessing a data byte located in a data block opened via the DB register.

The variable OB121_FLT_REG contains the error source as per the error code:

- ▷ The errored address (in the event of a read/write error)
- ▷ The errored area (in the event of a range error)
- ▷ The errored number of the block or the timer/counter

23.1.2 Access Errors

The operating system calls OB 122 if a runtime access error is detected.

The error code in variable OB122_SW_FLT is described in detail in section 23.2.1, "Error Filter". The contents of variable OB122_MEM_AREA correspond to the contents of variable OB121_RESERVED_1 (see above under "Programming Errors").

23.2 Synchronous Error Handling

The following system functions are provided for handling synchronous errors:

- ▷ SFC 36 MSK_FLT
  Mask synchronous errors (disable OB call)
- ▷ SFC 37 DMSK_FLT
  Unmask synchronous errors (reenable OB call)
- ▷ SFC 38 READ_ERR
  Read error register

The operating system enters the synchronous error in the diagnostic buffer without regard to the use of system functions SFC 37 to SFC 39. The parameters for these system functions are listed in Table 23.2.

23.2.1 Error Filters

The error filters are used to control the system functions for synchronous error handling. In the programming error filter, one bit stands for each programming error detected; in the access error filter, one bit stands for each access error detected. When you define an error filter, you set the bit that stands for the synchronous error you want to mask, unmask or query. The error filters returned by the system functions show a "1" for synchronous errors that are still masked or which have occurred.

Table 23.2 SFC Parameters for Synchronous Error Handling

SFC	Parameter Name	Declaration	Data Type	Contents, Description
36	PRGFLT_SET_MASK	INPUT	DWORD	New (additional) programming error filter
	ACCFLT_SET_MASK	INPUT	DWORD	New (additional) access error filter
	RET_VAL	OUTPUT	INT	W#16#0001 = The new filter overlaps the existing filter
	PRGFLT_MASKED	OUTPUT	DWORD	Complete programming error filter
	ACCFLT_MASKED	OUTPUT	DWORD	Complete access error filter
37	PRGFLT_RESET_MASK	INPUT	DWORD	Programming error filter to be reset
	ACCFLT_RESET_MASK	INPUT	DWORD	Access error filter to be reset
	RET_VAL	OUTPUT	INT	W#16#0001 = The new filter contains bits that are not set (in the current filter)
	PRGFLT_MASKED	OUTPUT	DWORD	Remaining programming error filter
	ACCFLT_MASKED	OUTPUT	DWORD	Remaining access error filter
38	PRGFLT_QUERY	INPUT	DWORD	Programming error filter to be queried
	ACCFLT_QUERY	INPUT	DWORD	Access error filter to be queried
	RET_VAL	OUTPUT	INT	W#16#0001 = The query filter contains bits that are not set (in the current filter)
	PRGFLT_CLR	OUTPUT	DWORD	Programming error filter with error messages
	ACCFLT_CLR	OUTPUT	DWORD	Access error filter with error messages

Table 23.3 Programming Error Filter

Bit	Error Code	Description
1	B#16#21	BCD conversion error (pseudo-tetrade detected during conversion)
2	B#16#22	Area length error on read (address not within area limits)
3	B#16#23	Area length error on write (address not within area limits)
4	B#16#24	Area length error on read (wrong area in area pointer)
5	B#16#25	Area length error on write (wrong area in area pointer)
6	B#16#26	Invalid timer number
7	B#16#27	Invalid counter number
8	B#16#28	Address error on read (bit address <>0 in conjunction with byte, word or doubleword access and indirect addressing)
9	B#16#29	Address error on write (bit address <>0 in conjunction with byte, word or doubleword access and indirect addressing)
16	B#16#30	Write error, global data block (write-protected block)
17	B#16#31	Write error, instance data block (write-protected block)
18	B#16#32	Invalid number of a global data block (DB register)
19	B#16#33	Invalid number of an instance data block (DI register)
20	B#16#34	Invalid number of a function (FC)
21	B#16#35	Invalid number of a function block (FB)
26	B#16#3A	Called data block (DB) does not exist
28	B#16#3C	Called function (FC) does not exist
30	B#16#3E	Called function block (FB) does not exist

The programming error filter is shown in Table 23.3, the Error Code column shows the contents of variable OB121_SW_FLT in the start information for OB 121.

The access error filter is shown in Table 23.4; the Error Code column shows the contents of variable OB122_SW_FLT in the start information for OB 122.

The S7-400 CPUs distinguish between two types of access error: access to a non-existent module and invalid access attempt to an existing module. If a module fails during operation, a time-out is signaled approximately 150 μs after an access attempt. At the same time, that module is marked "non-existent" and an I/O access error (PZF) is reported on every subsequent attempt to access the module. The CPU also reports an I/O access error when an attempt is made to access a non-existent module, regardless of whether the attempt was direct (via the I/O area) or indirect (via the process image).

The error filter bits not listed in the tables are not relevant to the handling of synchronous errors.

23.2.2 Masking Synchronous Errors

System function SFC 36 MSK_FLT disables synchronous error OB calls via the error filters. A "1" in the error filters indicates the synchronous errors for which the OBs are not to be called (the synchronous errors are "masked"). The masking of synchronous errors in the error filters is in addition to the masking stored in the operating system's memory. SFC 36 returns a function value indicating whether a (stored) masking already exists on at least one bit for the masking specified at the input parameters (W#16#0001).

SFC 36 returns a "1" in the output parameters for all currently masked errors.

If a masked synchronous error event occurs, the respective OB is not called and the error is entered in the error register. The Disable applies to the current priority class (priority level). For example, if you were to disable a synchronous error OB call in the main program, the synchronous error OB would still be called if the error were to occur in an interrupt service routine.

23.2.3 Unmasking Synchronous Errors

System function SFC 37 DMSK_FLT enables the synchronous error OB calls via the error filters. You enter a "1" in the filters to indicate the synchronous errors for which the OBs are once again to be called (the synchronous errors are "unmasked"). The entries corresponding to the specified bits are deleted in the error register. SFC 37 returns W#16#0001 as function value if a (stored) masking already exists on at least one bit for the masking specified at the input parameters .

SFC 37 returns a "1" in the output parameters for all currently masked errors.

If an unmasked synchronous error occurs, the respective OB is called and the event entered in the error register. The Enable applies to the current priority class (priority level).

Table 23.4 Access Error Filter

Bit	Error Code	Description
3	B#16#42	I/O access error on read S7-300: Module does not exist or does not acknowledge S7-400: An existing module does not acknowledge after first access operation (time-out)
4	B#16#43	I/O access error on write S7-300: Module does not exist or does not acknowledge S7-400: An existing module does not acknowledge after first access operation (time-out)
5	B#16#44	S7-400 only: I/O access error on attempt to read from non-existent module (PZF) or on repeated access to modules which do not acknowledge
6	B#16#45	S7-400 only: I/O access error on attempt to write to non-existent module (PZF) or on repeated access to modules which do not acknowledge

Table 23.5 Parameters for SFC 44 REPL_VA

SFC	Parameter Name	Declaration	Data Type	Contents, Description
44	VAL	INPUT	DWORD	Substitute value
	RET_VAL	OUTPUT	INT	Error information

23.2.4 Reading the Error Register

System function SFC 38 READ_ERR reads the error register. You must enter a "1" in the error filters to indicate the synchronous errors whose entries you want to read. SFC 38 returns W#16#0001 as function value if a (stored) masking already exists on at least one bit for the masking specified at the input parameters

SFC 38 returns a "1" in the output parameters for the selected errors when these errors occurred, and deletes these errors in the error register when they are queried. The synchronous errors that are reported are those in the current priority class (priority level).

23.2.5 Entering a Substitute Value

SFC 44 REPL_VAL allows you to enter a substitute value in accumulator 1 from within a synchronous error OB. Use SFC 44 when you can no longer read any values from a module (for instance when a module is defective). When you program SFC 44, OB 122 ("access error") is called every time an attempt is made to access the module in question. When you can SFC 344, you can load a substitute value into the accumulator; the program scan is then resumed with this replacement value. Table 23.5 lists the parameters for SFC 44.

You may call SFC 44 in only one synchronous error OB (OB 121 or OB 122).

23.3 Asynchronous Errors

Asynchronous errors are errors which can occur independently of the program scan. When an asynchronous error occurs, the operating system calls one of the organization blocks listed below:

OB 80 Timing error

OB 81 Power supply error

OB 82 Diagnostic interrupt

OB 83 Insert/remove module interrupt

OB 84 CPU hardware fault

OB 85 Program execution error

OB 86 Rack failure

OB 87 Communication error

The OB 82 call (diagnostic interrupt) is described in detail in section 23.4, "System Diagnostics".

The asynchronous error OB call can be disabled and enabled with system functions SFC 39 DIS_IRT and SFC 40 EN_IRT, and delayed and enabled with system functions SFC 41 DIS_AIRT and SFC 42 EN_AIRT.

Timing error

The operating system calls organization block OB 80 when one of the following errors occurs:

▷ Cycle monitoring time exceeded

▷ OB request error (the requested OB is still executing or an OB was requested too frequently within a given priority class)

▷ Time-of-day interrupt error (TOD interrupt time past because clock was set forward or after transition to RUN)

If no OB 80 is available and a timing error occurs, the CPU goes to STOP. The CPU also goes to STOP when the OB is called a second time in the same program scan cycle.

Power supply error

The operating system calls organization block OB 81 when one of the following errors occurs:

▷ At least one backup battery in the central controller or in an expansion unit is empty

▷ No battery voltage in the central controller or in an expansion unit

▷ 24 V supply failed in central controller or in an expansion unit

OB 81 is called for coming and leaving events. If there is no OB 81, the CPU continues functioning when a power supply error occurs.

Insert/remove module interrupt

The operating system monitors the module configuration once per second. An entry is made in the diagnostic buffer and in the system status list each time a module is inserted or removed in RUN, STOP or STARTUP mode. In addition, the operating system calls organization block OB 83 if the CPU is in RUN mode. If there is no OB 83, the CPU goes to STOP on an insert/remove module interrupt.

As much as a second can pass before the insert/remove module interrupt is generated. As a result, it is possible that an access error or an error relating to the updating of the process image could be reported in the interim between removal of a module and generation of the interrupt.

If a suitable module is inserted into a configured slot, the CPU automatically parameterizes that module, using data records already stored on that CPU. Only then is OB 83 called.

CPU hardware error

The operating system calls organization block OB 84 when an interface error (MPI network, PROFIBUS DP) occurs or disappears. If there is no OB 84, the CPU goes to STOP on a CPU hardware error.

Program execution error

The operating system calls organization block OB 85 when one of the following errors occurs:

▷ Start request for an organization block which has not been loaded

▷ Error occurred while the operating system was accessing a block (for instance no instance data block when a system function block (SFB) was called)

▷ I/O access error while executing a full process image update

If there is no OB 85, the CPU goes to STOP on a program execution error.

Rack failure

The operating system calls organization block OB 86 when it detects the failure of a rack (power failure, line break, defective IM), a subnet, or a distributed I/O station. OB 86 is called for both coming and leaving errors.

In multiprocessor mode, OB 86 is called in all CPUs when a rack fails.

If there is no OB 86, the CPU goes to STOP when a rack failure occurs.

Communication error

The operating system calls organization block OB 87 when a communication error occurs. Some examples of communication errors are:

▷ Invalid frame identification or frame length error in conjunction with global communication

▷ Sending of diagnostic entries not possible

▷ Clock Synchronization error

▷ GD status cannot be entered in a data block

If there is no OB 87, the CPU goes to STOP when a communication error is detected.

23.4 System Diagnostics

23.4.1 Diagnostic Events and Diagnostic Buffer

System diagnostics is the detection, evaluation and reporting of errors which occur in the programmable controller. Examples are errors in the user program or module failures, but also wirebreaks on signaling modules. These *diagnostic events* may be:

▷ Diagnostic interrupts from modules with diagnostic capability

▷ System errors and CPU operating mode transitions

▷ User messages issued via system functions.

The modules with diagnostic capability distinguish between programmable and non-programmable diagnostic events. Programmable diagnostic events are reported only when you have enabled diagnostics by parameterizing the proper parameters. Non-programmable diagnostic events are always reported, regardless of whether or not diagnostics were enabled. When a diagnostic event occurs,

▷ the fault LED on the CPU lights up

▷ the diagnostic event is forwarded to the CPU's operating system

▷ a diagnostic interrupt is generated if the respective parameters were set to enable diagnostic interrupts (by default, diagnostic interrupts are disabled).

Table 23.6 Parameters for SFC 52 WR_USMSG

SFC	Parameter Name	Declaration	Data Type	Contents, Description
52	SEND	INPUT	BOOL	When "1": Sending is enabled
	EVENTN	INPUT	WORD	Event ID
	INFO1	INPUT	ANY	Additional information 1 (one word)
	INFO2	INPUT	ANY	Additional information 2 (one doubleword)
	RET_VAL	OUTPUT	INT	Error information

All diagnostic events reported to the CPU operating system are entered in a *diagnostic buffer* in the order in which they occurred and with date and time stamp. The diagnostic buffer is a battery-backed memory area on the CPU whose contents are retained even when a memory reset is executed. The diagnostic buffer is a ring buffer, and its size depends on the CPU used. When the diagnostic buffer is full, the oldest entry is overwritten by the newest.

You can read out the contents of the diagnostic buffer at any time using a programming device. In the CPU's *System Diagnostics* parameter block you can specify you want extended diagnostic buffer entries (in addition, all OB calls). You can also specify whether the last diagnostic entry before the CPU goes to STOP should be sent to a particular node on the MPI bus.

23.4.2 Writing User Entries in the Diagnostic Buffer

System function SFC 52 WR_USMSG writes a user entry in the diagnostic buffer, and can also send that entry to all nodes that have logged on the MPI bus. Table 23.6 lists the parameters for SFC 52.

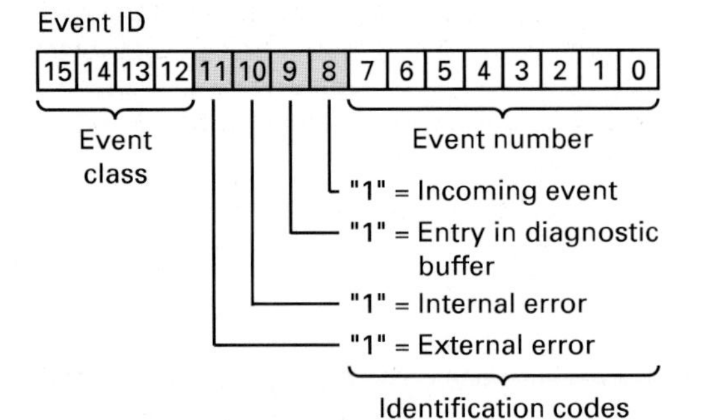

Figure 23.1 Event ID for Diagnostic Buffer Entries

The entry in the diagnostic buffer corresponds in format to that of a system event, for instance the start information for an organization block. Within the framework of the rules governing the event ID (EVENTN parameter) and the additional information (parameters INFO1 and INFO2), you can include any desired information.

The event ID is identical to the first two bytes of the buffer entry (Figure 23.1). Permissible for a user entry are event classes 8 (diagnostic entries for signal modules), 9 (standard user events), and A and B (arbitrary user events).

Additional information 1 corresponds to bytes 7 and 8 of the buffer entry (one word), additional information 2 to bytes 9 to 12 (one doubleword). The contents of both variables is arbitrary.

Set SEND to "1" to indicate that the diagnostic buffer entry is to be sent to the participating nodes. Even when transmission is not possible (for example when no nodes are logged on or when the Send buffer is full), the entry in the diagnostic buffer is still made (when bit 9 of the event ID is set).

23.4.3 Evaluating a Diagnostic Interrupt

When a diagnostic interrupt is coming or leaving, the operating system interrupts the current program scan and calls organization block OB 82. If OB 82 has not been programmed, the CPU goes to STOP when a diagnostic interrupt is generated. You can disable and enable OB 82 with system functions SFC 39 DIS_IRT and SFC 40 EN_IRT, and delay and enable it with system functions SFC 41 DIS_AIRT and SFC 42 EN_AIRT.

The first byte of the start information contains B#16#39 for a coming diagnostic interrupt and B#16#38 for a leaving diagnostic interrupt. The sixth byte is the address identifier (B#16#54 for

Table 23.7 Parameters for SFC 51 RDSYSST

SFC	Parameter Name	Declaration	Data Type	Contents, Description
51	REQ	INPUT	BOOL	When "1": Initiate job
	SZL_ID	INPUT	WORD	Sublist ID
	INDEX	INPUT	WORD	Type or number of the sublist object
	RET_VAL	OUTPUT	INT	Error information
	BUSY	OUTPUT	BOOL	When "1": Read still in progress
	SZL_HEADER	OUTPUT	STRUCT	Length and number of data records read
	DR	OUTPUT	ANY	Array for the data records read

an input, B#16#55 for an output); the INT variable which follows the address identifier contains the address of the module which sent the diagnostic interrupt. The next four bytes contain the diagnostic information provided by the module.

You can utilize system function SFC 59 RD_REC (read data record) in OB 82 to obtain detailed error information. The diagnostic information are consistent until OB 82 is exited, that is to say, the information remains "frozen". When OB 82 is exited, the diagnostic interrupt is acknowledged on the module.

A module's diagnostic data are in data records DS 0 and DS 1. Data record DS 0 contains four bytes of diagnostic data describing the current state of the module. The contents of these four bytes are identical to those of bytes 8 to 11 of the start information for OB 82. Data record DS 1 contains the four bytes in data record DS 0 as well as the module-specific diagnostic data.

23.4.4 Reading the System Status List

The system status list (SZL) describes the current status of a programmable controller. The contents of the system status list can be read out using information functions, but cannot be modified. You can read a section of the list (a so-called sublist) with system function SFC 51 RDSYSST. The sublists are virtual lists, that is, they are assembled by the CPU operating system only on request. The parameters for SFC 51 are listed in Table 23.7.

The Read operation is initiated with REQ = "1", and BUSY = "0" informs you when the operation has terminated. The operating system can process several asynchronous Read operations quasi-simultaneously, the actual number depending on the CPU used. Should SFC 51 return a function value indicative of a lack of resources (W#16#8085), simply reinitiate the Read.

The contents of parameters SZL_ID and INDEX are CPU-dependent. Parameter SZL_HEADER has the data type STRUCT, with variables LENGTHDR (data type WORD) and N_DR (WORD) as components. LENGTHDR contains the length of a data record, N_DR the number of data records read.

Specify the variable or data area in which SFC 51 is to enter the data record in the DR parameter. For example, P#DB200.DBX0.0 WORD 256 makes available an area of 256 data words in data block DB 200, beginning at DBB 0. If the area provided is too small, as many data records as possible will be entered. Only complete data records are transferred. However, the area must be capable of accommodating at least one data record.

Appendix

This section of the book contains useful supplements to the LAD programming language, an overview of the contents of the STEP 7 Block Libraries and a function overview of all LAD elements.

- You can also provide blocks with LAD program with **block protection**. For this purpose, you use the source-oriented Editor in the STL programming language.
- You can use a further function of STL, **indirect addressing** to transfer data areas in the LAD programming language; the addresses of these data areas are then not calculated until runtime. The accompanying diskette contains an **"Example Message Frame"** showing how to set up and transfer data areas.
- The scope of supply of STEP 7 includes **Block Libraries** with loadable functions and function blocks and with block headers and interface descriptions for system blocks (SFCs and SFBs).
- A **function overview** of all LAD functions completes the book.

On the **diskette** accompanying the book you will find the file LAD-BOOK.ARJ, that contains an archive library. You retrieve the library under the SIMATIC Manager with FILE → RETRIEVE. Select the archive (the diskette) from the dialog field displayed. You define the designation directory in the next dialog field. In general, libraries are located under ...\STEP7\S7LIBS; but you can choose any other directory, for example ...\STEP7\S7PROJ, which normally contains the projects. In the final dialog field "Retrieve – Options" deactivate the option "Restore full path".

The library LAD-book contains eight programs which are essentially illustrative examples of LAD representation. Two extensive examples show the programming of functions, function blocks and local instances (Conveyor Example) and the handling of data (Message Frame Example). The memory requirements are approximately 1.93 Mbytes.

To try out an example, set up a project that corresponds to your hardware configuration and copy the program, including the symbol table, from the library to the project. Now you can test the example on-line.

24 LAD Supplements

24.1 Block Protection

The keyword KNOW_HOW_PROTECT represents block protection. You cannot view, print or modify a block with this attribute. The Editor only displays the block header and the declaration table with the block parameters. In source-oriented input, you can protect any block yourself with KNOW_HOW_PROTECT. This means that no-one, including yourself, can view the compiled block (keep the source file in a safe place!).

You can only enter the KNOW_HOW_PROTECT block protection source-oriented with STL. To do so, proceed as follows:

• You create the block under LAD in the usual way. Later, this block will be overwritten in the user program *Blocks* by the block with the keyword. If you want to retain the (original) block (strongly recommended when entering block protection), you can store the block in, for example, a (user-created) library before entering the keyword. You can also store your entire user program in this way.

• Create a source container. If there is no source container *Source Files* under the S7 program (on the same level as the user program *Blocks*), you must create one: Mark the S7 program and insert the symbol *Source Files* with INSERT → S7 SOFTWARE → SOURCE FILE CONTAINER.

• Generate an STL source for the block. Change to the Editor (via the taskbar, for example, or open any block in *Blocks* and then close it again) and select the menu point FILE → GENERATE SOURCE FILE. In the displayed dialog form, set your project, mark the source container *Source Files*, assign a name for the source file under "Object Name". Terminate with "OK". The next dialog form shows you all blocks in the container *Blocks*; you select the block(s) from which you want to create a source file. Terminate with "OK".

• Open the source file (for example, with a double-click on the source file symbol in the SIMATIC Manager or with the Editor and FILE → OPEN). You now see the ASCII source of your LAD block. If you have previously selected several blocks, these blocks will be arranged in order in the source file.

The entries for a code block are in the following order:

▷ Keyword for the block type (FUNCTION, FUNCTION_BLOCK, ORGANIZATION_BLOCK) with specification of the address. This may be followed by the block title (starting with TITLE=...) and the block comment (starting with //...).

▷ Block attributes (depending on whether you have filled in fields in the Properties page for the block and, if so, how many).

▷ Variable declaration (several sections with the keywords VAR_xxx, ..., END_VAR); depending on whether you have declared block-local variables and, if so, which ones.

▷ Program, starts with BEGIN and ends with the keyword for the block end (for example, END_FUNCTION_BLOCK).

The entries for a data block are in the following order:

▷ Keyword for the block type (DATA_BLOCK) with specification of the address.

▷ Block attributes (depending on whether you have filled in fields in the Properties page for the block and, if so, how many).

▷ Variable declaration (starting with STRUCT and ending with END_STRUCT); in instance data blocks, the address of the associated function block is found at this location.

▷ Variable initialization, starting with BEGIN and ending with END_DATA_BLOCK.

• You enter the keyword KNOW_HOW_PROTECT in the source file, in its own line in each case, following the block attributes and before the variable declaration. If you have created a source from several blocks, enter the keywords in all selected blocks. Finally, store the source file.

• Compile the source file with the menu point FILE → COMPILE. The compiler creates a block with the entered block attributes (in the creation language STL in the case of code blocks; the creation language is not significant here since KNOW_HOW_PROTECT means the block can no longer be viewed or printed out). The compiled (new) block is in the user program *Blocks* and replaces the (old) block with the same number.

24.2 Indirect Addressing

With the programming language STL, you have a method of accessing operands whose addresses are not calculated until runtime. This is also possible to a limited degree in LAD: You can wait until runtime to define which data areas you want to copy with SFC 20 BLKMOV.

However, first some useful information on pointers.

24.2.1 Pointers: General

For indirect addressing, you require a data format that contains the bit address as well as the byte address and, if applicable, the operand area. This data format is a *pointer*. A pointer is also used to point to an operand.

There are three types of pointers:

▷ Area pointers; these are 32 bits long and contain a specific operand or its address

▷ DB pointers; these are 48 bits long and in addition to the area pointer they also contain the number of the data block

Table 24.1 Area Coding in the Area Pointer

Area		Coding		
P	Peripheral I/O	0	0	0
I	Inputs	0	0	1
Q	Outputs	0	1	0
M	Memory bits	0	1	1
DBX	Global data	1	0	0
DIX	Instance data	1	0	1
L[1)]	Temporary local data	1	1	0
V[2)]	Temporary local data of the preceding block	1	1	1

[1)] not with area-crossing addressing
[2)] used in transferring block parameters

▷ ANY pointers; these are 80 bits long and, in addition to the DB pointer, they contain further information such as the data type of the operand

24.2.2 Area Pointer

The area pointer contains the operand address and, if applicable, the operand area. Without operand area, it is an *area-internal* pointer; if the pointer also contains an operand area, it is referred to as an *area-crossing* pointer.

The notation for constant representation is as follows:

P#y.x for an area-internal pointer
e.g. P#22.0

P#Zy.x for an area-crossing pointer
e.g. P#M22.0

with x = bit address, y = byte address and Z = area. You specify the operand ID (Table 24.1) as the area. The assignment of bit 31 differentiates the two pointer types (Figure 24.1).

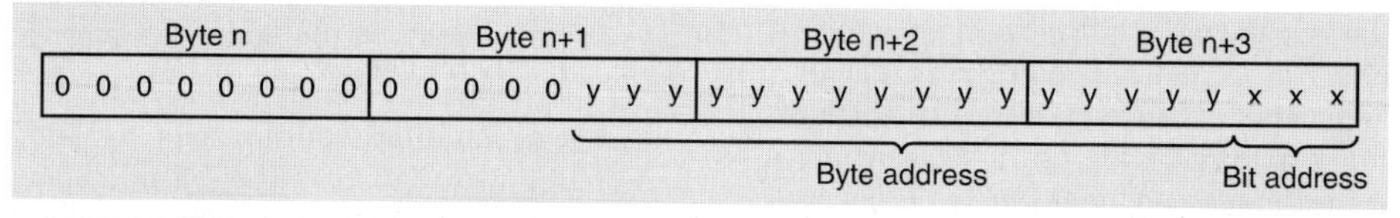

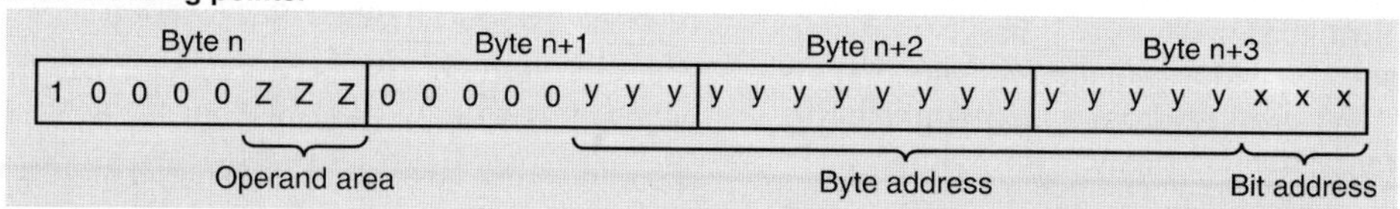

Figure 24.1 Structure of an Area Pointer

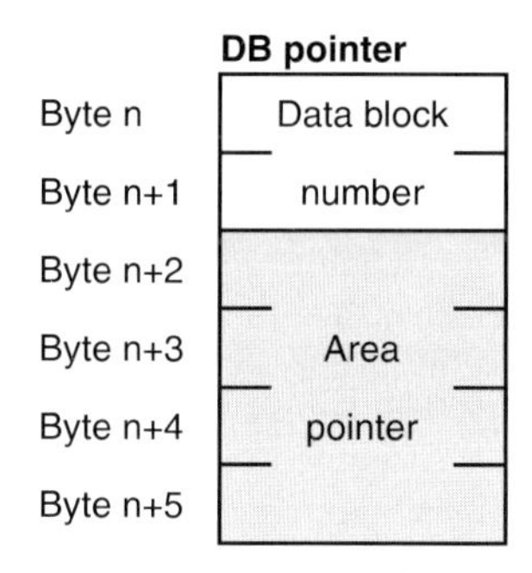

Figure 24.2 Structure of a DB Pointer

The area pointer has, in principle, a bit address that must always be specified even with digital operands; in the case of digital operands, specify 0 as the bit address. Example: You can use area pointer P#M22.0 to address memory bit M 22.0, but also memory byte MB 22, memory word MW 22 or memory doubleword MD 22.

24.2.3 DB Pointer

In addition to the area pointer, a DB pointer also contains a data block number as a positive INT number (Figure 24.2). It specifies the data block if the area pointer contains the operand areas global data or instance data. In all other cases, the first two bytes contain zero.

The notation of the pointer is familiar to you from full addressing of data operands. Here also, the data block and the data operand are specified, separated by a dot:

P#DataBlock.DataOperand

Example: P#DB 10.DBX 20.5

You can apply this pointer to a block parameter of parameter type POINTER in order to point to a data operand. The Editor uses this pointer type internally to transfer actual parameters.

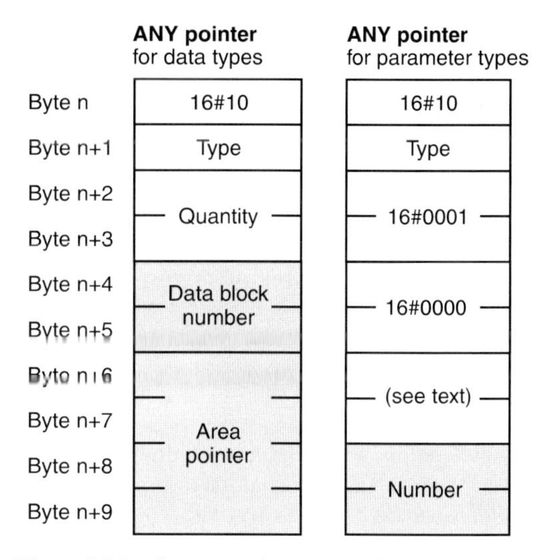

Figure 24.3 Structure of an ANY Pointer

24.2.4 ANY Pointer

In addition to the DB pointer the ANY pointer also contains the data type and a repetition factor. This makes it also possible to point to a data area.

The ANY pointer is available in two variants: For variables with data types and for variables with parameter types. If you point to a variable with a data type, the ANY pointer contains a DB pointer, the type and a repetition factor. If the ANY pointer points to a variable with a parameter type, it contains only the number instead of the DB pointer in addition to the type (Figure 24.3). In the case of a timer or counter function, the type is repeated in byte (n+6); byte (n+7) contains B#16#00. In all other cases, these two bytes contain the value W#16#0000.

Table 24.2 Data Type Coding in ANY Pointers

Data Type	Type	Data Types	Type	Parameter Types	Type
BOOL	01	REAL	08	BLOCK_FB	17
BYTE	02	DATE	09	BLOCK_FC	18
CHAR	03	TOD	0A	BLOCK_DB	19
WORD	04	TIME	0B	BLOCK_SDB	1A
INT	05	S5TIME	0C	COUNTER	1C
DWORD	06	DT	0E	TIMER	1D
DINT	07	STRING	13		

The first byte of the ANY pointer contains the syntax ID; in STEP 7 it is always 10_{hex}. The type specifies the data type of the variables for which the ANY pointer applies. Variables of elementary data types, DT and STRING receive the type shown in Table 24.2 and quantity 1.

If you apply a variable of data type ARRAY or STRUCT (also UDT) at an ANY parameter, the Editor generates an ANY pointer to the field or the structure. This ANY pointer contains the ID for BYTE (02_{hex}) as the type, and the byte length of the variable as the quantity. The data type of the individual field or structure components is not significant here. Thus, an ANY pointer points with double the number of bytes to a WORD field. Exception: A pointer to a field consisting of components of data type CHAR is also created with CHAR type (03_{hex}).

You can apply an ANY pointer at a block parameter of parameter type ANY if you want to point to a variable or an operand area. The constant representation for data types is as follows:

P#[DataBlock.]Operand Type Quantity

Examples:

▷ P#DB 11.DBX 30.0 INT 12
Area with 12 words in DB 11 from DBB 30

▷ P#M 16.0 BYTE 8
Area with 8 bytes from MB 16

▷ P#E 18.0 WORD 1
Input word IW 18

▷ P#E 1.0 BOOL 1
Inputs I 1.0

In the case of parameter types, you write the pointer as follows:

L#Number Type Quantity

Examples:

▷ L#10 TIMER 1
Timer function T 10

▷ L#2 COUNTER 1
Counter function C 2

The Editor then applies an ANY pointer that agrees in type and quantity with the specifications in the constant representation. Please note that the operand address in the ANY pointer must also be a bit address for data types.

Specification of a constant ANY pointer makes sense if you want to access a data area for which you have not declared any variables. In principle, you can also apply variables or operands at an ANY parameter. For example, the representation "P#I 1.0 BOOL 1" is identical with "I 1.0" or the corresponding symbol address.

If you do not specify any defaults when declaring an ANY parameter at a function block, the Editor assigns 10_{hex} to the syntax ID and 00_{hex} to the remaining bytes. The Editor then represents this (empty) ANY pointer (in the data view) as follows: P#P0.0 VOID 0.

24.2.5 "Variable" ANY Pointer

When copying with SFC 20, you specify at the source and destination parameters either an absolute addressed area (for example, P#DB127.DBX0.0 BYTE 32) or a variable. In both cases, the source area and the destination area are fixed during programming (variable indexing is not possible even with field components). The following method is available for modifying, at runtime, a data area created at a block parameter of the ANY type:

You create a variable of data type ANY in the temporary local data and use this to initialize an ANY parameter. The Editor then does not generate an ANY pointer (as it would if you created a different variable), but takes the ANY variable into the temporary local data as an ANY pointer to a source or destination area. The ANY variable in the temporary local data is structured in the same way as an ANY pointer; you can now modify the individual entries at runtime.

This procedure functions not only in the case of SFC 20 BLKMOV but also with block parameters of type ANY in other blocks.

The diskette accompanying the book has an example of the "variable" ANY pointer in the program "Message Frame Example".

24.3 Brief Description of "Message Frame Example"

This example is broken down as follows:

▷ Message frame data, shows how to handle data structures

▷ Time-of-day check, shows how to handle system blocks and standard blocks

▷ Prepare message frame, shows the use of SFC 20 BLKMOV with fixed addresses

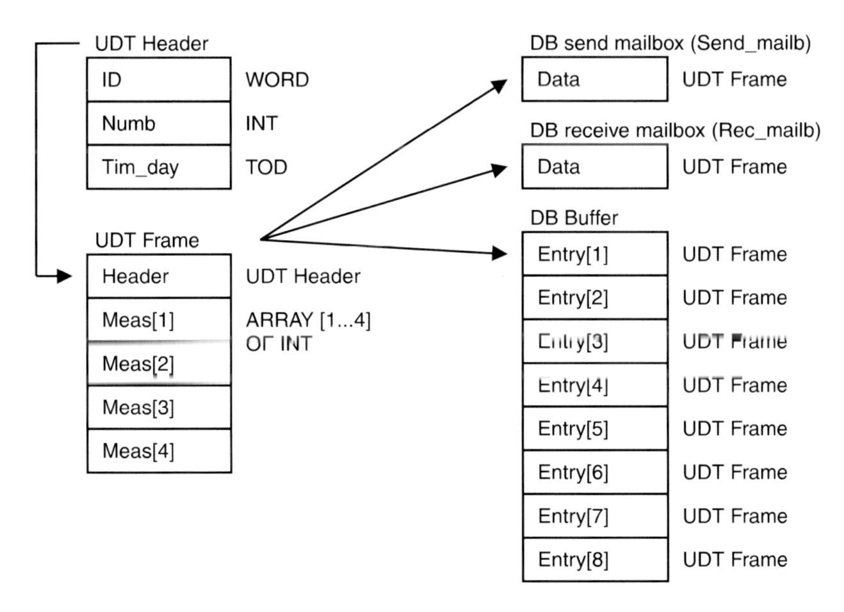

Figure 24.4
Data Structure for the Message Frame Data Example

- ▷ Copy data area indirect, shows a function "copy indirect" when using "variable" ANY pointers
- ▷ Save message frame, shows the use of "indirect copying"

Message frame data

The example shows how you can define frequently occurring data structures as your own data type and how you use this data type when declaring variables and parameters.

We construct a data store for incoming and outgoing message frames: A send mailbox with the structure of a message frame, a receive mailbox with the same structure and a (receive) ring buffer that is to provide intermediate storage for incoming message frames (Figure 24.4). Since the data structure of the message frame occurs frequently we will define it as a user-defined data type (UDT) *Frame*. The message frame contains a frame header whose structure we also want to give a name to. The send mailbox and the receive mailbox are to be data blocks that each contain a variable with the structure of *Frame*. Finally, there is also a ring buffer, a data block with a field consisting of eight components that are also of the data structure *Frame*.

UDT 51 User-defined data type Header
UDT 52 User-defined data type Frame
DB 61 Send mailbox
DB 62 Receive mailbox
DB 63 Ring buffer
DB 64 Data block for measured values

Time-of-day check

The examples shows how to handle system and standard blocks (evaluating errors, copying from the library, renaming).

The function *Time-of-day check* is to output the time-of-day in the integral CPU real-time clock as a function value. For this purpose, we require the system function SFC 1 READ_CLK that reads the date and the time-of-day of the real-time clock in the DATE_AND_TIME or DT data format. Since we only want to read the time-of-day, we also require the IEC function FC 8 DT_TOD. This function fetches the time-of-day in the TIME_OF_DAY or TOD format from the DT data format (Figure 24.5).

Error evaluation

The system functions signal an error via the binary result BR and via the function value RET_VAL. An error exists if the binary result BR = "0"; the function value is then also negative (bit 15 is set). The IEC standard functions signal an error only via the binary result. Both types of error evaluation are shown in the example. If an error exists, an invalid value is output for the time-of-day. In addition, the binary result is affected. After the *Time-of-day check* function has been called, you can therefore also use the binary result to see if an error has occurred.

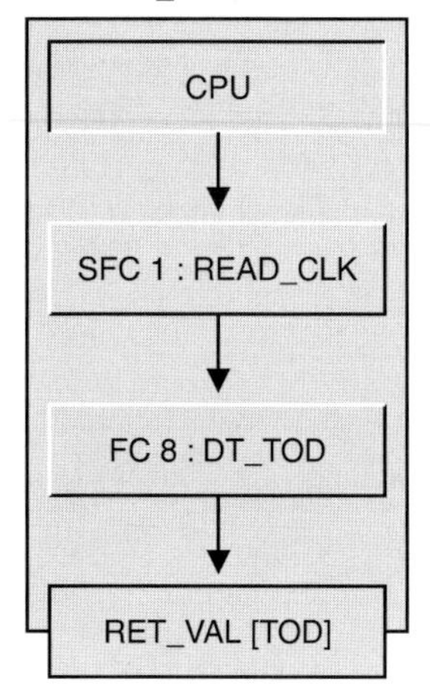

Figure 24.5
Example: Checking the Time-of-Day

Off-line programming of system functions

Before saving the entered block, the system functions SFC 1 and the standard function FC 8 must be included in the off-line user program. Both functions are included in the scope of supply of STEP 7. You will find these functions in the block libraries supplied. (For the system functions integrated into the CPU, the library contains only has an interface description instead of the system functions program. The function can be called off-line via this interface description; the interface description is not transferred to the CPU. The loadable functions such as the IEC functions are available in the library as executable programs.)

Select the library "StdLib30" with FILE → OPEN under the SIMATIC Manager and open the library "Built In". Under *Blocks*, you will find here all interface descriptions for the system functions. If you still have the project window of your project open, you can arrange both windows side-by-side with WINDOW → ARRANGE → VERTICALLY and "drag" the selected system functions with the mouse into your program (mark the SFC with the mouse, keep the mouse key pressed, drag to *Blocks* or into its opened window and release). Copy standard function FC 8 in the same way. You will find this in the library "IEC". FC 8 is a loadable function; it therefore occupies user memory, in contrast to SFC 1.

If a standard function block is called under "Libraries" in the Program Element Catalog using the Editor, it is automatically copied into *Blocks* and entered in the Symbol Table.

Renaming standard functions

You can rename a loadable standard function. Mark the standard function (for example, FC 8) in the project window and click once (again) on the designator. A frame and the name appear and you can specify a new address (for example, FC 98). If you now press the F1 key while the standard function (renamed to FC 98) is marked, you will nevertheless receive the on-line help for the original standard function (FC 8).

If an identically addressed block exists when copying, a dialog box appears where you can choose between overwriting and renaming.

Symbol address

You can assign names to the system functions and the standard functions in the Symbol Table so that you can also access these functions symbolically. You can assign these names freely within the permissible definition for block names. In the example, the block name in each case is selected as a symbol name (for better identification).

Prepare message frame

The data block *Send_mailb* is to be filled with the data for a message frame. We use a function block that has the ID and the consecutive number stored in its instance data block. The net data are stored finally in a global data block; they are copied into the send mailbox with the system function BLKMOV. We use the function *CPU_Time* to take the time-of-day from the CPU's real-time clock (Figure 24.6).

The first network in the function block FB *Generate_Frame* transfers the ID stored in the instance data block to the frame header. The consecutive number is incremented by +1 and is also transferred to the frame header.

The second network contains the call of the function *CPU_Time* that takes the time-of-day from the real-time clock and enters it in the frame header in the format TIME_OF_DAY.

In the subsequent networks, you will see a method of copying selected variables at runtime with the system function SFC 20 BLKMOV and

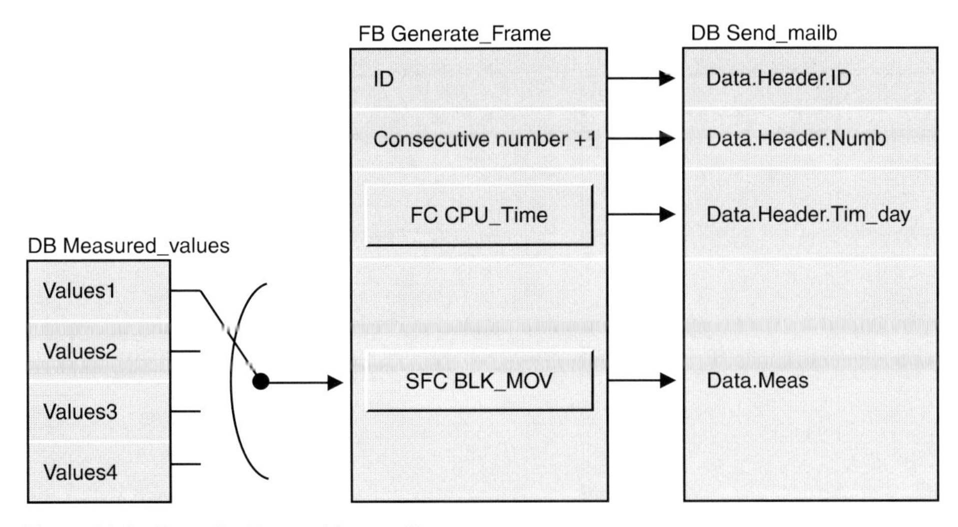

Figure 24.6 Example: Prepare Message Frame

without using indirect addressing. It is therefore also not necessary to know the absolute address and the structure of the variables. The principle is extremely simple: The desired copy function is selected with comparison functions. The numbers 1 to 14 are permissible as selection criteria.

The FB *Generate_Frame* is programmed in such a way that it is called via a signal edge for generating a message frame.

Indirect copying of a data area

The example shows the preparation and use of a "variable" ANY pointer with LAD program elements.

The function *CopyData* copies a data area whose address and length you can set freely via block parameters. The individual block parameters correspond to the individual elements of an ANY pointer (see Section 24.2.4 "ANY pointer"). The specifications at the block parameters must correspond to valid values; they are not checked (a copy error is signaled by SFC 20 BLKMOV via its function value that is transferred to the function value of the function *CopyData*).

Essential elements are the two temporary variables *SoPointer* and *DesPointer* of data type ANY. They contain the ANY pointers for the system function SFC 20 BLKMOV. *SoPointer* points to the source area of the data to be transferred and *DesPointer* points to the destination area. Figure 24.7 shows the structure of the variable *SoPointer; DesPointer* is structured identically. The individual bytes, words and doublewords of the ANY variables are accessed via their absolute addresses.

Save message frame

The example shows the use of the function *CopyData* (copying a data area with programmable address).

A message frame in the data block *Rec_mailb* is to be written to the next location in the data block *Buffer*. The block-local variable *Entry* determines the location in the ring buffer; the value of this location is used to calculate the address in the ring buffer Figure 24.8).

The *Entry* has a value range from 0 to 7. In the first network, a comparator determines whether *Entry* is less than 7. If this is the case, *Entry* is incremen-

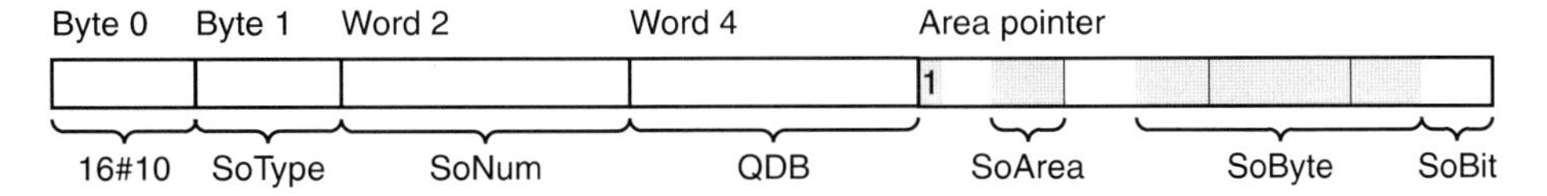

Figure 24.7 Structure of the Variable SoPointer

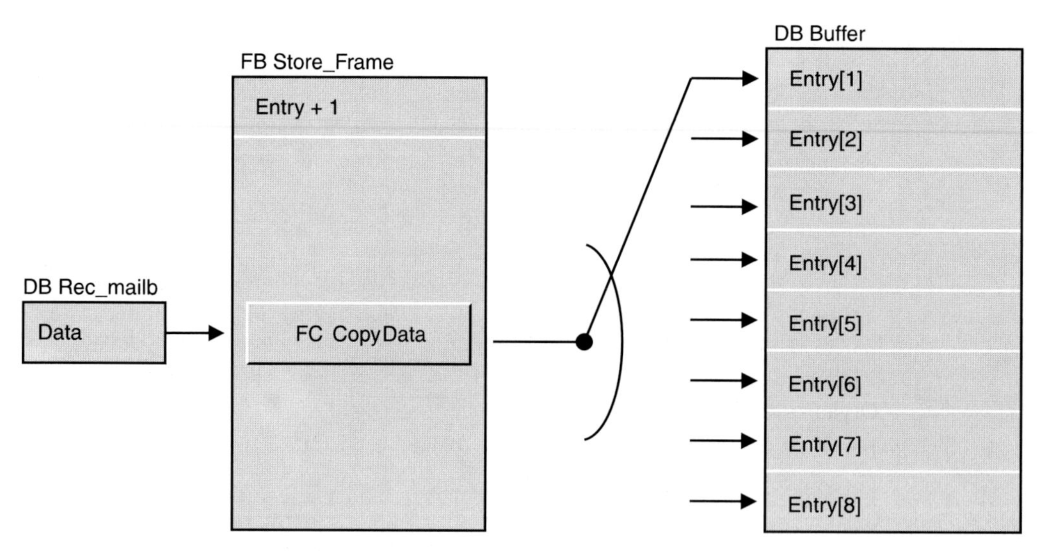

Figure 24.8 Example: Save Message Frame

ted by 1 in the next network, otherwise it is set to zero. *Entry* multiplied by 16 gives the absolute byte address of the next entry in the ring buffer (the data structure *Frame* consists of 16 bytes).

The function *CopyData* that copies the message frame from the receive mailbox (data block DB 62) into the ring buffer (data block DB 63) is called in Network 3.

25 Block Libraries

The library StdLib30 contains the following library programs and is included in the scope of supply of the STEP 7 Standard Software Version 3:

- Std OBs — Organization blocks
- Built In — Integrated system blocks
- IEC — IEC functions
- FB Lib 1 — S5 converter functions
- FB Lib 2 — Further standard functions
- PID Control — Closed-loop control functions
- Net DP — DP functions

You can copy blocks or interface descriptions into Version 3 projects from the library programs described. If you want to transfer blocks or interface descriptions into a Version 2 project, you must use the library "stdlibs".

25.1 Std Obs Library

(Prio = Default priority class)

OB	Prio	Designation
1	1	Main program
10	2	Time-of-day interrupt 0
11	2	Time-of-day interrupt 1
12	2	Time-of-day interrupt 2
13	2	Time-of-day interrupt 3
14	2	Time-of-day interrupt 4
15	2	Time-of-day interrupt 5
16	2	Time-of-day interrupt 6
17	2	Time-of-day interrupt 7
20	3	Time-delay interrupt 0
21	4	Time-delay interrupt 1
22	5	Time-delay interrupt 2
23	6	Time-delay interrupt 3
30	7	Watchdog interrupt 0 (5 s)
31	8	Watchdog interrupt 1 (2 s)
32	9	Watchdog interrupt 2 (1 s)
33	10	Watchdog interrupt 3 (500 ms)
34	11	Watchdog interrupt 4 (200 ms)
35	12	Watchdog interrupt 5 (100 ms)
36	13	Watchdog interrupt 6 (50 ms)
37	14	Watchdog interrupt 7 (20 ms)
38	15	Watchdog interrupt 8 (10 ms)
40	16	Hardware interrupt 0
41	17	Hardware interrupt 1
42	18	Hardware interrupt 2
43	19	Hardware interrupt 3
44	20	Hardware interrupt 4
45	21	Hardware interrupt 5
46	22	Hardware interrupt 6
47	23	Hardware interrupt 7
60	25	Multicomputing interrupt
80	26	Time error
	28	Time error (startup)
81	26	Power supply fault
	28	Power supply fault (startup)
82	26	Diagnostics interrupt
	28	Diagnostics interrupt (startup)
83	26	Insert/remove-module interrupt
	28	Insert/remove-module interrupt (startup)
84	26	CPU hardware fault
	28	CPU hardware fault (startup)
85	26	Priority class error
	28	Priority class error(startup)
86	26	DP error
	28	DP error (startup)
87	26	Communications error
	28	Communications error (startup)
90	29	Background processing
100	27	Complete restart
101	27	Restart
121	–	Programming error
122	–	I/O access error

25.2 Built In Library

IEC timers and IEC counters

SFB	Name	Designation
0	CTU	Up counter
1	CTD	Down counter
2	CTUD	Up/down counter
3	TP	Pulse
4	TON	On delay
5	TOF	Off delay

Copy and block functions

SFC	Name	Designation
20	BLKMOV	Copy data area
21	FILL	Preassign data area
22	CREAT_DB	Create data block
23	DEL_DB	Delete data block
24	TEST_DB	Test data block
25	COMPRESS	Compress memory
44	REPL_VAL	Enter replacement value

Program control

SFC	Name	Designation
43	RE_TRIGR	Retrigger cycle time monitoring
46	STP	Change to STOP state
47	WAIT	Wait delay time

Drum

SFB	Name	Designation
32	DRUM	Drum

CPU clock and operating hours counter

SFC	Name	Designation
0	SET_CLK	Set clock
1	READ_CLK	Read clock
2	SET_RTM	Set operating hours counter
3	CTRL_RTM	Control operating hours counter
4	READ_RTM	Read operating hours counter
48	SNC_RTCB	Synchronize slave clocks
64	TIME_TCK	Read system time

Interrupt events

SFC	Name	Designation
28	SET_TINT	Set time-of-day interrupt
29	CAN_TINT	Cancel time-of-day interrupt
30	ACT_TINT	Activate time-of-day interrupt
31	QRY_TINT	Check time-of-day interrupt
32	SRT_DINT	Start time-delay interrupt
33	CAN_DINT	Cancel time-delay interrupt
34	QRY_DINT	Check time-delay interrupt
35	MP_ALM	Initiate multicomputing interrupt
36	MSK_FLT	Mask synchronous error
37	DMSK_FLT	Demask synchronous error
38	READ_ERR	Read event status register
39	DIS_IRT	Disable asynchronous error
40	EN_IRT	Enable asynchronous error
41	DIS_AIRT	Delay asynchronous error
42	EN_AIRT	Enable asynchronous error

Data record transfer

SFC	Name	Designation
55	WR_PARM	Write dynamic parameter
56	WR_DPARM	Write predefined parameter
57	PARM_MOD	Parameterize module
58	WR_REC	Write data record
59	RD_REC	Read data record

System diagnostics

SFC	Name	Designation
6	RD_SINFO	Read start information
51	RDSYSST	Read SYS ST part list
52	WR_USMSG	Entry in the diagnostics buffer

Create block-related messages

SFB	Name	Designation
33	ALARM	Messages with acknowledgment display
34	ALARM_8	Messages without accompanying values
35	ALARM_8P	Messages with accompanying values
36	NOTIFY	Messages without acknowledgment display
37	AR_SEND	Send archive data

SFC	Name	Designation
9	EN_MSG	Enable messages
10	DIS_MSG	Disable messages
17	ALARM_SQ	Messages that can be acknowledged
18	ALARM_S	Messages that are always acknowledged
19	ALARM_SC	Determine acknowledgment status

Update the process image

SFC	Name	Designation
26	UPDAT_PI	Update process-image inputs
27	UPDAT_PO	Update process-image outputs
79	SET	Set I/O bit field
80	RSET	Reset I/O bit field

Address modules

SFC	Name	Designation
5	GADR_LGC	Determine logical address
49	LGC_GADR	Determine slot
50	RD_LGADR	Determine all logical addresses

Distributed I/O

SFC	Name	Designation
7	DP_PRAL	Initiate hardware interrupt
13	DPNRM_DG	Read diagnostics data
14	DPRD_DAT	Read slave data
15	DPWR_DAT	Write slave data

Global data communications

SFC	Name	Designation
60	GD_SND	Send GD packet
61	GD_RCV	Receive GD packet

Communications via unconfigured connections

SFC	Name	Designation
65	X_SEND	Send data externally
66	X_RCV	Receive data externally
67	X_GET	Read data externally
68	X_PUT	Write data externally
69	X_ABORT	Abort external connection
72	I_GET	Read data internally
73	I_PUT	Write data internally
74	I_ABORT	Abort internal connection

Communications via configured connections

SFC	Name	Designation
62	CONTROL	Check communications status

SFB	Name	Designation
8	USEND	Uncoordinated send
9	URVC	Uncoordinated receive
12	BSEND	Block-oriented send
13	BRCV	Block-oriented receive
14	GET	Read data from partner
15	PUT	Write data to partner
16	PRINT	Write data to printer
19	START	Initiate complete restart in the partner
20	STOP	Set partner to STOP
21	RESUME	Initiate restart in the partner
22	STATUS	Check status of partner
23	USTATUS	Receive status of partner

Integrated functions CPU 312/314/614

SFC	Name	Designation
63	AB_CALL	Call assembler block

SFB	Name	Designation
29	HS_COUNT	High-speed counter
30	FREQ_MES	Frequency meter
38	HSC_A_B	Control "Counter A/B"
39	POS	Control "Positioning"
41	CONT_C	Continuous closed-loop control
42	CONT_S	Step-action control
43	PULSEGEN	Generate pulse

25.3 IEC Library

Date and time functions

FC	Name	Designation
3	D_TOD_DT	Combine DATE and TOD to DT
6	DT_DATE	Extract DATE from DT
7	DT_DAY	Extract day-of-the-week from DT
8	DT_TOD	Extract TOD from DT
33	S5TI_TIM	Convert S5TIME to TIME
40	TIM_S5TI	Convert TIME to S5TIME
1	AD_DT_TM	Add TIME to DT
35	SB_DT_TM	Subtract TIME from DT
34	SB_DT_DT	Subtract DT from DT

Comparisons

FC	Name	Designation
9	EQ_DT	Compare DT for equal to
28	NE_DT	Compare DT for not equal to
14	GT_DT	Compare DT for greater than
12	GE_DT	Compare DT for greater than or equal to
23	LT_DT	Compare DT for less than
18	LE_DT	Compare DT for less than or equal to
10	EQ_STRNG	Compare STRING for equal to
29	NE_STRNG	Compare STRING for not equal to
15	GT_STRNG	Compare STRING for greater than
13	GE_STRNG	Compare STRING for greater than or equal to
24	LT_STRNG	Compare STRING for less than
19	LE_STRNG	Compare STRING for less than or equal to

String functions

FC	Name	Designation
21	LEN	Length of a STRING
20	LEFT	Left section of a STRING
32	RIGHT	Right section of a STRING
26	MID	Middle section of a STRING
2	CONCAT	Concatenate STRINGs
17	INSERT	Insert STRING
4	DELETE	Delete STRING
31	REPLACE	Replace STRING
11	FIND	Find STRING
16	I_STRNG	Convert INT to STRING
5	DI_STRNG	Convert DINT to STRING
30	R_STRNG	Convert REAL to STRING
38	STRNG_I	Convert STRING to INT
37	STRNG_DI	Convert STRING to DINT
39	STRNG_R	Convert STRING to REAL

Math functions

FC	Name	Designation
22	LIMIT	Limiter
25	MAX	Maximum selection
27	MIN	Minimum selection
26	SEL	Binary selection

25.4 FB Lib 1 Library

Floating-point arithmetic

FC	Name	Designation
61	GP_FPGP	Convert fixed-point to floating-point
62	GP_GPFP	Convert floating-point to fixed-point
63	GP_ADD	Add floating-point numbers
64	GP_SUB	Subtract floating-point numbers
65	GP_MUL	Multiply floating-point numbers
66	GP_DIV	Divide floating-point numbers
67	GP_VGL	Compare floating-point numbers
68	GP_RAD	Find the square root of a floating-point number

Signal functions

FC	Name	Designation
69	MLD_TG	Clock pulse generator
70	MLD_TGZ	Clock pulse generator with timer function
71	MLD_EZW	Initial value single blinking wordwise
72	MLD_EDW	Initial value double blinking wordwise
73	MLD_SAMW	Group signal wordwise
74	MLD_SAM	Group signal
75	MLD_EZ	Initial value single blinking
76	MLD_ED	Initial value double blinking
77	MLD_EZWK	Initial value single blinking (wordwise) memory bit
78	MLD_EZDK	Initial value double blinking (wordwise) memory bit
79	MLD_EZK	Initial value single blinking memory bit
80	MLD_EDK	Initial value double blinking memory bit

Integrated functions

FC	Name	Designation
81	COD_B4	BCD-binary conversion 4 decades
82	COD_16	Binary-BCD conversion 4 decades
83	MUL_16	16-bit fixed-point multiplier
84	DIV_16	16-bit fixed-point divider

Basic functions

FC	Name	Designation
85	ADD_32	32-bit fixed-point adder
86	SUB_32	32-bit fixed-point subtractor
87	MUL_32	32-bit fixed-point multiplier
88	DIV_32	32-bit fixed-point divider
89	RAD_16	16-bit fixed-point square root extractor
90	REG_SCHB	Bitwise shift register
91	REG_SCHW	Wordwise shift register
92	REG_FIFO	Buffer (FIFO)
93	REG_LIFO	Stack (LIFO)
94	DB_COPY1	Copy data area (direct)
95	DB_COPY2	Copy data area (indirect)
96	RETTEN	Save scratchpad memory (S5 155U)
97	LADEN	Load scratchpad memory (S5 155U)
98	COD_B8	BCD-binary conversion 8 decades
99	COD_32	Binary-BCD conversion 8 decades

Analog functions

FC	Name	Designation
100	AE_460_1	Analog input module 460
101	AI_460_2	Analog input module 460
102	AI_463_1	Analog input module 463
103	AE_463_2	Analog input module 463
104	AE_464_1	Analog input module 464
105	AE_464_2	Analog input module 464
106	AE_466_1	Analog input module 466
107	AE_466_2	Analog input module 466
108	RLG_AA1	Analog output module
109	RLG_AA2	Analog output module
110	PER_ET1	ET 100 distributed I/O
111	PER_ET2	ET 100 distributed I/O

Math functions

FC	Name	Designation
112	SINUS	Sine
113	COSINUS	Cosine
114	TANGENS	Tangent
115	COTANG	Cotangent
116	ARCSIN	Arc sine
117	ARCCOS	Arc cosine
118	ARCTAN	Arc tangent
119	ARCCOT	Arc cotangent
120	LN_X	Natural logarithm
121	LG_X	Logarithm to base 10
122	B_LOG_X	Logarithm to any base
123	E_H_N	Exponential function with base e
124	ZEHN_H_N	Exponential function with base 10
125	A2_H_A1	Exponential function with any base

25.5 FB Lib 2 Library

FC	Name	Designation
80	TONR	Retentive on delay
81	IBLKMOV	Transfer data area indirect
82	RSET	Reset process-image bitwise
83	SET	Set process-image bitwise
84	ATT	Enter value in table
85	FIFO	Output first value in table
86	TBL_FIND	Find value in table
87	LIFO	Output last table value
88	TBL	Execute table operation
89	TBL_WRD	Copy value from the table
90	WSR	Store data item
91	WRD_TBL	Combine table element
92	SHRB	Shift bit to bit-shift register
93	SEG	Bit pattern for 7-segment display
94	ATH	ASCII-hexadecimal conversion
95	HTA	Hexadecimal-ASCII conversion
96	ENCO	Least-significant set bit
97	DECO	Set bit in the word
98	BCDCPL	Generate ten's complement
99	BITSUM	Count set bits
100	RSETI	Reset PIQ bytewise
101	SETI	Set PI bytewise
102	DEV	Calculate standard deviation
103	CDT	Correlated data tables
104	TBL_TBL	Table combination
105	SCALE	Scale values
106	UNSCALE	Descale values

25.6 PID Control Library

FB	Name	Designation
41	CONT_C	Continuous control
42	CONT_S	Step control
43	PULSGEN	Generate pulse

25.7 DP Library

FC	Name	Designation
1	DP_SEND	Send data
2	DP_RECV	Receive data
3	DP_DIAG	Diagnostics
4	DP_CTRL	Control

26 Function Set

The LAD programming language provides you with the functions shown in the overview below.

26.1 Basic Functions

26.1.1 Binary Checks and Combinations

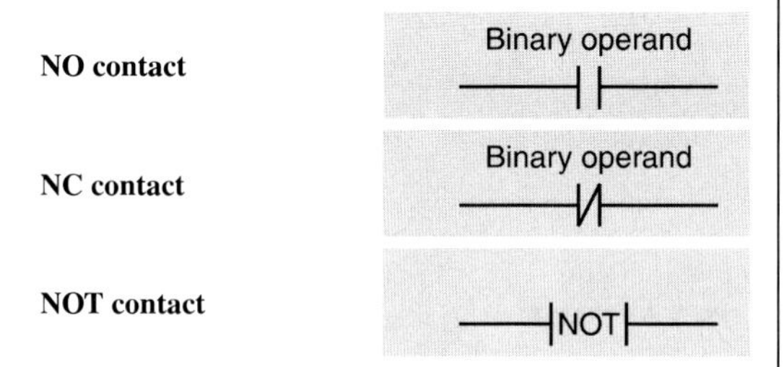

26.1.2 Memory Functions

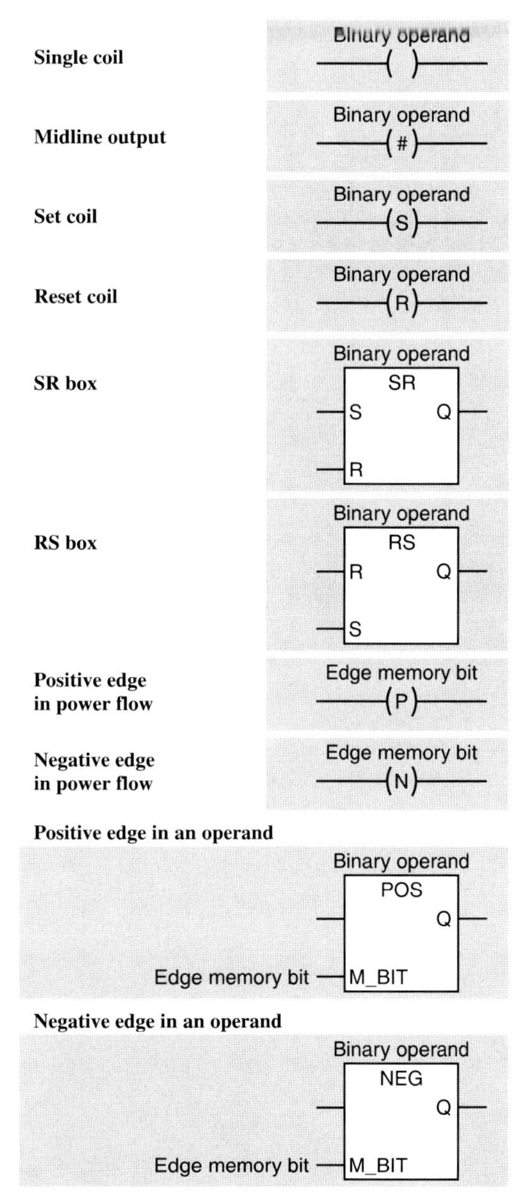

26.1.3 Transfer Functions

MOVE box

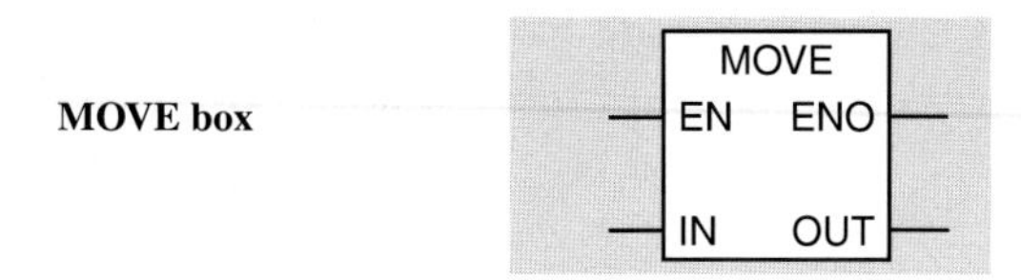

26.1.4 Timer Functions

Timer box

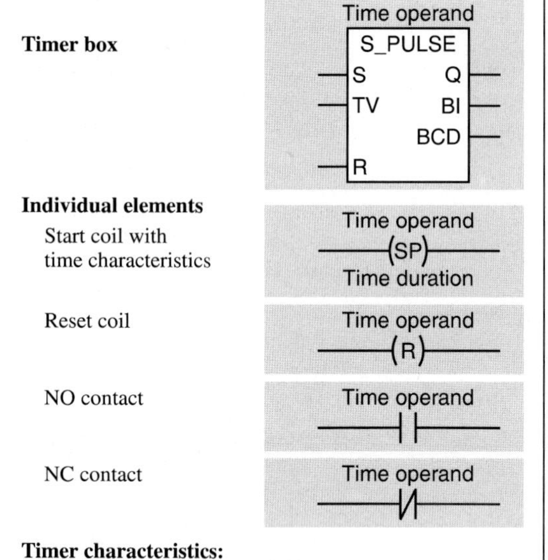

Individual elements

Start coil with time characteristics

Reset coil

NO contact

NC contact

Timer characteristics:

S_PULSE	SP	Pulse
S_PEXT	SE	Extended pulse
S_ODT	SD	ON delay
S_ODTS	SS	Stored ON delay
S_OFFDT	SF	OFF delay

26.1.5 Counter Functions

Counter box

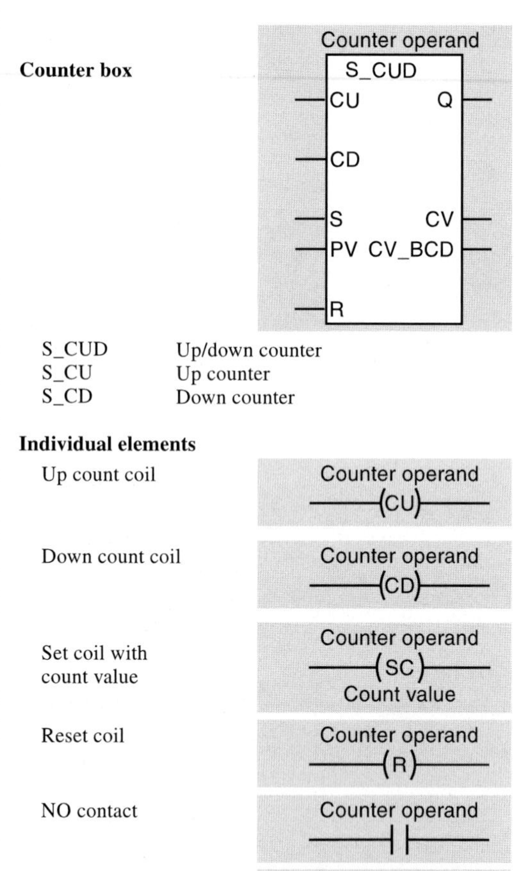

S_CUD	Up/down counter
S_CU	Up counter
S_CD	Down counter

Individual elements

Up count coil — Counter operand —(CU)—

Down count coil — Counter operand —(CD)—

Set coil with count value — Counter operand —(SC)— Count value

Reset coil — Counter operand —(R)—

NO contact — Counter operand

NC contact — Counter operand

26.2 Digital Functions

26.2.1 Comparison Functions

Comparison box

CMP ==I
IN1
IN2

Compare for	according to		
	INT	DINT	REAL
equal to	==I	==D	==R
not equal to	<>I	<>D	<>R
greater than	>I	>D	>R
greater than or equal to	>=I	>=D	>=R
less than	<I	<D	<R
less than or equal to	<=I	<=D	<=R

26.2.2 Arithmetic Functions

Arithmetic box

ADD_I
EN ENO
IN1 OUT
IN2

Calculation	according to		
	INT	DINT	REAL
Addition	ADD_I	ADD_DI	ADD_R
Subtraction	SUB_I	SUB_DI	SUB_R
Multiplication	MUL_I	MUL_DI	MUL_R
Division	DIV_I	DIV_DI	DIV_R
Modulus	–	MOD_DI	–

26.2.3 Math Functions

Math box

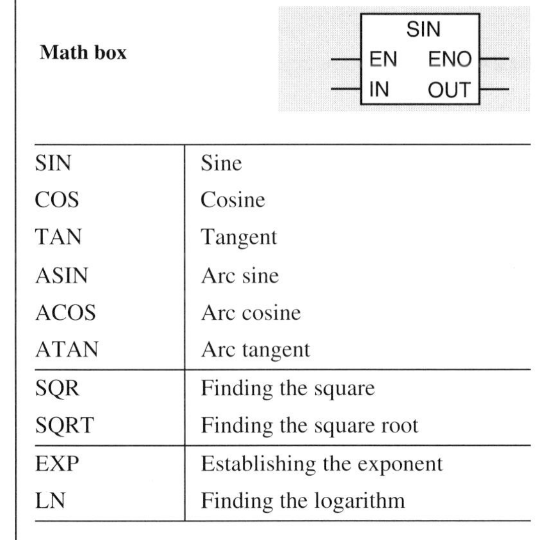

SIN	Sine
COS	Cosine
TAN	Tangent
ASIN	Arc sine
ACOS	Arc cosine
ATAN	Arc tangent
SQR	Finding the square
SQRT	Finding the square root
EXP	Establishing the exponent
LN	Finding the logarithm

26.2.4 Conversion Functions

Conversion box

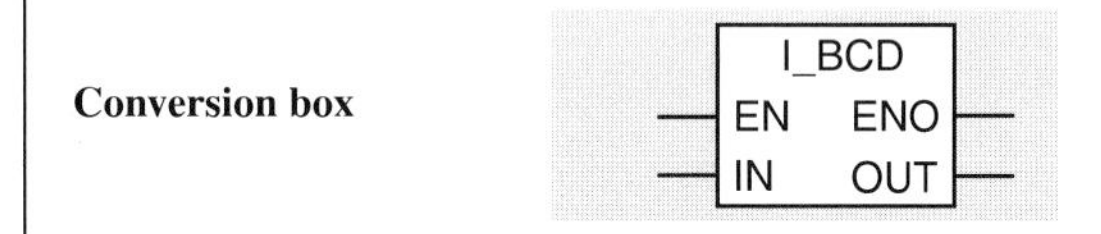

I_DI	Conversion of INT to DINT
I_BCD	Conversion of INT to BCD
DI_BCD	Conversion of DINT to BCD
DI_R	Conversion of DINT to REAL
BCD_I	Conversion of BCD to INT
BCD_DI	Conversion of BCD to DINT
	Conversion of REAL to DINT with rounding
CEIL	to next higher number
FLOOR	to next lower number
ROUND	to next whole number
TRUNC	without rounding
INV_I	INT one's complement
INV_DI	DINT one's complement
NEG_I	INT negation
NEG_DI	DINT Negation
NEG_R	REAL negation
ABS	REAL absolute-value generation

26.2.5 Shift Functions

Shift box

SHL_W: EN, IN, N → ENO, OUT

SHL_W	Shift word left
SHL_DW	Shift doubleword left
SHR_W	Shift word right
SHR_DW	Shift doubleword right
SHR_I	Shift word with sign
SHR_DI	Shift doubleword with sign
ROL_DW	Rotate left
ROR_DW	Rotate right

26.2.6 Word Logic

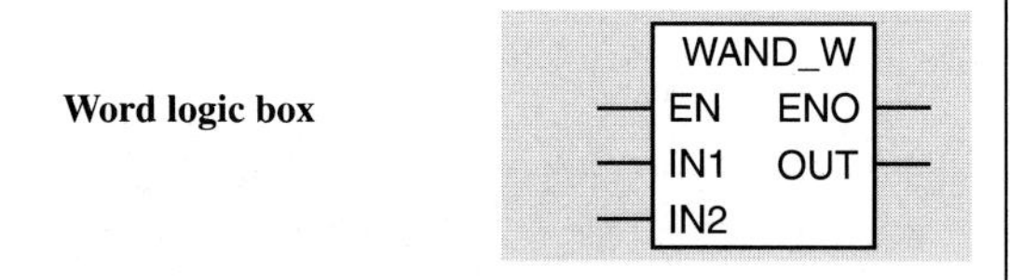

WAND_W	AND word
WOR_W	OR word
WXOR_W	Exclusive OR word
WAND_DW	AND doubleword
WOR_DW	OR doubleword
WXOR_DW	Exclusive OR doubleword

26.3 Program Flow Control

26.3.1 Status Bits

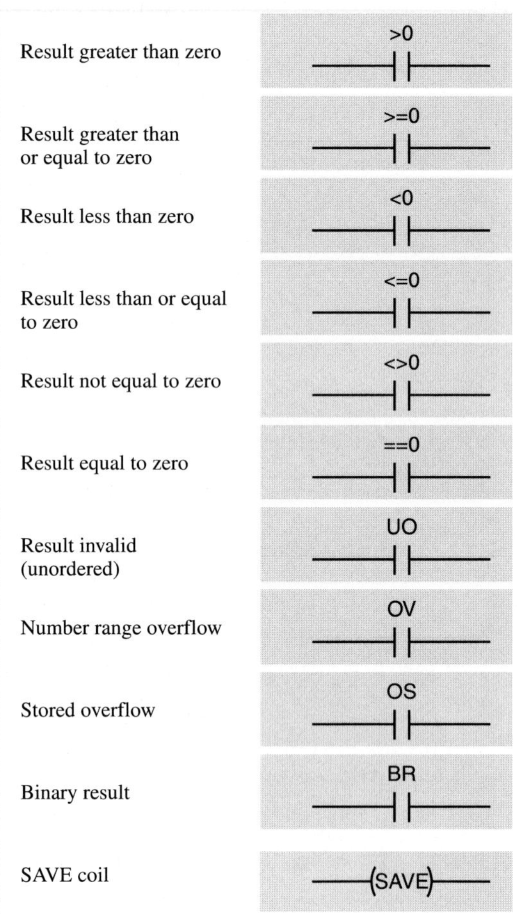

26.3.2 Jump Functions

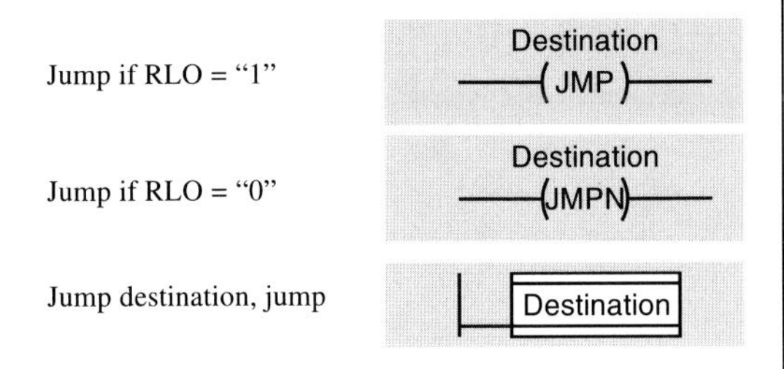

26.3.3 Master Control Relay

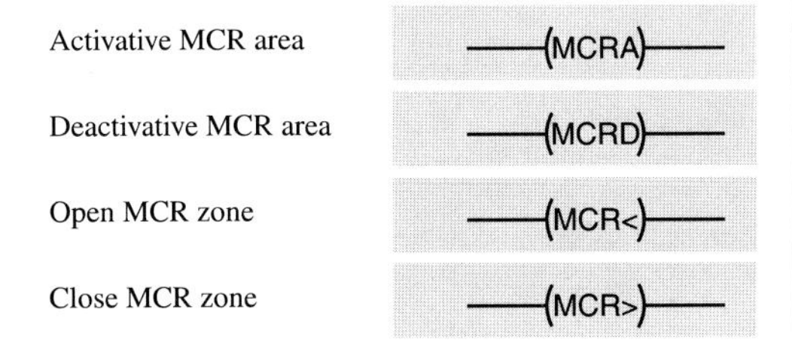

26.3.4 Block Functions

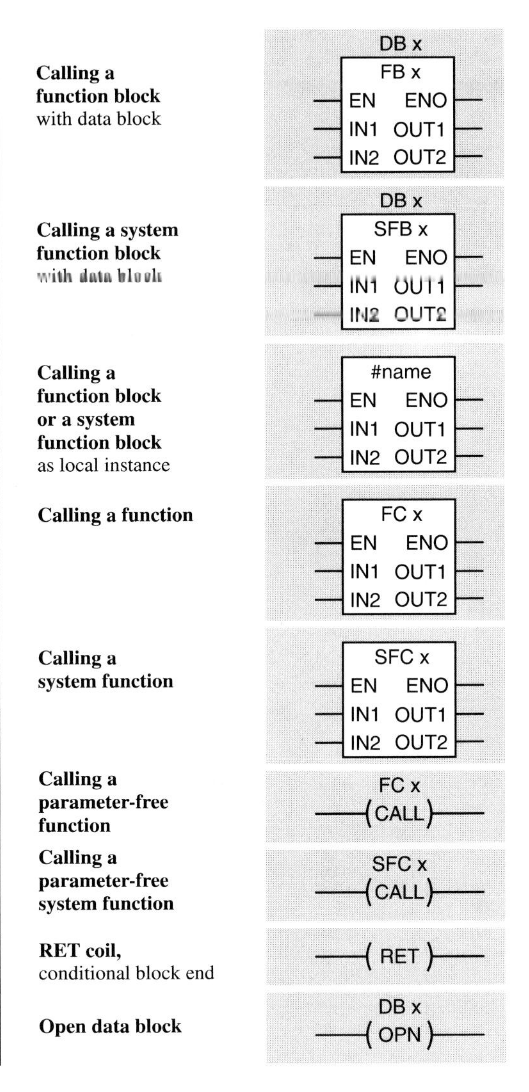

Index

H

I

J

L

M

N

O

P

R

S

T

U

V

W

Abbreviations

AI	Analog input
AO	Analog output
AS	Automation system
AS-I	Actuator-sensor-interface
CFC	Continuous function chart
CP	Communications processor
CPU	Central processing unit
DB	Data block
DI	Digital input
DO	Digital output
DP	Distributed I/O
EPROM	Erasable programmable read-only memory
FB	Function block
FBD	Function block diagram
FC	Function call
FEPROM	Flash erasable programmable read-only memory
FM	Function module
IM	Interface module
LAD	Ladder diagram
MCR	Master Control Relay
MPI	Multipoint interface
OB	Organization block
OP	Operator panel
PG	Programming device
PLC	Programmable controller
PS	Power supply
RAM	Random access memory
RLO	Result of logic operation
SCL	Structured control language
SDB	System data block
SFB	System function block
SFC	System function call
SM	Signal module
SSL	System status list
STL	Statement list
UDT	User-defined data type
VAT	Variable table

Berger, Hans

Automating with STEP 7 in STL

SIMATIC S7-300/400 Programmable Controllers

⊔ Book and Disk

1998, approx. 316 pages, approx.
100 illustrations, approx. 90 tables,
1 disk $3^1/2''$, 17.3 cm × 25 cm, hardcover
ISBN 3-89578-093-6
approx. DEM 114.00/öS 832.00/sFr 101.00

SIMATIC S7 programmable controllers are used to implement industrial control systems for machines, manufacturing plants and industrial processes. The relevant open-loop and closed-loop control tasks can be solved using the STEP 7 programming software, which has been developed on the basis of STEP 5, with its various programming languages.
This book describes elements and applications of the instruction-oriented STL (statement list) programming language for use with both SIMATIC S7-300 and SIMATIC S7-400. It is aimed at all users of SIMATIC S7 programmable controllers. First-time users will be introduced to the field of programmable logic control whereas advanced users will learn about specific applications of SIMATIC S7 programmable controllers.
The enclosed diskette contains all programming examples described in the book – and a few extra examples – also intended as exercises. The examples can be viewed, modified and tested using STEP 7.

Contents

Principle of Operation of a Programmable Controller · System Overview: SIMATIC S7 and STEP 7 · STL Programming Language · Basic Functions · Digital Functions · Program Sequence Control · User Program Execution · Use of Variables.

Bezner, Heinrich

Dictionary of Power Engineering and Automation

Part 1 German/English

3rd revised and enlarged edition, 1993, 579 pages, 14.8 cm × 22 cm,
hardcover
ISBN 3-8009-4118-X
DEM 98.00/öS 715.00/sFr 96.00

Part 2 English/German

3rd revised and enlarged edition, 1993, 511 pages, 14.8 cm × 22 cm,
hardcover
ISBN 3-8009-4119-8
DEM 98.00/öS 715.00/sFr 96.00

CD-ROM Version

Edition 1996
System requirements
- Microsoft® Windows™ 3.x, Windows 95
- Personal computer with at least a 386 microprocessor and
- min. 4 MB RAM
- CD-ROM drive
- VGA or higher resolution video adapter
- Mouse or adequate pointing device

ISBN 3-89578-045-6
Price for single user
DEM 229.00/öS 1672.00/sFr 203.00
Price for multi-user on request.

This publication covers terms essentially from the following fields: Power generation, transmission and distribution; drives, switchgear and installation technology, power electronics; measuring and analysis technology, test engineering; automation technology and process control.
With over 65,000 entries in the German-English part of the dictionary and 52,000 in the English-German part, as well as with its many cross-references to VDE codes and standards (DIN, VDI, IEC, BS, ANSI, CEE, ISO), this dictionary is a reliable reference source for all those who read, prepare or process power engineering and automation texts in either language.